Edited by
Oleg G. Okhotnikov

Semiconductor Disk Lasers

Related Titles

Gywat, O., Krenner, H. J., Berezovsky, J.

Spins in Optically Active Quantum Dots

Concepts and Methods

2010

ISBN: 978-3-527-40806-1

Saleh, B. E. A., Teich, M. C.

Fundamentals of Photonics

2006

ISBN: 978-0-471-35832-9

Buus, J., Amann, M.-C., Blumenthal, D. J.

Tunable Laser Diodes and Related Optical Sources

2005

ISBN: 978-0-471-20816-7

Fukuda, M.

Optical Semiconductor Devices

1999

ISBN: 978-0-471-14959-0

Bimberg, D., Grundmann, M., Ledentsov, N. N.

Quantum Dot Heterostructures

1999

ISBN: 978-0-471-97388-1

Edited by
Oleg G. Okhotnikov

Semiconductor Disk Lasers

Physics and Technology

WILEY-VCH Verlag GmbH & Co. KGaA

The Editor

Prof. Dr. Oleg G. Okhotnikov
Optoelectronics Research Center
Tampere University
Finland
deg.okhotnikov@tut.fi

Cover Image
Spieszdesign, Neu-Ulm
Germany

All books published by Wiley-VCH are carefully produced. Nevertheless, authors, editors, and publisher do not warrant the information contained in these books, including this book, to be free of errors. Readers are advised to keep in mind that statements, data, illustrations, procedural details or other items may inadvertently be inaccurate.

Library of Congress Card No.: applied for

British Library Cataloguing-in-Publication Data
A catalogue record for this book is available from the British Library.

Bibliographic information published by the Deutsche Nationalbibliothek
The Deutsche Nationalbibliothek lists this publication in the Deutsche Nationalbibliografie; detailed bibliographic data are available on the Internet at http://dnb.d-nb.de.

© 2010 WILEY-VCH Verlag GmbH & Co. KGaA, Weinheim

All rights reserved (including those of translation into other languages). No part of this book may be reproduced in any form – by photoprinting, microfilm, or any other means – nor transmitted or translated into a machine language without written permission from the publishers. Registered names, trademarks, etc. used in this book, even when not specifically marked as such, are not to be considered unprotected by law.

Composition   Thomson Digital, Noida, India
Printing and Bookbinding   betz-druck GmbH, Darmstadt
Cover Design   Adam Design, Weinheim

Printed in the Federal Republic of Germany
Printed on acid-free paper

ISBN: 978-3-527-40933-4

Contents

Preface *XI*
List of Contributors *XIII*

1	**VECSEL Semiconductor Lasers: A Path to High-Power, Quality Beam and UV to IR Wavelength by Design** *1*	
	Mark Kuznetsov	
1.1	Introduction *1*	
1.2	What Are VECSEL Semiconductor Lasers *2*	
1.2.1	History of VECSELs: Semiconductor Lasers, Optical Pumping, and External Cavity *2*	
1.2.2	Basic Principles of Operation: VECSEL Structure and Function *6*	
1.2.3	Basic Properties of VECSEL Lasers: Power Scaling, Beam Quality, and Intracavity Optical Elements *9*	
1.2.3.1	Power Scaling *9*	
1.2.3.2	Beam Quality *10*	
1.2.3.3	Laser Functional Versatility Through Intracavity Optical Elements *11*	
1.2.4	VECSEL Wavelength Versatility Through Materials and Nonlinear Optics *12*	
1.2.4.1	Wavelength Versatility Through Semiconductor Materials and Structures *12*	
1.2.4.2	Wavelength Versatility Through Nonlinear Optical Conversion *14*	
1.3	How Do You Make a VECSEL Laser *16*	
1.3.1	Semiconductor Gain Medium and On-Chip Bragg Mirror *16*	
1.3.1.1	Semiconductor Gain Design for VECSELs *16*	
1.3.1.2	On-Chip Multilayer Laser Bragg Mirror *21*	
1.3.1.3	Semiconductor Wafer Structure *22*	
1.3.2	Optical Cavity: Geometry, Mode Control, and Intracavity Elements *24*	
1.3.3	Optical and Electrical Pumping *29*	
1.3.4	VECSEL Laser Characterization *33*	
1.4	Demonstrated Performance of VECSELs and Future Directions *38*	

Semiconductor Disk Lasers. Physics and Technology. Edited by Oleg G. Okhotnikov
Copyright © 2010 WILEY-VCH Verlag GmbH & Co. KGaA, Weinheim
ISBN: 978-3-527-40933-4

1.4.1	Demonstrated Power Scaling and Wavelength Coverage	38
1.4.2	Commercial Applications	45
1.4.3	Current and Future Research Directions	48
1.4.4	Future of VECSEL Lasers: Scalable Power with Beam Quality from UV to IR	54
	References	57
2	**Thermal Management, Structure Design, and Integration Considerations for VECSELs**	**73**
	Stephane Calvez, Jennifer E. Hastie, Alan J. Kemp, Nicolas Laurand, and Martin D. Dawson	
2.1	Introduction	73
2.2	VECSEL Structure Design	74
2.2.1	Material System Selection	74
2.2.2	Gain	76
2.2.3	Mirrors	79
2.2.4	Subcavity Designs	80
2.2.5	Growth	81
2.2.6	Structure Characterization	82
2.2.7	Laser Cavity	83
2.3	Thermal Management	84
2.3.1	Introduction: Why Is Thermal Management Important?	84
2.3.2	Thermal Management Strategies in VECSELs	85
2.3.3	Modeling of Heat Flow in VECSELs: Guidelines	87
2.3.4	The Thin Device and Heat Spreader Approaches at 1 and 2 μm	87
2.3.5	Important Parameters: The Thermal Conductivity of the Mirror Structure, Submount, and Heat Spreader	89
2.3.6	Power Scaling of VECSELs	91
2.3.7	Wavelength Versatility	94
2.4	Laser Performance and Results	96
2.4.1	Power	96
2.4.2	Efficiency	100
2.4.3	Tuning	103
2.5	Integration	105
2.5.1	Microchip	105
2.5.2	Pump Integration	107
2.5.3	Fiber-Tunable VECSELs	109
2.6	Conclusions	111
	References	111
3	**Red Semiconductor Disk Lasers by Intracavity Frequency Conversion**	**119**
	Oleg Okhotnikov and Mircea Guina	
3.1	Introduction	119
3.2	SDL with Frequency Doubling	121

3.2.1	General Principle of Frequency Doubling	121
3.2.2	Power Scaling of SDLs	124
3.3	SDL Frequency Doubled to Red	126
3.3.1	Dilute Nitride Heterostructures for 1.2 µm Light Emission	127
3.3.2	Plasma-Assisted MBE Growth of Dilute Nitrides	128
3.3.3	Design and Characteristics of Dilute Nitride Gain Media	130
3.3.4	Performance of 1220 nm SDL	133
3.3.5	SDL Intracavity Light Conversion to Red–Orange	136
3.4	Conclusions	139
	References	139
4	**Long-Wavelength GaSb Disk Lasers**	**143**
	Benno Rösener, Marcel Rattunde, John-Mark Hopkins, David Burns, and Joachim Wagner	
4.1	Introduction	143
4.2	The III-Sb Material System	144
4.3	Epitaxial Layer Design and Growth of III-Sb Disk Laser Structures	146
4.3.1	Basic Structural Layout	147
4.3.2	Sample Growth and Post-Growth Analysis	149
4.3.3	Epitaxial Design of In-Well-Pumped SDLs	153
4.3.4	Sb-Based Active Regions on GaAs/AlGaAs DBRs	155
4.4	High-Power 2.X µm Disk Lasers	157
4.4.1	Initial Experiments	157
4.4.2	Sb-Based SDLs Using Intracavity Heat Spreaders	158
4.4.3	In-Well-Pumped Sb-Based Semiconductor Disk Lasers	163
4.4.4	Sb-Based Semiconductor Disk Lasers on GaAs Substrates	166
4.5	Tunable, Single-Frequency Lasers	167
4.5.1	Tunability	168
4.5.2	Single-Frequency Operation	172
4.5.3	Experimental Results of a 2.3 µm Single-Frequency SDL	176
4.6	Disk Lasers At and Above 3 µm Wavelength	179
4.7	Conclusions	179
	References	181
5	**Semiconductor Disk Lasers Based on Quantum Dots**	**187**
	Udo W. Pohl and Dieter Bimberg	
5.1	Introduction	187
5.2	Size Quantization in Optical Gain Media	187
5.2.1	Quantum Dots in Lasers	189
5.2.2	Species of Quantum Dots	191
5.2.3	Energies of Confined Charge Carriers	191
5.2.4	Quantum Dot Lasers	196
5.2.4.1	Edge-Emitting Quantum Dot Lasers	196
5.2.4.2	Surface-Emitting Quantum Dot Lasers	198

5.3	Development of Disk Lasers Based on Quantum Dots 200
5.3.1	Concepts of Gain Structures 200
5.3.2	Adjustment of Quantum Dot Emission Wavelength 202
5.3.2.1	Tuning of Stranski–Krastanow Quantum Dots 203
5.3.2.2	Tuning of Submonolayer Quantum Dots 204
5.3.3	Characteristics of Quantum Dot Disk Lasers 205
5.3.3.1	Disk Lasers with Stranski–Krastanow Quantum Dots 205
5.3.3.2	Disk Lasers with Submonolayer Quantum Dots 207
5.4	Conclusions 207
	References 208

6 Mode-Locked Semiconductor Disk Lasers 213
Thomas Südmeyer, Deran J.H.C Maas, and Ursula Keller

6.1	Introduction 213
6.1.1	Ultrafast Lasers 213
6.1.2	Ultrafast Semiconductor Lasers 215
6.1.3	Application Areas 216
6.2	SESAM Mode Locking of Semiconductor Disk Lasers 219
6.2.1	Macroscopic Key Parameters of a SESAM 220
6.2.1.1	Nonlinear Optical Reflectivity 220
6.2.1.2	Temporal SESAM Response 225
6.2.2	Pulse Formation 227
6.2.2.1	Model for the Pulse Shaping 229
6.2.2.2	Mode-Locking Stability and the Importance of Gain and SESAM Saturation 231
6.2.2.3	Importance of Group Delay Dispersion 232
6.2.3	SESAM Designs 233
6.2.3.1	SESAM Structure for Field Enhancement Control 233
6.2.3.2	Comparison of Quantum Well and Quantum Dot SESAMs 236
6.3	Mode Locking Results 239
6.3.1	Introduction 239
6.3.2	Mode-Locked VECSELs with High Average Output Power 240
6.3.2.1	Power Scaling of Mode-Locked VECSELs 240
6.3.2.2	Experimental Results 242
6.3.2.3	Outlook 243
6.3.3	VECSEL Mode Locking at High Repetition Rates 244
6.3.3.1	Mode Locking with Similar Area on Gain and Absorber (1:1 Mode Locking) 245
6.3.3.2	Mode-Locked VECSELs with up to 50 GHz 245
6.3.3.3	Outlook 246
6.3.4	Femtosecond Mode-Locked VECSELs 246
6.3.4.1	Introduction 246
6.3.4.2	Mode Locking Results 247
6.3.5	Electrically Pumped Mode-Locked VECSELs 247

6.4	Mode-Locked Integrated External-Cavity Surface-Emitting Laser (MIXSEL) *248*	
6.4.1	Introduction *248*	
6.4.2	Integration Challenges *249*	
6.4.3	Results *251*	
6.4.4	Outlook *252*	
6.5	Summary and Outlook *254*	
	References *256*	
7	**External-Cavity Surface-Emitting Diode Lasers** *263*	
	Aram Mooradian, Andrei Shchegrov, Ashish Tandon, and Gideon Yoffe	
7.1	Introduction *263*	
7.2	Device Design and Performance *264*	
7.3	Mode Control, Cavity Design, and Thermal Lensing *270*	
7.4	High-Power Arrays and Multielement Devices *274*	
7.4.1	Design of the Chip *276*	
7.5	Carrier Dynamics *286*	
7.5.1	Mode Locking *287*	
7.6	Nonlinear Optical Conversion with Surface-Emitting Diode Lasers: Design and Performance *287*	
7.6.1	Visible Laser Sources: Applications and Requirements *287*	
7.6.2	Cavity Design Optimization and Trade-Offs for Second Harmonic Generation with Surface-Emitting Diode Lasers *289*	
7.6.3	Nonlinear Crystals Used in Intracavity Frequency Conversion *291*	
7.6.4	Low-Noise, High Mode Quality, Continuous-Wave Visible Laser Sources for Instrumentation Applications *294*	
7.6.5	Compact Visible Sources Scalable to Array Architecture *297*	
	References *301*	

Index *305*

Preface

Semiconductor disk lasers (SDLs), also known as vertical-external-cavity surface-emitting lasers (VECSELs), are a relatively new type of quantum electronic device. This book presents impressive state-of-the-art achievements in the field. The unique feature of SDL systems is their successful combination of advanced mode control using extended cavities and power scalability. SDLs incorporate the positive traits inherent in the disk laser concept, that is, low thermal lensing and excellent heat dissipation properties, which are further enhanced by the principal advantage of semiconductor lasers – the versatility afforded by the wide portfolio of available semiconductor quantum-confined compounds, including quantum dot media. Among various attractive features of the SDL concept are the potential for high repetition rate short pulse generation and efficient intracavity frequency conversion, combined with high-power, diffraction-limited output beams.

We should, however, expect further progress in both mid-infrared and visible SDLs, the latter still waiting for powerful short-wavelength pump diodes. It is not necessary to be especially expert or farsighted to predict a bright future for SDL technology; the demand for these lasers already exists, and hence will be launched in practice.

It is my honor and pleasure to introduce this book, written by internationally renowned experts in the field actively engaged in various aspects of SDL development. This is the first time researchers have joined forces to overview their achievements in a single volume and to speculate on the future perspectives of SDL technology.

The book includes the basic concepts, technology, and some applications of semiconductor disk lasers. It is useful for scientists and engineers working in the field as well as for graduate students interested in pursuing careers related to vertical-cavity lasers.

I wish to thank all the authors for their contribution to this edition. A special thanks goes to the staff of Wiley-VCH Verlag GmbH & Co. for their drive to keep the book on track for publication.

September 2009 *Oleg G. Okhotnikov*
Tampere

Semiconductor Disk Lasers. Physics and Technology. Edited by Oleg G. Okhotnikov
Copyright © 2010 WILEY-VCH Verlag GmbH & Co. KGaA, Weinheim
ISBN: 978-3-527-40933-4

List of Contributors

Dieter Bimberg
Technische Universität Berlin
Institut für Festkörperphysik
Hardenbergstr. 36
10623 Berlin
Germany

David Burns
University of Strathclyde
Institute of Photonics
106 Rottenrow
Glasgow G4 0NW
UK

Stephane Calvez
University of Strathclyde
Institute of Photonics
106 Rottenrow
Glasgow G4 0NW
UK

Martin D. Dawson
University of Strathclyde
Institute of Photonics
106 Rottenrow
Glasgow G4 0NW
UK

Mircea Guina
Tampere University of Technology
Optoelectronics Research Centre
Korkeakoulunkatu 3
33720 Tampere
Finland

Jennifer E. Hastie
University of Strathclyde
Institute of Photonics
106 Rottenrow
Glasgow G4 0NW
UK

John-Mark Hopkins
University of Strathclyde
Institute of Photonics
106 Rottenrow
Glasgow G4 0NW
UK

Alan J. Kemp
University of Strathclyde
Institute of Photonics
106 Rottenrow
Glasgow G4 0NW
UK

Ursula Keller
ETH Zurich
Institute of Quantum Electronics
Physics Department
Wolfgang-Pauli-Strasse 16
8093 Zurich
Switzerland

Mark Kuznetsov
Axsun Technologies
1 Fortune Drive
Billerica, MA 01821
USA

Nicolas Laurand
University of Strathclyde
Institute of Photonics
106 Rottenrow
Glasgow G4 0NW
UK

Deran J.H.C. Maas
ETH Zurich
Institute of Quantum Electronics
Physics Department
Wolfgang-Pauli-Strasse 16
8093 Zurich
Switzerland

Aram Mooradian
1240 Heartwood Drive
Rohnert Park, CA 94928
USA

Oleg Okhotnikov
Tampere University of Technology
Optoelectronics Research Centre
Korkeakoulunkatu 3
33720 Tampere
Finland

Udo W. Pohl
Technische Universität Berlin
Institut für Festkörperphysik
Hardenbergstr. 36
10623 Berlin
Germany

Marcel Rattunde
Fraunhofer-Institut für Angewandte
Festkörperphysik
Tullastrasse 72
79108 Freiburg
Germany

Benno Rösener
Fraunhofer-Institut für Angewandte
Festkörperphysik
Tullastrasse 72
79108 Freiburg
Germany

Andrei Shchegrov
Spectralus Corporation
2953 Bunker Hill Lane, Suite 400
Santa Clara, CA 95054
USA

Thomas Südmeyer
ETH Zurich
Institute of Quantum Electronics
Physics Department
Wolfgang-Pauli-Strasse 16
8093 Zurich
Switzerland

Ashish Tandon
Stion Corporation
6321 San Ignacio Avenue
San Jose, CA 95119
USA

Joachim Wagner
Fraunhofer-Institut für Angewandte
Festkörperphysik
Tullastrasse 72
79108 Freiburg
Germany

Gideon Yoffe
Kaiam Corporation
39655 Eureka Drive
Newark, CA 94560
USA

1
VECSEL Semiconductor Lasers: A Path to High-Power, Quality Beam and UV to IR Wavelength by Design
Mark Kuznetsov

1.1
Introduction

Since its invention and demonstration in 1960, several types of laser have been developed, such as solid-state, semiconductor, gas, excimer, and dye lasers [1]. Today, lasers are used in a wide range of important applications, particularly in optical fiber communication, optical digital recording (CD, DVD, and Blu-ray), laser materials processing, biology and medicine, spectroscopy, imaging, entertainment, and many others. A number of properties enable the application of lasers in these diverse areas, each application requiring a particular combination of these properties. Some of the most important laser properties are laser emission wavelength; output optical power; method of laser excitation, whether by optical pumping or electrical current injection; laser power consumption and efficiency; high-speed modulation or short pulse generation ability; wavelength tunability; output beam quality; device size; and so on. Thus, optical fiber communication [2], a major application that enables modern Internet, commonly requires lasers with emission wavelengths in the 1.55 μm low-loss band of glass fibers and with single-transverse mode output beams for coupling into single-mode optical fibers. Typically, a given laser type excels in some of these properties, while exhibiting shortcomings in others. For example, by using different material compositions and structures, the most widely used semiconductor diode laser [3–12] can cover a wide range of wavelengths from the ultraviolet (UV) to the mid-IR, can be advantageously driven by diode current injection, and is very compact and efficient. However, the good beam quality, that is, single-transverse mode near-circular beam operation, can be typically achieved in semiconductor lasers only for output powers below 1 W. Much higher power levels are achievable from semiconductor lasers only with large aspect ratio highly multimoded poor quality optical beams. On the other hand, the solid-state lasers [13, 14], including fiber lasers [15], can emit hundreds of watts of output power with excellent beam quality, however, their emission wavelengths are restricted to discrete values of electronic transitions in ions, such as the classic 1064 nm wavelength of the Nd:YAG laser, making them

---

Semiconductor Disk Lasers. Physics and Technology. Edited by Oleg G. Okhotnikov
Copyright © 2010 WILEY-VCH Verlag GmbH & Co. KGaA, Weinheim
ISBN: 978-3-527-40933-4

inapplicable for applications requiring specific inaccessible wavelengths. For example, the 488 nm excitation wavelength required for many fluorescent labels in biomedical applications [16], such as the green fluorescent protein (GFP), is not accessible by direct solid-state laser transitions. Therefore, the 488 nm wavelength biomedical applications have required in the past the use of large and inefficient Ar gas lasers, which serendipitously have the required emission wavelength. This explains the large variety of laser types used today, where one or another type fits a given application with its beneficial properties, while carrying the baggage of its undesirable properties.

It is therefore useful and important to develop a laser that exhibits simultaneously the application required and desired laser properties, such as emission wavelength, optical power, beam quality, efficiency, compact size, and so on. Vertical-external-cavity surface-emitting laser (VECSEL) [17–24], also called optically pumped semiconductor laser (OPSL) or semiconductor disk laser (SDL), is a relatively new laser family that uniquely combines many of these desirable laser properties simultaneously, and because of this, it is becoming the laser of choice for a wide range of laser applications. This chapter describes VECSEL lasers and their history; discusses how they are made and characterized; explains how VECSEL structure enables their basic properties; and indicates key applications enabled by this unique combination of properties. Other chapters in this book address in more detail the various aspects and applications of this remarkable new class of lasers.

1.2
What Are VECSEL Semiconductor Lasers

1.2.1
History of VECSELs: Semiconductor Lasers, Optical Pumping, and External Cavity

Vertical-external-cavity surface-emitting lasers were developed in the mid-1990s [17, 18] to overcome a key problem with conventional semiconductor lasers: how to generate watt-level and higher optical powers with fundamental transverse mode circular optical beam quality. The versatile semiconductor diode lasers are very widely used because of their numerous advantageous properties, such as size, efficiency, electrical current laser excitation and modulation, and wide wavelength coverage. Using GaN, GaAs, InP, and GaSb semiconductor material systems, for example, these lasers can access 0.4, 0.8, 1.5, and 2.0 μm emission wavelength regions. However, obtaining lasers with both high optical power and good beam quality simultaneously has always been a difficult task, although it is key for many important scientific and commercial laser applications. Such combination is required, for example, for efficient nonlinear optical second harmonic generation [14, 25].

The conventional semiconductor lasers have two major configurations: edge-emitting [3–6] and surface-emitting lasers [9–11] (see Figure 1.1). The edge-emitting

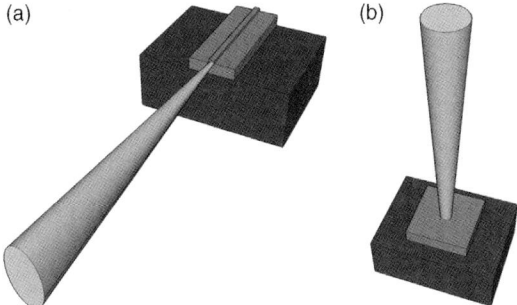

Figure 1.1 (a) Semiconductor edge-emitting laser. (b) Semiconductor vertical-cavity surface-emitting laser (VCSEL).

lasers use a waveguide to confine light to the plane of the semiconductor chip and emit light from the edge of the chip (Figure 1.1a). Output beam cross section is typically about one by several microns, with the wider dimension in the plane of the chip. Such small waveguide dimensions are required for single-transverse mode operation, but result in the asymmetric and strong angular divergence of the laser beam. Laser output power is typically limited by the required excess heat dissipation from the chip active region or catastrophic optical damage at the semiconductor surface [9, 12]. Scaling up laser output power requires wider waveguides with larger area beams: this improves heat dissipation by reducing active stripe thermal impedance and avoids catastrophic optical damage by decreasing beam optical intensity. In this way, up to several hundred milliwatts of output power is achievable in a single-transverse mode waveguide configuration [9, 12]. For still wider waveguides, of the order of a 100 μm, single-stripe edge-emitting lasers can emit tens of watts of output power, but the waveguide is then highly multimoded in the plane of the chip, and output beam is very elongated with a very large, 100: 1, aspect ratio. Multiple stripe semiconductor laser bars can emit hundreds of watts, but again with a highly multimoded output beam [9, 12].

In contrast, vertical-cavity surface-emitting lasers [10, 11] have laser cavity axis and emit light perpendicular to the plane of the laser chip (Figure 1.1b). Such lasers can emit circular fundamental transverse mode beam with powers up to several milliwatts and beam diameter of several microns. With circular cross section and larger beam size, the laser output beam is also symmetrical and has much smaller divergence than for edge-emitting lasers. Again, the required heat dissipation limits the output power and the scaling to higher powers demands larger active areas. But for output beam diameters greater than about 10 μm, laser output beam quickly becomes multimoded, and uniform current injection over such large areas is difficult with edge injection through transparent contact layers. Arrays of semiconductor lasers have been a typical path to high output power [12, 25]. In short, surface-emitting lasers have good fundamental mode circular beams, but at powers of only a few milliwatts, while edge-emitting lasers can emit up to several hundred milliwatts but with elliptical beam profile. For still higher powers, both laser types emit highly transverse multimoded output beams. In short, high power and good beam quality cannot be achieved simultaneously with conventional edge- or surface-emitting semiconductor lasers.

Two things become clear from the above description of semiconductor lasers. First, scaling up optical power to watt and higher levels with circular output beams requires beam diameters of tens and possibly hundreds of microns, which can be satisfied only by surface-emitting laser geometry. Second, good beam quality with fundamental transverse mode operation requires strong transverse mode control of the laser cavity. Such transverse mode control can be provided by optical cavity elements external to the laser chip, which assure that fundamental transverse mode of the laser cavity, the desired operating laser mode, has diameter approximately equal to the gain region diameter. In this way we arrive to the concept of vertical-external-cavity surface-emitting laser.

When the beam diameter of a surface-emitting laser becomes tens of microns large and the laser cavity is extended by an external optical element, the issue of laser excitation acquires additional importance. Injecting carriers uniformly across a wide area is difficult in the traditional diode current injection [10]; this requires a thick doped semiconductor current spreading layer. Such a doped layer has strong free carrier absorption inside the extended laser cavity, which can degrade laser threshold and efficiency. One possible solution to this problem is the use of optical pumping, which can inject excitation carriers uniformly across a wide area without using intracavity lossy doped regions. Simple and efficient semiconductor diode pump lasers with multimode beams and very high powers have been developed and are available for pumping solid-state and fiber lasers. VECSEL lasers have been made with both types of excitation, optical pumping and diode current injection. To emphasize their distinction from the common semiconductor diode lasers that use electrical pumping, optically pumped VECSELs are frequently referred to as OPSLs or optically pumped semiconductor lasers.

External optical cavity elements had been used previously with semiconductor lasers. For edge-emitting lasers, external reflectors provide a longer laser cavity for pulse repetition rate control in mode locking [27] and for inserting intracavity optical elements, such as spectral-filtering gratings [27]. There had also been attempts to stabilize transverse modes of surface-emitting lasers using external spherical mirrors [28].

Optical pumping of semiconductor lasers has a long history, where optical pumping had been used not only for characterization of novel semiconductor laser structures but also for generation of higher output powers or for short pulse generation. As early as in 1973, pulsed operation was demonstrated with optically pumped edge-emitting GaAs semiconductor lasers [29]. Later, surface-emitting thin-film InGaAsP lasers [30] were used to generate gain-switched picosecond pulses in the 0.83–1.59 μm wavelength range using dye laser pumping. Using an external optical cavity for pulse repetition rate and transverse mode control, optically pumped mode locking was demonstrated with a CdS platelet laser [31]. High peak power was observed in an external-cavity GaAs platelet laser pumped by a Ti:sapphire laser [32].

Using diode laser pumping, low-power 10 mW CW operation was demonstrated with GaAs VCSEL lasers [33]; in external cavity, however, such lasers emitted only 20 µW [34]. A diode-laser-pumped surface-emitting optical amplifier was demonstrated at 1.5 µm using InGaAs–InGaAlAs multiquantum well structures [35]. Using 77 K low temperature operation and a Nd:YAG pump laser, 190 mW continuous output power was obtained from an external-cavity InGaAs–InP surface emitting laser [36]. In a similar configuration, an external-cavity GaAs VCSEL laser at 77 K has demonstrated CW output power of 700 mW using a 1.8 W krypton–ion pump laser [37]. To obtain high power from a diode-laser-pumped semiconductor laser, specially designed edge-emitting InGaAs–GaAs laser structures were used to generate as much as 4 W average power [38, 39], however the beams were strongly elongated with aspect ratios between 10 and 50 to 1. These works had demonstrated the potential capabilities of the optically pumped semiconductor lasers; however, the goal of a high-power compact and efficient diode-pumped room-temperature laser with circular diffraction-limited beam profile had remained elusive prior to OPS-VECSEL demonstration in 1997 [17]. What enabled the appearance of the modern VECSEL lasers is the availability of sophisticated custom-designed multilayered bandgap-engineered semiconductor structures, modern high-power multimode semiconductor pump lasers, and thermal designs for efficient heat dissipation from the active semiconductor chip.

Figure 1.2 shows basic configuration of an optically pumped VECSEL. A thin active semiconductor chip, containing gain region and multilayer high-reflectivity mirror, is placed on a heat sink and is excited by an incident optical pump beam. Laser cavity consists of the on-chip mirror and an external spherical mirror, which defines the laser transverse mode and also serves as the output coupler. Typical laser beam diameters on the gain chip range between 50 and 500 µm; VECSELs have been made

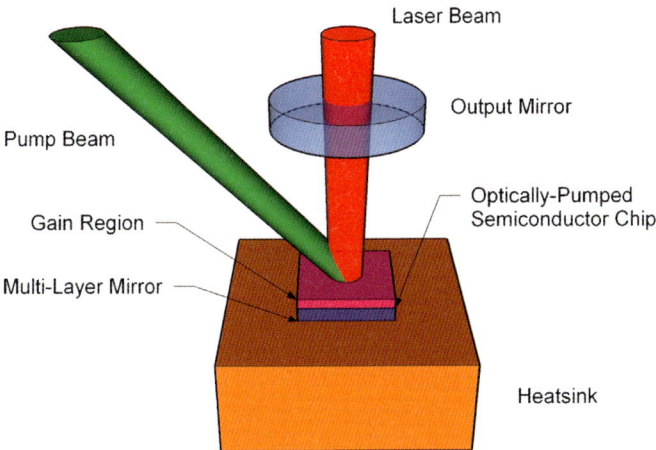

Figure 1.2 Optically pumped semiconductor vertical-external-cavity surface-emitting laser (VECSEL).

with output powers ranging from 20 mW to 20 W and higher. Optically pumped VECSEL can be thought of as a brightness or mode converter, converting a high-power low-quality multimode pump beam with poor spatial and spectral brightness into a high-power high-quality fundamental transverse mode laser output beam with the desired spatial and spectral properties. In this way an optically pumped VECSEL is similar to solid-state and fiber lasers [13–15], which similarly act as brightness or mode converters. Indeed, an optically pumped VECSEL can be thought of as a solid-state laser, where the gain medium, instead of the traditional active ions in a transparent host material, uses bandgap-engineered semiconductor structures to achieve the desired laser absorption and emission properties. Just as evolution of semiconductor lasers to high power and good beam operation has arrived at the VECSEL laser configuration, diode pumped solid-state DPSS lasers have arrived at the very similar solid-state disk laser configuration [40, 41], which has demonstrated kilowatt-level output powers. In such a solid-state disk laser, with a geometry similar to that in Figure 1.2, a thin solid-state gain medium, such as a Yb:YAG crystal, with a thin-film high-reflectivity mirror coating is placed directly on a heat sink with external spherical mirror stabilizing the cavity transverse mode and diode optical pumping providing laser excitation. An important benefit of using semiconductors, in contrast to other solid-state gain media, is that the on-chip multilayer mirror can be made of alternating different composition semiconductor layers and can be grown in the same epitaxial growth step as the gain region itself. Externally deposited mirror on the semiconductor laser chip can also be used. Because of their similarity to the solid-state disk lasers, VECSELs have also been referred to as semiconductor disk lasers or SDLs. Optically pumped VECSELs form a hybrid between traditional semiconductor and solid-state lasers, hence the interest in these lasers has come from both of these laser communities. For high-power good beam quality operation with wavelength versatility, such optically pumped VECSEL lasers have many significant advantages compared to both the traditional semiconductor diode lasers and the traditional solid-state lasers, including disk lasers.

1.2.2
Basic Principles of Operation: VECSEL Structure and Function

Basic operating principles of VECSEL lasers are illustrated in Figure 1.3. The key element of the laser is the semiconductor chip, which contains both a multilayer laser mirror and a gain region; Figure 1.3 shows the conduction and valence band energy levels across the semiconductor layers and explains the functions of the various layers. For optically pumped operation, incident pump photons with higher photon energy are absorbed in separate pump-absorbing layers that also serve as the quantum well barriers. The excited carriers, electrons and holes, then diffuse to the smaller bandgap quantum wells that provide gain to the optical wave, emitting lasing photons with lower photon energy. These separate pump absorption and quantum well laser emission layers allow independent optimization of the pump absorption and laser gain properties. For optically pumped VECSEL operation,

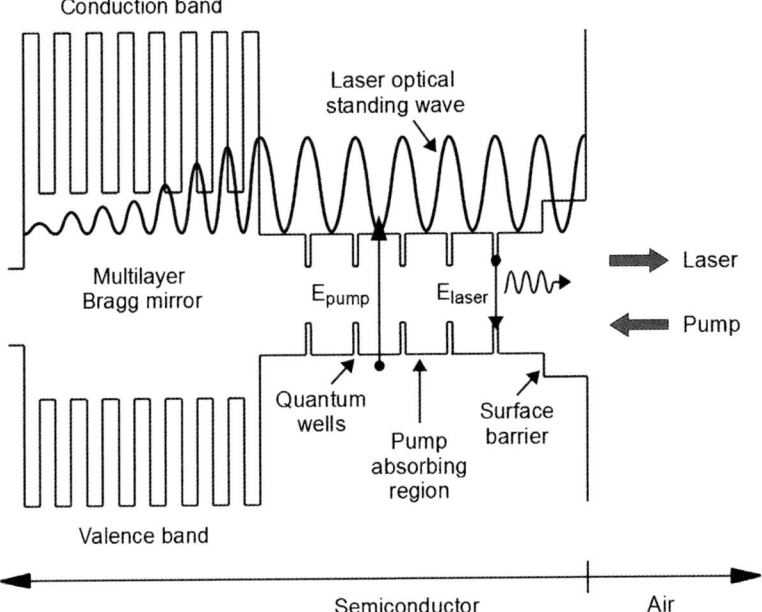

Figure 1.3 Operating principles of optically pumped VECSELs.

semiconductor layers are typically undoped, thus significantly simplifying semiconductor wafer growth and eliminating free carrier absorption of the doped regions. For electrically pumped operation, p- and n-doped regions are used to form a p–n junction for diode current carrier injection, but this also results in optical losses inside the laser cavity. A higher bandgap surface barrier window layer on the chip prevents carriers from diffusing to the semiconductor–air interface, where they could recombine nonradiatively and thus deplete laser gain. Optical wave of the laser mode back-reflecting from the on-chip laser cavity mirror sets up an intracavity standing wave inside the chip. Quantum wells have to be placed near the antinodes of this standing wave in order to provide efficient gain to the laser. This is the so-called resonant periodic gain (RPG) arrangement [42]; one or more closely spaced wells can be placed near a given standing wave antinode. Typically, gain region thickness covers several periods of this laser mode standing wave.

Incident pump photons have higher energy than the emitted laser photons, the difference of the two photon energies is the quantum defect. This quantum defect is one of the major contributors to the overall laser operating efficiency; this pump–laser photon energy difference, together with contributions from other lasing inefficiencies, has to be dissipated as heat from the device active region. Heat dissipation from the VECSEL active semiconductor chip is provided by heat spreaders connected to heat sinks: either a soldered heat spreader below the mirror structure or a transparent heat spreader above the surface window of the chip, or possibly both (Chapter 2).

Good heat dissipation and heat sinking are critical for high-power operation of all semiconductor lasers. Without these, temperature of the active region would rise and excited carriers would escape thermally from the quantum wells into the barrier region, thus depleting laser gain and turning the laser off in a thermal rollover process. Such thermal rollover is typically the dominant mechanism that limits output power in VECSEL lasers [43]. Smaller quantum defect produces less excess heat, but typically also implies smaller energy difference, or confinement energy, between electron and hole states in the wells and the barriers, making it easier for electrons to escape thermally from the wells into the barriers, and thus making lasers more sensitive to the temperature rise. Optimization of the quantum defect and electron confinement energy is required for high-power room temperature device operation.

Optical absorption in semiconductors is very strong for pump photon energies above the bandgap, of the order of $10^4\,cm^{-1} = 1\,\mu m^{-1}$. This means that $\sim$63% of pump light is absorbed on a single pass through 1 μm thick semiconductor absorbing layer, 86% is absorbed in 2 μm. In most cases, single-pass pump absorption is sufficient; a pump-reflecting mirror can be included on the chip if double-pass absorption is desired. A very simple pump-focusing optics can be used, since multimode pump light does not have a chance to diverge in a few microns before it is absorbed; no depth of focus is required for pump optics and high brightness is not required of the multimode pump sources. Compare this with $\sim$7 cm^{-1} absorption in Yb:YAG, a typical active medium in solid-state disk lasers. Such a thousand times weaker absorption requires a much thicker absorbing region, 100–300 μm, and, in addition, multiple pump beam passes for efficient pump absorption, with correspondingly complex pump optics to handle divergent multimode pump beams on multiple absorption passes [40, 41]. In an attempt to reduce quantum defect and improve efficiency in VECSEL lasers, in-well, rather than barrier, optical pumping has also been used for these devices [44–48]. Another important advantage of optically pumped semiconductor gain medium is its spectrally broad absorption and hence tolerance of broad pump wavelength variation. Essentially, any pump wavelength is useful that is shorter than the absorber region bandgap wavelength. Therefore, tight wavelength selection and temperature control of pump diode lasers are not required, unlike the case for solid-state and fiber lasers. Diode laser pumping of VECSELs also offers fast direct VECSEL modulation capability via pump laser current modulation, since VECSEL semiconductor gain medium has short, sub-nanosecond, carrier lifetimes, as compared with microseconds to milliseconds lifetimes of typical solid-state gain media.

Since laser optical axis is perpendicular to the surface of the gain chip and quantum well gain layers are very thin, the single-pass optical gain is at most only a few percent. This means that external output coupling mirror transmission should also be of the order of a few percent and the on-chip mirror reflectivity should be as high as possible, say greater than 99.9%. Intracavity losses should also be kept very low, less than a percent, in order to maintain efficient laser operation.

1.2.3
Basic Properties of VECSEL Lasers: Power Scaling, Beam Quality, and Intracavity Optical Elements

Basic configuration of VECSEL lasers enables their many key advantageous properties, such as power scaling, beam quality, and laser functional versatility; in this section, we describe these connections between VECSEL laser structure and device functionality.

1.2.3.1 Power Scaling

One of the key important properties of VECSEL lasers is their output power scalability: efficient research and commercial optically pumped devices have been demonstrated with power levels between 10 mW and 60 W, a range of almost four orders of magnitude, while maintaining good beam quality. Such efficient power scalability is enabled by the laser mode and pump spot-size scalability on the VECSEL semiconductor laser chip. Since output power of semiconductor lasers is typically limited by heat dissipation and optical intensity-induced damage, increasing beam diameter in a VECSEL helps on both accounts, distributing heat and optical power over larger beam area. For well-designed heat sinking with thin semiconductor chips, heat flow from the laser active region into heat sink is essentially one-dimensional. Therefore, increasing beam area is essentially equivalent to operating multiple lasing elements in parallel, without changing thermal or optical intensity regime of the individual lasing elements. In this scenario, both output laser power and pump power scale linearly with the active area. VECSELs have been operated with on-chip beam diameters between 30 and 900 µm, which scale the beam area and potentially output power by a factor of 900. For such power scaling, the same semiconductor wafer and chip structure can be used, adjusting only the laser cavity optics and pump laser arrangement. Optical pumping allows simple uniform excitation of such widely scalable mode areas. In contrast, with electrical pumping, uniform carrier injection over hundreds of microns wide area is extremely difficult, typically leaving a weakly pumped region in the center of the active area and making power scaling of electrically pumped devices very challenging. Direct optical coupling of pump diode chips into the VECSEL chip is possible with relatively simple pump lens arrangements; alternatively, multi-mode fiber-coupled pump diode sources can also be used. If pump power available from a single pump diode is limited, multiple pump diodes can be used with multiple pump beams incident on a single VECSEL chip from different angles. When heat dissipation from a single semiconductor chip becomes the limiting factor, further scaling of optical power is possible by arranging multiple semiconductor gain chips within a single VECSEL laser cavity and reflecting the laser beam sequentially from these reflecting vertical amplifier chips [22, 49–52]. All of these factors combined make it possible to scale optically pumped VECSEL power by the demonstrated four orders of magnitude, and potentially more in the future.

1.2.3.2 Beam Quality

Another critically important property of VECSEL lasers is their beam quality: VECSELs operate with a circular beam, fundamental transverse TEM_{00} mode, and essentially diffraction-limited low beam divergence with $M^2 \sim 1.0-1.3$. Here, beam spatial quality parameter M^2 [9] indicates how much faster a laser beam diverges angularly in the two transverse directions as compared with a single-transverse mode diffraction-limited beam, which has $M^2 = 1$. Several factors contribute to this beam quality in VECSELs. Most important, VECSEL laser external-cavity optics defines and stabilizes the circular fundamental laser transverse mode; such optical elements and their stabilization effect are not available with the more conventional edge- and surface-emitting semiconductor lasers. Using pump and laser cavity optics, VECSELs have independent control allowing matching of the pump spot size and the laser fundamental transverse mode size. If the pump spot is too small, compared to the fundamental mode size, laser threshold will be high because of the lossy unpumped regions encountered by the laser mode. If the pump spot is too large, higher order transverse laser modes with a larger transverse extent will be excited, causing multimode laser operation and thus degraded beam quality. Optimally adjusted pump spot size gives preferentially higher gain to the fundamental laser mode, while giving excess loss from the unpumped regions to the spatially wider higher order transverse modes; this stabilizes the fundamental transverse mode operation. Large VECSEL laser beam and pump spot sizes on the chip, tens to hundreds of microns, as compared with just a few microns for edge-emitting semiconductor lasers, contribute to the ease of alignment and thus to the mechanical stability of the VECSEL laser cavity, and thus also to the stability of its fundamental transverse mode operation. Such a large VECSEL beam diameter also conveniently has very low divergence angles, as compared with extremely fast divergence of micron-sized beams of the edge-emitting, and even surface-emitting, lasers.

The second important factor contributing to spatial beam quality and stability of VECSEL lasers is the negligible thermal lensing in the thin, 2–8 μm, VECSEL semiconductor chip [22] when proper heat spreading/heat sinking is used. Thermal lensing and other beam phase profile distortions are caused by thermally induced refractive index gradients in the laser gain material. In VECSELs, thin semiconductor active region with good heat sinking implies that optical path length thermal distortions and hence beam profile changes and distortions are negligible. In contrast, such thermal lensing is much stronger in solid-state lasers that require much longer, hundreds of microns, gain path length. As a result, thermal lensing typically forces optimal solid-state laser operation only within a narrow range of pump and output powers where the thermal gradients and the resulting thermal intracavity lens produce laser mode size consistent with the pumping profile. VECSEL semiconductor lasers operate efficiently and with excellent beam quality across a wide range of operating power regimes from near to high above threshold. There is some trade-off between output power and beam quality in VECSELs: a multimode laser beam can better overlap the pump spot and thus produce somewhat higher output power. Thus, when a few transverse modes can be tolerated with a somewhat degraded $M^2 \sim 3-4$ beam parameter, VECSELs have been operated in such regime with higher efficiency and higher output power than for single-transverse mode regime [22].

1.2.3.3 Laser Functional Versatility Through Intracavity Optical Elements

The external optical cavity of semiconductor VECSELs, which controls the laser transverse modes, can be viewed as mechanically cumbersome, making these lasers more complex and requiring assembly and alignment as compared with the simple integrated edge-emitting and surface-emitting semiconductor lasers. On the other hand, such an external cavity gives tremendous versatility to VECSEL device configurations and functions. Flexible VECSEL laser cavities, such as linear two-mirror cavity, three-mirror V-shaped cavity, and four-mirror Z-shaped cavity [18, 20, 49–54], allow flexible insertion of intracavity optical elements. Such intracavity functional elements are very difficult or impossible to use with integrated semiconductor devices. We have already discussed one example of such extended cavity versatility – inserting multiple gain elements in the cavity in series for power scaling of VECSELs.

One important option allowed by the external cavity is the insertion of intracavity spectral filters, such as etalons [55, 56], Brewster's angle birefringent filters [54, 57, 58], volume gratings (Chapter 7) [59], or high-reflectivity gratings [60], to control longitudinal spectral modes of the laser and possibly to select a single longitudinal lasing mode. Tuning a birefringent filter by rotation then achieves tunable VECSEL operation; greater than 20 nm tuning range with multiwatt output was demonstrated with this approach [61].

VECSEL external cavity also allows the insertion of intracavity saturable absorber elements to achieve laser passive mode locking with picosecond and subpicosecond pulse generation (Chapter 6) [20]. In this case, the length of the external cavity also allows control of the pulse repetition rates, with rates as high as 50 GHz demonstrated with short cavities [62]. External cavity optics also allows different beam spot sizes on the gain and absorber elements, which controls optical intensity of the beam spots and is typically required to achieve mode locking [20, 63, 64].

The open cavity of VECSELs allows placement of transparent intracavity heat spreaders in direct contact with the laser gain element without thermally resistive laser mirrors in the path of heat dissipation (Chapter 2) [53, 65–68]. Since thermal management is critical for high-power VECSEL operation, the possibility of using such heat spreaders tremendously broadens laser design options with chip gain, mirror, and substrate materials that do not allow effective heat removal through the on-chip mirror. Another option for VECSELs allowed by the external cavity is the microchip laser regime [69–72]. Here, an imperfectly heat-sunk semiconductor VECSEL chip produces an intracavity thermal lens that stabilizes laser transverse modes in a short external plane–plane laser cavity.

Low intracavity loss of VECSELs, combined with their wide gain bandwidth, allows insertion of intracavity absorption cells, such as gas cells, for intracavity laser absorption spectroscopy (ICLAS) [73, 74]. Laser output spectrum then reflects the absorption spectrum of the intracavity absorption cell. With intracavity real absorption length of the order of 1 m, equivalent laser intracavity absorption path length as long as 130 km has been produced [75], allowing sensitive measurements of extremely weak absorption lines.

VECSEL output power coupling with only a few percent of transmission implies that its intracavity laser power is 20–100 times higher than the output power. Availability of such high intracavity power, together with high beam quality, allows very efficient nonlinear optical operation, such as second harmonic generation, by inserting nonlinear optical crystals inside the external laser cavity (Chapter 3) [14, 22, 24, 53, 54, 58, 76]. Using intracavity second harmonic generation, VECSELs have provided efficient laser output at wavelengths not accessible by other laser materials and techniques (Chapter 3).

Several examples discussed here show the tremendous versatility of VECSELs allowed by its external cavity and the various intracavity functional elements; these allow VECSELs to operate in a wide variety of operating regimes with a correspondingly large variety of laser applications.

To summarize this section, some of the key properties of VECSELs that make them so useful, namely, output power scaling, beam quality, and functional device versatility, follow from the unique structure of these devices, including semiconductor chip, external laser cavity, and optical pumping configuration.

1.2.4
VECSEL Wavelength Versatility Through Materials and Nonlinear Optics

One of the most important properties of VECSEL lasers is their wavelength versatility: VECSELs have been made with output wavelengths ranging from 244 nm [77] and 338 nm [24, 78] in the UV; through the 460–675 nm range of blue, green, yellow, orange, and red in the visible (Chapter 3) [22, 24, 52–54, 58, 79]; through the 0.9–2.4 µm in the near-infrared (NIR) (Chapter 4) [18, 23, 67, 68, 80, 81]; to the 5 µm in the mid-IR [82–85]. In principle, any wavelength in this range is accessible by design. This is simply not possible with any other laser type, not with the power, beam quality, and efficiency available from VECSELs. In this section, we discuss how such wavelength versatility of VECSELs is made possible by using different semiconductor materials and structures, in combination with the use of efficient nonlinear optical conversion.

1.2.4.1 Wavelength Versatility Through Semiconductor Materials and Structures

Compound semiconductor materials have different bandgap energies, and thus different photon emission wavelengths, for different material compositions. Controlling bandgaps of multiple layers of semiconductor structures, including the quantum well photon-emitting layers, allows control of laser emission wavelength by design. Such bandgap control of semiconductor heterostructures is required for VECSEL structures as in Figure 1.3, which also includes mirror layers, pump absorbing layers, and so on. Over more than 40 years of compound semiconductor technology, several semiconductor material systems have been developed that allow reliable growth of such bandgap-engineered multilayer structures. Starting substrates for VECSEL structures are binary semiconductor wafers; lattice constants of the multiple layers of an epitaxially grown structure have to be closely matched to the substrate to avoid large strain and the resulting growth of crystalline defects that destroy laser operation. Epitaxial growth of ternary, quaternary, and even quinary semiconductor alloys has been developed, which allows independent control of semiconductor layer bandgap energy while maintaining lattice match to the substrate. Using group III–V semiconductor GaAs substrate material system with its ternary (e.g., InGaP, AlGaAs, InGaAs, GaAsP, GaAsSb), quaternary (e.g., InGaNAs, InAlGaAs), and quinary (e.g., InAlGaAsP) alloys, VECSEL lasers have been demonstrated with emission wavelengths in the 660–1300 nm wavelength range [18, 23, 68, 79, 86–88]. InP-based material system using quaternary alloys (e.g., InGaAsP, InGaAlAs) allows VECSEL lasers to access the 1500–1600 nm optical fiber communication wavelength regions [23, 80, 89–93]. Starting with GaSb substrate with ternary (e.g., GaInSb, AlAsSb) and quaternary (e.g., GaInAsSb, GaAlAsSb) alloys, VECSELs with emission wavelengths in the 2.0–2.3 µm range have been demonstrated (Chapter 4) [23, 48, 81, 94]. Recently, group IV–VI semiconductor PbTe/PbEuTe and PbSe/PbEuTe-based material systems have been used to demonstrate VECSELs in the 4.5–5 µm wavelength range [82–85]. VECSELs have also been fabricated in the GaN/InGaN material system [95, 96], which opens the 400 nm wavelength region for direct VECSEL operation.

It is important to note that a lattice-matched semiconductor material system for VECSELs must allow not only the desired emission wavelength but also the creation of very high reflectivity on-chip Bragg mirror. To achieve such high reflectivity, high refractive index contrast is required between the high- and low-index mirror materials. High index contrast is available in the GaAs material system. However, such contrast is poor in the InP material system, requiring thicker mirrors with more layers and correspondingly larger thermal impedance, which is detrimental to laser operation. InP material system is important because of its access to the 1500–1600 nm telecom wavelength emission range. Improved VECSEL laser performance at these wavelengths has been achieved by bonding or fusing InP-based gain region wafers with high-reflectivity GaAs-based Bragg mirror wafers [97, 98]. No lattice matching is required for this wafer fusion approach, thus broadening the choices available for laser emission wavelength materials and mirror materials.

The use of dimensionally, or quantum, confined semiconductor active regions [4, 8] allows further control of the VECSEL emission wavelengths. By adjusting the thickness of quantum wells (2D) or diameter of quantum wires (1D) and dots (0D), as well as the composition of confining barrier layers, the quantum-confined electron and hole energy levels are shifted and VECSEL designer acquires the additional fine control of laser emission wavelength (Chapter 5) [4, 8, 18, 87, 99–102]. Using a controlled amount of strain [4, 8, 18, 87] in the quantum-confined light-emitting regions further expands the range of available material compositions and thus emission wavelengths for a semiconductor material system. Thus, for example, the strained InGaAs on GaAs material system is used very successfully for light emission in the 950–1150 nm wavelength region [43, 68, 87].

Such diversity and flexibility of alloyed compound semiconductor material systems allows designing VECSELs with direct laser emission in the wide 0.6–5 µm wavelength range [18, 23, 79]; many of these emission wavelengths have already been demonstrated, both in research and in commercial devices.

Semiconductor gain media, especially with engineered quantum-confined quantum well and quantum dot structures, can have very large gain bandwidths, from tens to more than a hundred nanometers in wavelength [4–9]. Using intracavity tunable filters, tunable VECSELs have been demonstrated, for example, with 80 nm near 2.0 μm [103], 30 nm near 1175 nm [58], 33 nm near 975 nm [51], 30 nm near 850 [66], and 10 nm near 674 nm [79]. Such tunability is useful for such laser applications as spectroscopy. At the same time, the use of intracavity tunable filters allows to set VECSEL output wavelength to a specific precise stable value required for a given application. This ability to set VECSEL output wavelengths has proven very valuable in certain applications, such as exciting marker fluorescent proteins in biological imaging applications, where specific excitation wavelengths, for example, 488 nm, are most efficient for a given marker protein [16]. Using intracavity filters, VECSELs have also been operated in the single longitudinal mode, or single-frequency regime [104], important for some applications, such as spectroscopy. In this way, VECSELs, by design, access not only a wide output wavelength range but also very specific desired wavelengths at arbitrary locations within that range.

An optically pumped laser is not very useful if a suitable pump laser is not easily available; this is frequently the case with solid-state lasers, which require pump lasers with a very narrow, only several nanometers wide, range of pumping wavelengths. An important factor that makes the wide VECSEL emission wavelength range possible is the wide pump wavelength range, tens to hundreds of nanometers wide, allowed by optically pumped VECSELs. In practice, this means that a desired emission wavelength can be achieved in VECSELs while using widely available efficient pump diode lasers at commonly available pump wavelengths. For example, the 808 nm wavelength pump lasers, the standard pumping wavelength for Nd:YAG solid-state lasers and thus wavelength where pump diodes are easily available, have been used for VECSEL lasers emitting in the 920–1300 nm wavelength range [18, 67, 68, 87]. The common 980 nm pump lasers, the standard wavelength for pumping Er-doped fiber amplifiers, have been used to pump 1550 nm VECSELs [97]. Pump lasers at 790, 808, 830, and 980 nm have been used to pump 2.0–2.3 μm lasers [81, 94, 105, 106]. Pump wavelength of 1.55 μm has been used for pumping 5 μm lasers [82–85]. Thus, pump wavelength versatility is an important contributor to the emission wavelength versatility in VECSELs.

1.2.4.2 Wavelength Versatility Through Nonlinear Optical Conversion

Perhaps the biggest contributor to the wavelength versatility of VECSEL lasers, as well as to their commercial success so far, has been efficient nonlinear optical conversion possible with VECSELs. Nonlinear optical conversion uses a nonlinear optical crystal to generate light at harmonics, as well as sum or difference frequencies of the incoming light beams (Chapter 3) [14, 24, 25]. The most common is the second harmonic generation SHG process, where light is generated at twice the frequency or half the wavelength of the fundamental laser emission. For example, a very useful 488 nm wavelength visible blue output can be produced as a second harmonic of the fundamental 976 nm near-infrared laser emission. The nonlinear optical processes can be further cascaded to generate third, fourth, and so on harmonics of the fundamental input light frequency. Using this approach, fundamental wavelengths in the near-infrared between 0.8 and 1.3 μm, which are more readily accessible directly by semiconductor VCSELS, have been converted efficiently to the 0.24–0.65 μm ultraviolet and visible, including blue, green, yellow, orange, and red, wavelength range (Chapter 3) [22, 24, 53, 54, 57, 58, 78]. Nonlinear optical conversion has tremendously broadened the wavelength range accessible by VECSELs and made efficient light sources available at wavelengths that previously had been accessible only by inefficient gas lasers, such as Ar laser at 488 nm for fluorescent marker applications, or where no effective light sources had been available at all, such as 577 nm yellow wavelength for photocoagulation treatment in ophthalmology [57, 58].

Efficient nonlinear optical conversion requires high optical intensity; this is provided by high optical power and good beam quality, which allows focusing laser beam to small diameters. High optical power and good beam quality are precisely the fundamental properties of VECSEL lasers, which make them very efficient sources for nonlinear optical conversion. The efficiency of nonlinear optical conversion increases with the increase in optical intensity. Even though the fundamental output power of VECSELs is high, their intracavity power is still much higher. With typical output coupling mirror in VECSELs having a transmission of between 1 and 5%, intracavity laser power is remarkably 20–100 times higher than the output power. Thus, a 20 W output power VECSEL with output coupling transmission of 0.7% has an intracavity power of 2.8 kW [107], while beam quality $M^2 \sim 1.1$ is also very high. Such high VECSEL intracavity powers allow the use of very efficient intracavity nonlinear optical conversion (Chapter 3) [14, 24, 25], which because of the higher intracavity power can be much more efficient than similar conversion done outside the laser cavity. For intracavity SHG, for example, a nonlinear optical crystal is inserted inside the laser cavity and a dichroic laser output mirror has 100% reflectivity at the fundamental laser wavelength and 100% transmission at the second harmonic frequency; laser cavity output is then emitted at the second harmonic frequency (Chapter 3). Such nonlinear optical processes can be further cascaded inside the laser cavity to produce third and fourth harmonic laser output [108], producing, for example, 355 nm third harmonic UV radiation from the fundamental 1065 nm near-infrared laser emission. Alternatively, intracavity-doubled laser output can then be further doubled in frequency outside the laser cavity; in this way, 244 nm UV output has been produced from the 976 nm fundamental laser wavelength [77].

Another approach to extend wavelengths accessible by VECSEL lasers is to operate these lasers in a dual wavelength mode with intracavity nonlinear optical sum or difference frequency generation [109–113]. Also possible is intracavity sum frequency generation in a VECSEL laser with externally injected solid-state laser beam [114]. Using such intracavity difference frequency generation, VECSEL laser becomes a room temperature source of 4–20 μm wavelength mid- to far-infrared radiation. Still longer wavelength terahertz radiation, 0.1–2.0 THz or 150–3000 μm wavelength, can be generated using short-pulse mode-locked VECSELs [115]. In this approach, very short 260–480 fs optical pulses incident on a photoconductive antenna produce terahertz radiation with bandwidth inversely proportional to the pulse width. A similar short pulse-driven photoconductive antenna is used for time domain detection of the terahertz radiation [115].

In summary, VECSEL lasers can access an extraordinarily wide range of wavelengths from the UV, through the visible and near-infrared, to mid- and far-infrared, and even to the terahertz frequency range. Two key factors are the source of such a wide wavelength range: flexibility of semiconductor material systems and structures in combination with nonlinear optical conversion techniques. More important, in contrast with other laser systems, such as gas- and solid-state laser systems, which emit only at discrete wavelengths of existing electronic transitions of active ions, semiconductor VECSELs can generate light by design within this wide range at essentially arbitrary specific target wavelengths required for different applications.

1.3
How Do You Make a VECSEL Laser

Now that we have described what VECSEL lasers are, their basic operating principles and fundamental properties that make them so uniquely useful, in this section we outline the key elements in designing and making VECSELs. First, we describe the design of basic building blocks of VECSELs, see Figures 1.2 and 1.3: gain medium, on-chip Bragg mirror, laser optical cavity, and optical pumping arrangement. Finally, we describe VECSEL laser characterization. Note that here we describe the design principles of the more common optically pumped version of VECSELs; the electrically pumped VECSEL is described in detail in Chapter 7 of this book; the two differently pumped versions share most of the design principles that are not related to pumping.

1.3.1
Semiconductor Gain Medium and On-Chip Bragg Mirror

The main component of VECSEL lasers is the semiconductor laser chip, which includes the semiconductor gain medium and the laser-cavity multilayer Bragg mirror. First, we address the design of the semiconductor gain medium; this will tell us the key design and operational parameters of VECSEL lasers: the required gain level and number of quantum wells, laser mirror reflectivities required, laser threshold and operational pump powers, and laser output power and efficiency. We then describe the semiconductor mirror design and show an example of a full VECSEL semiconductor wafer structure.

1.3.1.1 Semiconductor Gain Design for VECSELs
To model VECSEL lasers, we use a simple analytical phenomenological model of semiconductor quantum well gain, which then gives us a very useful analytical description of VECSEL laser design and operation [18]. The model here does not

include thermal considerations, which are very important in the laser design and are considered in detail in Chapter 2 of this book.

Semiconductor quantum well gain g, cm^{-1}, has an approximately logarithmic dependence on well carrier density N, cm^{-3},

$$g = g_0 \ln(N/N_0), \tag{1.1}$$

where g_0 is the semiconductor material gain parameter and N_0 is the transparency carrier density. VECSEL laser threshold condition states that intracavity optical field is reproduced upon a round-trip inside the cavity:

$$R_1 R_2 T_{loss} \exp(2\Gamma g_{th} N_w L_w) = 1, \tag{1.2}$$

where R_1 and R_2 are the cavity mirror reflectivities, T_{loss} is the transmission factor due to round-trip cavity loss, g_{th} is the threshold material gain, N_w is the number of quantum wells in the gain medium, and L_w is the thickness of a quantum well. Longitudinal confinement factor Γ [116] of this resonant periodic gain structure characterizes overlap between the intracavity optical standing wave and the quantum wells spaced inside the active region. Carrier density N below threshold can be calculated from the incident pump power P_p:

$$N = \frac{\eta_{abs} P_p}{h\nu_p (N_w L_w A_p)} \tau(N). \tag{1.3}$$

Here η_{abs} is the pump absorption efficiency, $h\nu_p$ is the pump photon energy, A_p is the pump spot area, and τ is the carrier lifetime. Carrier lifetime is given by

$$\frac{1}{\tau(N)} = A + BN + CN^2, \tag{1.4}$$

where A, B, and C are the monomolecular, bimolecular, and Auger recombination coefficients. From Eqs (1.1)–(1.4), we derive simple expressions for the threshold carrier density N_{th} and the threshold pump power P_{th}:

$$N_{th} = N_0 (R_1 R_2 T_{loss})^{-(2\Gamma g_0 N_w L_w)^{-1}}, \tag{1.5}$$

$$P_{th} = N_{th} \frac{h\nu(N_w L_w A_p)}{\eta_{abs} \tau(N_{th})}. \tag{1.6}$$

VECSEL output power is then given by

$$P_{las} = (P_p - P_{th})\eta_{diff}, \tag{1.7}$$

where laser differential efficiency η_{diff} is

$$\eta_{diff} = \eta_{out} \eta_{quant} \eta_{rad} \eta_{abs}. \tag{1.8}$$

The components of the differential efficiency are the output efficiency η_{out}:

$$\eta_{out} = \frac{\ln(R_2)}{\ln(R_1 R_2 T_{loss})}, \tag{1.9}$$

Table 1.1 Laser and material parameters used in the OPS-VECSEL laser threshold and output power calculations.

Parameter	Description	Value	Units
g_0	Material gain coefficient	2000	cm^{-1}
N_0	Transparency carrier density	$1.7 \times 10^{+18}$	cm^{-3}
Γ	RPG longitudinal confinement factor	2.0	—
L_w	Quantum well thickness	8.0	nm
R_1	On-wafer mirror reflectivity	0.999	—
T_{loss}	Round-trip loss transmission factor	0.990	—
λ_{laser}	Laser wavelength	980	nm
λ_{pump}	Pump wavelength	808	nm
d_{pump}	Pump spot diameter	100	µm
η_{abs}	Pump absorption efficiency	0.85	—
A	Monomolecular recombination coefficient	$1.0 \times 10^{+7}$	s^{-1}
B	Bimolecular recombination coefficient	1.0×10^{-10}	$cm^3 s^{-1}$
C	Auger recombination coefficient	6.0×10^{-30}	$cm^6 s^{-1}$

where R_2 is the laser output mirror reflectivity; the quantum-defect efficiency η_{quant}:

$$\eta_{quant} = \frac{\lambda_{pump}}{\lambda_{laser}}, \tag{1.10}$$

given by the ratio of pump λ_{pump} and laser λ_{laser} wavelengths; and the radiative efficiency η_{rad}:

$$\eta_{rad} = \frac{BN_{th}}{A + BN_{th} + CN_{th}^2}. \tag{1.11}$$

To illustrate laser design and operation, we choose typical material and laser parameter values for a ~1 µm emission wavelength InGaAs/GaAs optically pumped VECSEL laser, as summarized in Table 1.1. Figure 1.4 shows the interplay between the laser design and operational parameters, as calculated using the above model. Figures 1.4a and b show the threshold pump power and laser output power, respectively, as a function of the number of quantum wells in the structure for several external mirror reflectivities $R_{ext} = R_2$. Because quantum wells are so thin, only ~8 nm, they provide only a small amount of gain to an optical wave propagating normal to the plane of the well. With on-chip mirror reflectivity of 99.9%, typical output coupling mirror reflectivities need also be high and range between 96 and 99%; intracavity laser loss is assumed here to be 1%. Multiple wells are required for lasing, with approximately 5–15 wells minimizing laser threshold depending on the output mirror reflectivity. Lower output mirror reflectivity provides higher output coupling but has higher threshold and requires larger number of quantum wells for operation. With 100 µm pump spot size here, threshold pump powers range between 100 and 300 mW. For smaller number of wells, threshold power rises very rapidly, whereas for larger number of wells, the threshold increases only slowly.

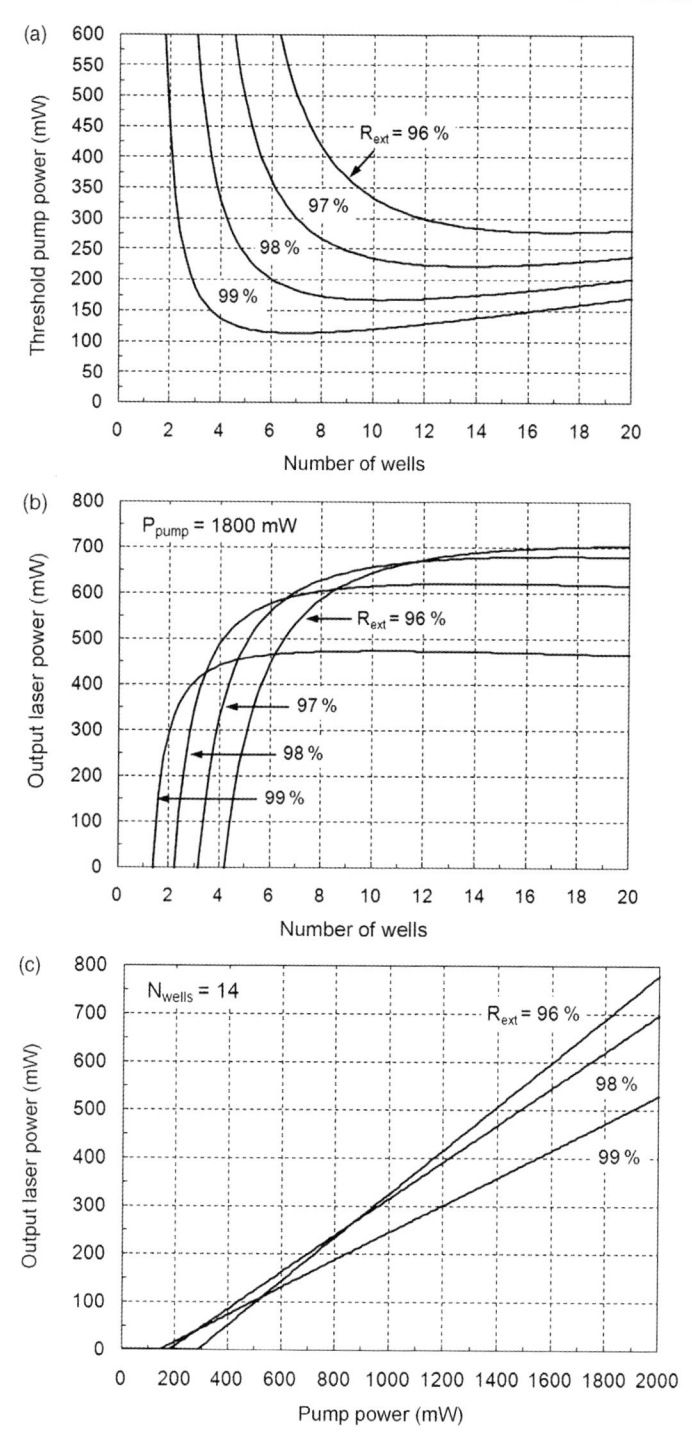

Figure 1.4 Calculated characteristics of the OPS-VECSEL lasers. (a) Threshold pump power versus the number of quantum wells. (b) Maximum output power versus the number of quantum wells. (c) Output power versus input power.

Figure 1.4b shows the calculated output laser power as a function of the number of quantum wells in the structure for several different external mirror reflectivities; the calculation assumes pump power of 1800 mW. Output power is maximized above 650 mW for the number of wells greater than 8–10 and external mirror reflectivities of 96–97%. Figure 1.4c shows the calculated output power of the laser as a function of the input pump power. Lower external mirror reflectivity increases output slope efficiency at the expense of the higher laser threshold.

Note that threshold power, Eq. (1.6), scales linearly with the pump, and laser, spot area; thus smaller pump spot areas are desired for lower thresholds. However, thermal impedance between VECSEL active region and heat sink increases with the decrease in pump area (Chapter 2), which leads to increasing active region temperatures, with the corresponding decrease in semiconductor gain and increase in laser threshold for such smaller pump areas. VECSEL lasers should be designed with the target output power in mind, and thus with approximate pump and thermal load levels. Given these power levels, pump spot size should be minimized such that thermal impedance and temperature rise are not too high; the number of quantum wells is then optimized for lower thresholds, and the output coupling mirror is optimized for highest output power at the available pump levels.

Here we have described a simple phenomenological analytical model of the semiconductor VECSEL; this model is useful to describe very simply and quickly the basic design and scaling principles of VECSELs, such as the number of quantum wells required, threshold pump levels and output power levels, output coupling optimization, and so on. This laser gain and power model should then be coupled with thermal models of VECSELs (Chapter 2) to define device thermal impedance and the desired pump/mode spot sizes, temperature rise of the active region, and the design wavelength offsets of the material photoluminescence PL peak and various cavity resonances [18]. Alternatively, detailed numerical microscopic models of semiconductor lasers can be used [117, 118]. Such models predict gain and emission properties of semiconductor materials without resorting to phenomenological description with adjustable model parameters; however such models are much more complex.

Multiple quantum wells required for laser gain are placed at the antinodes of the laser optical field standing wave, with none, one, or more closely spaced wells at each antinode, see Figure 1.3. Pump absorbing regions form the space between the antinodes. The number of quantum wells at the different antinodes is chosen such as to produce uniform quantum well excitation from the pump power that decays exponentially from the wafer surface as it is absorbed in the semiconductor. Placing quantum wells at the laser field antinodes resonantly enhances gain in this resonant periodic gain structure, as described by the confinement factor Γ. Such resonant periodic gain arrangement effectively eliminates spatial hole burning of the laser gain medium and enables simple single-frequency operation of these lasers, both with [55, 104] and sometimes without [119, 120] intracavity spectral filtering. Resonant gain enhancement, however, narrows the otherwise broad spectral bandwidth available from the laser gain medium. Such broad spectral bandwidth is desired, for example, for tunable laser operation or for ultrashort pulse generation. In this case, quantum wells can be displaced from their antinode positions in the structure to provide larger gain bandwidth at the expense of lower gain enhancement that comes with the lower confinement factor Γ [18]. Another structural factor that affects VECSEL gain bandwidth is the etalon formed between the on-chip laser mirror and the residual reflectivity at the chip surface. When this etalon is resonant, gain bandwidth is narrowed, laser gain is enhanced, and laser threshold is lowered. Designing this etalon to be antiresonant enhances gain bandwidth at the expense of the lower gain [20, 74].

In semiconductor lasers, strained quantum wells are frequently used, both because this allows access to a larger range of laser wavelengths and because of the improved operating characteristics of strained quantum well lasers [4, 8], such as lower threshold and improved temperature dependence. Because a large number of strained quantum wells are typically used in a VECSEL wafer structure, their total thickness can easily exceed the Matthews and Blakeslee strain critical thickness limit [4, 8], leading to strain relaxation via crystalline defect formation, which destroys laser operation because of strong nonradiative recombination at such "dark line" defects. Strain compensation [8, 18] must be used in this case, where layers of semiconductor material with the opposite sign of strain are introduced near the strained quantum wells such as to balance out net strain in the wafer structure. For example, compressively strained InGaAs quantum wells on GaAs are commonly used for ~1 μm laser emission. Here, using more than three strained quantum wells in the laser structure exceeds the critical thickness limit and requires strain compensation; tensile-strained GaAsP layers are typically used for strain compensation in this material system. Reliable semiconductor VECSEL operation has been obtained with such strain-compensated wafer structures [18] with over 35 000 h of lifetime data [54]. Using quantum dots in the laser active layers (Chapter 5) [99–102], instead of quantum wells, can provide further laser advantages, such as increased material gain bandwidth and improved temperature dependence [101]. VECSEL lasers utilizing quantum dot active regions are described in detail in Chapter 5.

1.3.1.2 On-Chip Multilayer Laser Bragg Mirror

The second critical component of the VECSEL semiconductor chip is the multilayer Bragg mirror that serves as one of the mirrors of the laser cavity, see Figures 1.2 and 1.3. As we have seen from the laser designs in Figure 1.4, on-chip laser mirror reflectivity should be very high, of the order of 99.9%, to keep laser thresholds low and output differential efficiency high. Since this mirror also forms a thermal barrier between the active region and the heat sink, another requirement is that mirror thermal impedance be low so that the active region has its heat dissipated efficiently and its temperature rise is limited. To produce efficient on-chip mirrors, multiple quarter-wave layer pairs of semiconductor materials with a high refractive index contrast are required. This yields high reflectivity with fewest layer pairs and thus lowest thickness and thermal impedance; good thermal conductivity of the mirror materials is also important. In addition, mirror materials should be nonabsorbing at the laser and, possibly, pump wavelengths. For optically pumped VECSELs, the mirror layers can be undoped, which significantly simplifies their epitaxial growth;

electrically pumped VECSELs require complex doping profiles in the mirror layers (Chapter 7). To achieve more efficient pump absorption in the pump absorbing regions, a pump light reflecting mirror can be included with the on-chip laser mirror; this, however, detrimentally increases the overall mirror thermal impedance. Higher thermal impedance of the on-chip laser mirrors can be counteracted somewhat by using front-side transparent heat spreaders (Chapter 2).

In the GaAs material system near 1 µm wavelength, high index contrast lattice-matched mirror materials, GaAs and AlAs, are available; a 30-pair mirror of such materials has the desired reflectivity with the mirror thickness of about 4.5 µm [18]. In contrast to GaAs, InP material system near 1.55 µm does not have such high contrast materials available; the InGaAsP/InP mirrors here require 48 quarter-wave pairs to achieve the desired reflectivity [121], with the resulting higher thermal impedance and significantly lower demonstrated output powers. Several alternatives have been explored for improved mirrors in this case. Metamorphic, or non-lattice-matched, semiconductor mirror materials have been used, for example, GaAs/AlAs mirrors on InP substrate, as well as hybrid metal-enhanced metamorphic mirrors [122, 123]. Dielectric mirrors [124] have also been used in VECSELs with optical pumping since no current injection is required in this case. But metamorphic and dielectric materials have poor thermal conductivity and such mirrors still have higher than desired thermal impedance. A novel way to overcome this material limitation is to use wafer fusion [97, 98]. In this approach, lattice-matched laser mirrors are grown on one substrate in a semiconductor material system with available high index contrast materials, while the gain medium is grown on a different substrate in another material system with the desired output wavelength range. The laser mirror and laser gain wafers with different lattice constants are then fused together; this is again much simpler for optically pumped structures where no current injection is required across the fusion interface. A dramatically improved output power performance, from 0.8 to 2.6 W CW, of 1.3 and 1.57 µm VECSELs was demonstrated using such an approach with InP-based gain wafer fused with GaAs/AlAs-based mirror wafer [97, 98].

1.3.1.3 Semiconductor Wafer Structure

Figure 1.5 shows an example of the full semiconductor window-on-substrate wafer structure of a 980 nm VECSEL. It follows the design principles outlined above and has been used in the first demonstration of high-power VECSEL lasers [18]. Starting from GaAs substrate, first the output window and then the gain region are grown, followed on top by the on-chip Bragg mirror structure. The active region contains 14 $In_{0.16}Ga_{0.84}As$ quantum wells of 8.0 nm thickness with 1.15% compressive strain. Each quantum well is paired with a 25.7 nm thick $GaAs_{0.90}P_{0.10}$ strain compensating layer with 0.36% tensile strain such that the net averaged strain of the structure is zero. These strain-compensated quantum wells are placed at the consecutive anti-nodes of the laser standing wave using $Al_{0.08}Ga_{0.92}As$ spacers that serve as pump absorbing layers and are designed for optical pumping at 808 nm. High-reflectivity Bragg mirror at the top of the structure consists of 30 pairs of $Al_{0.8}Ga_{0.2}As/GaAs$ quarter-wave thick mirror layers. The total mirror thickness is 4.5 µm and it has a

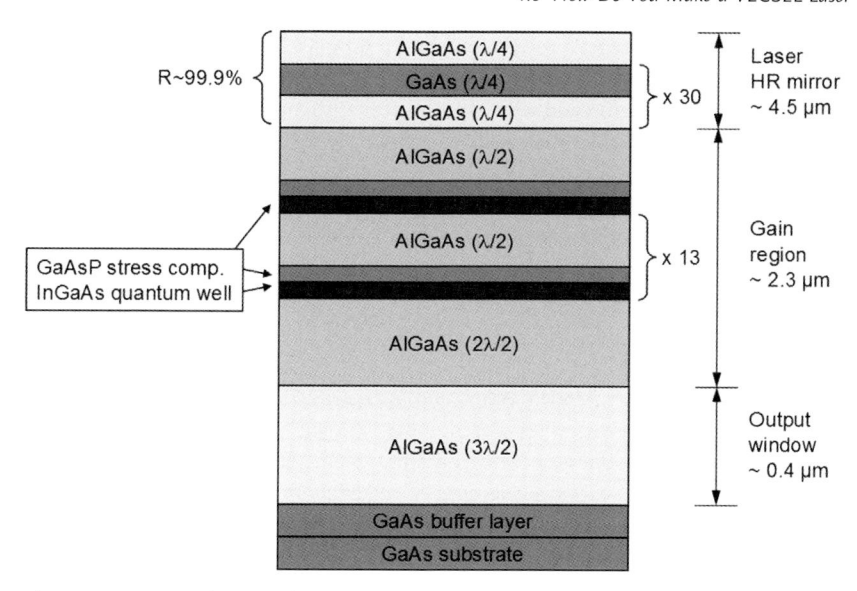

Figure 1.5 Semiconductor wafer structure of a 980 nm InGaAs/GaAs VECSEL laser.

thermal impedance of $21°\,\text{K}\,\text{W}^{-1}$ for a $100 \times 100\,\mu\text{m}^2$ laser spot size [18]. VECSEL semiconductor wafer structures require very good epitaxial growth control of the layer compositions, thicknesses and strains; such control is available with the modern metal-organic vapor-phase epitaxy (MOVPE) [18, 57, 125, 126] and molecular beam epitaxy (MBE) [127–129] semiconductor growth techniques.

Optically pumped VECSELs require very little processing after wafer growth; no lithographic processing is required. To use such semiconductor chip in a laser, it is first metallized, thinned, and soldered mirror side down onto a diamond heat spreader; subsequently, the thinned GaAs substrate is removed by selective wet chemical etching [18, 130] and the exposed window surface is antireflection (AR) coated. This AR coating serves to eliminate pump reflection loss at the OPS chip surface. It also strongly reduces chip surface intracavity reflection at the laser wavelengths, thus weakening the subcavity etalon formed by the on-chip mirror and the surface reflection, which otherwise can limit tuning bandwidth of the laser. This AR coating can also be made with semiconductor layers grown epitaxially on the semiconductor wafer together with the other structure layers [107]. Diamond heat spreader with the soldered OPS chip is in turn soldered onto a thermoelectrically temperature-controlled heat sink. Copper heat sink is typically used; however, using diamond in place of copper heat sink [107] can further decrease the chip thermal impedance, limit its temperature rise, and produce higher output powers.

The VECSEL window-on-substrate wafer layer arrangement leaves no extraneous substrate material, with its excess thermal impedance, between the mirror and heat spreader and gives excellent high-power laser operation. An alternative is the mirror-on-substrate wafer structure, where mirror layers are grown first on the substrate, followed by the active region and the output window [18]. Such mirror-on-substrate structures require back side substrate thinning, metallization, and soldering to heat sink. Handling thin semiconductor chips is very difficult; as a result, the residual substrate thickness is 15–30 μm with the corresponding thermal impedance and severe limitation to high-power laser operation [18]. Using front side transparent heat sinks with such structures, using materials such as sapphire [65], silicon carbide SiC [66], or single-crystal diamond [67], is very effective in addressing the thermal impedance problem and has demonstrated high-power laser operation (Chapter 2) [131].

1.3.2
Optical Cavity: Geometry, Mode Control, and Intracavity Elements

We next outline external optical cavity configurations used for VECSEL lasers; other than on-chip laser mirror, optical elements forming these cavities are external to the laser chip. VECSEL optical cavities allow control of the laser fundamental transverse mode operation as well as the insertion of various intracavity elements: saturable absorbers for laser passive mode locking, optical filters for laser wavelength selection and tuning, nonlinear optical crystals for intracavity second harmonic generation, and so on. Such optical cavities also allow combining of multiple gain elements in series for higher-power laser operation.

VECSEL lasers were first demonstrated with simple two mirror cavities [18, 65], as illustrated in Figures 1.2 and 1.6a. Later, as additional optical functional elements were added to the lasers, more complex three-mirror V-shaped cavities emerged. In one version of the V-shaped cavity, as illustrated in Figure 1.6b, the OPS gain chip serves as one end mirror of the cavity [57, 66, 97, 132], with the laser output taken variously either through the other end mirror [66, 97] or through the folding mirror [57, 132] of the cavity. In another version of the V-shaped cavity, as illustrated in Figure 1.6c, the OPS gain chip serves as the folding mirror in the middle of the cavity [20, 62, 133]. Four-mirror Z-shaped cavities, as shown in Figure 1.6d, give even more flexibility in placing intracavity functional elements and manipulating laser beam size at these elements [53, 131, 134]. Even more complex multimirror cavities have been used with two (Figure 1.6e) [49, 50] and three (Figure 1.6f) [135] active OPS gain chips in the cavity. Going beyond the above linear cavity configurations, a planar ring cavity has been used for a passively mode-locked VECSEL [136], and a nonplanar ring laser cavity has been used for a VECSEL-based ring laser gyro [137]. A diffractive unstable optical resonator has also been used with VECSELs where Gaussian beam output was extracted from a hard-edged outcoupling aperture [138].

A two-mirror, stable, plane–curved optical cavity (Figure 1.6a) [139] of length L_c and with curved mirror radius R_c has fundamental TEM_{00} laser mode beam $1/e^2$ diameters w_1 on the planar semiconductor chip and w_2 on the output spherical mirror given by

$$w_1^2 = \frac{4\lambda_{\text{laser}} L_c}{\pi} \sqrt{(R_c - L_c)/L_c}, \tag{1.12}$$

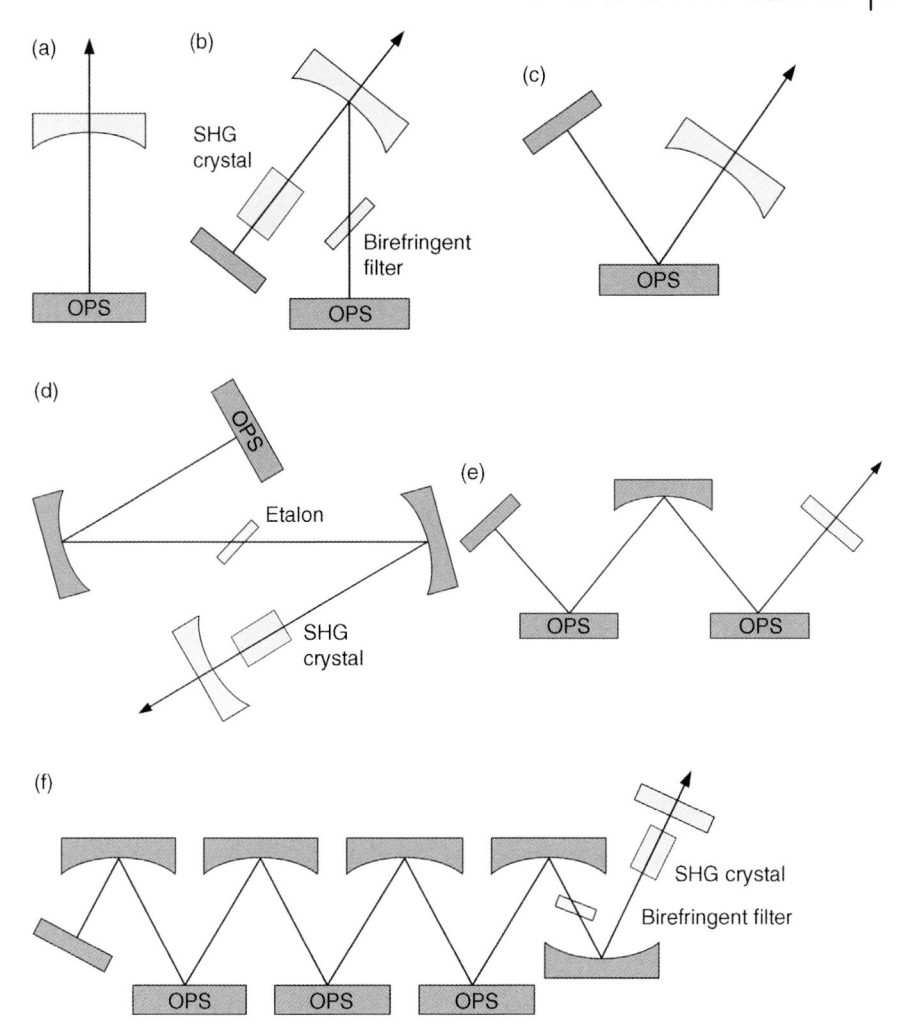

Figure 1.6 VECSEL laser cavities. (a) Two-mirror linear cavity. (b) Three-mirror V-shaped cavity for second harmonic generation (SHG). (c) Three-mirror V-shaped cavity for passive mode locking. (d) Four-mirror Z-shaped cavity. (e) VECSEL cavity with two gain chips. (f) VECSEL cavity with three gain chips.

$$w_2^2 = \frac{4\lambda_{\text{laser}} R_c}{\pi} \sqrt{L_c/(R_c - L_c)}. \tag{1.13}$$

Figures 1.7a and b illustrate variation of the laser mode diameters on the two cavity end mirrors as a function of the cavity length. Mode diameters on the chip between 100 and 200 μm can be easily achieved in this simple cavity for cavity lengths less than

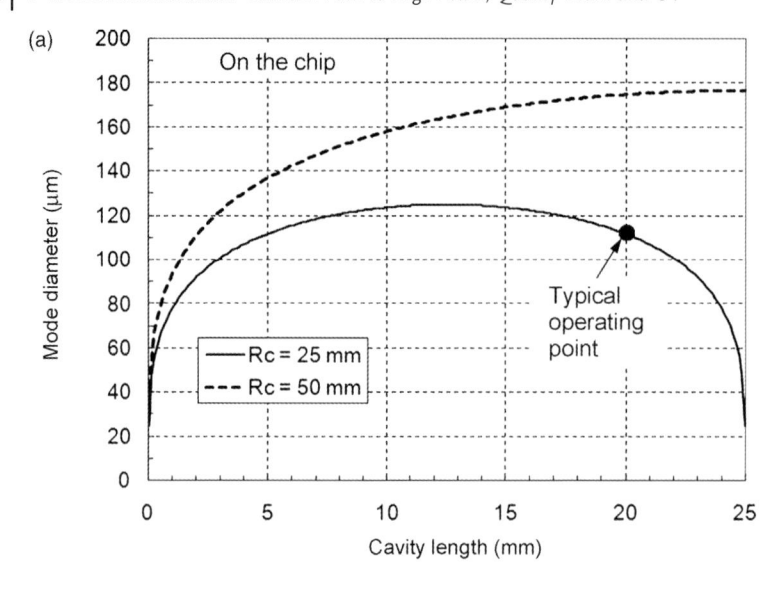

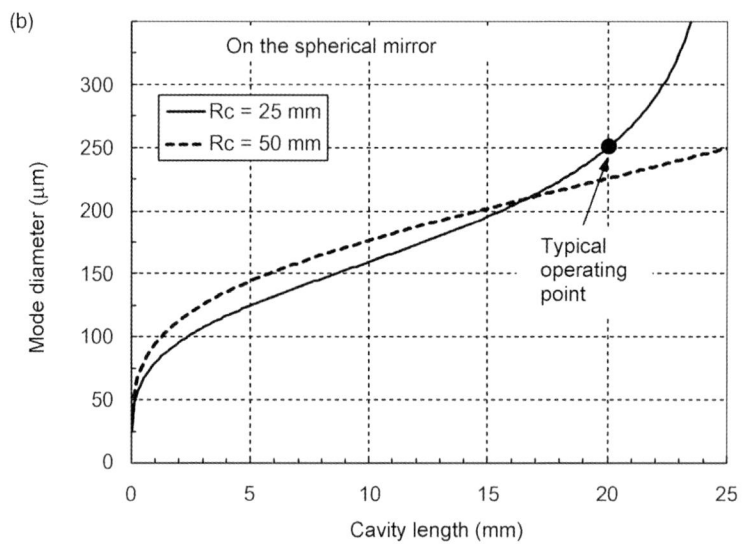

Figure 1.7 Mode spot $1/e^2$ diameters of the planar–spherical cavity (a) on the semiconductor chip w_1 and (b) on the output spherical mirror w_2 for spherical mirror radii of curvature of $R_c = 25$ and 50 mm and $\lambda_{laser} = 980$ nm.

25 mm. Pump spot size should be of the order of the laser mode size on the chip to provide efficient gain aperturing for fundamental transverse mode selection. Such two-mirror cavity VECSEL lasers have also been operated in unstable resonator regime with cavity lengths longer than mirror radius of curvature [18]. In this case, laser transverse mode is stabilized by gain aperturing with strong optical loss outside the pumped spot on the chip.

Another compact version of the two-mirror VECSEL cavity is the microchip laser configuration [69–71, 140], where a short plane–plane laser cavity is transverse mode stabilized by a thermal lens formed in the gain medium due to temperature gradients within the pump spot. Here, intracavity diamond heat spreader was used, with its planar outer surface coated with high-reflectivity output coupling mirror [70]; on-chip mirror is the second mirror of the cavity. Such arrangement allows very compact cavities; array laser operation was demonstrated for such microchip VECSELs [70]. Since thermally optimized VECSEL chip mounting, with on-chip mirror soldered on heat sink without intervening substrate, produces negligible thermal lensing [22], for microchip laser operation sufficient thermal impedance by design is required between the VECSEL chip and the heat sink. Microchip VECSELs operate optimally only within a well-defined window of pump powers and pump spot sizes. Intracavity thermal lens and a microchip mode of laser operation are also used in electrically pumped VECSELs (Chapter 7). Another version of a compact microchip VECSEL laser cavity does not rely on thermal lensing in the semiconductor, but instead uses spherical microlenses, or micromirrors, etched directly into outer surface of diamond heat spreaders in contact with the OPS chip surface [141–144]; arrays of such microchip lasers have also been demonstrated [143]. A potential compact laser cavity for VECSELs is the single-transverse mode optical resonator cavity [145], where shaping of the nonplanar resonator mirror can make all higher order transverse modes fundamentally unstable. Unlike other optically pumped VECSEL laser implementations, this resonator would require some lithographic processing of the semiconductor chip.

Inserting additional intracavity functional optical elements in the VECSEL typically requires additional mirrors for the laser cavity. Thus, a three-mirror V-shaped laser cavity was used for VECSEL passive mode locking [62], as in Figure 1.6c, with a flat SESAM semiconductor saturable absorber mirror at one end of the cavity, OPS gain chip as the folding mirror in the middle of the cavity, and a spherical output coupling mirror at the other end of the cavity. This cavity was only 3 mm long and demonstrated mode locking at 50 GHz high pulse repetition rate. The V-shaped laser cavity arrangement also allows controlling relative mode spot sizes on the gain and saturable absorber elements, with saturation intensity conditions of passive mode locking typically requiring smaller beam area on the absorber than the gain element (Chapter 6) [20].

An important enabling spatial flexibility of VECSEL laser cavities is the ability to inject pump light at various angles to the OPS chip without concern for pump beam divergence or specific beam angle. Thus, for passive mode locking in Ref. [62], pump beam was injected at 45° to the chip with incident beam direction in the plane perpendicular to the plane of the picture in Figure 1.6c. This pump flexibility is enabled by the thin disk nature of the laser gain medium with pump absorption length of the order of only 1 μm. A similar three-mirror V-shaped laser cavity (Figure 1.6b) is used extensively for VECSELs with intracavity second harmonic generation [57, 97, 132]. A four-mirror double-folded Z-shaped laser cavity with intracavity optical elements, such as the one shown in Figure 1.6d, has been used, for example, for VECSELs with intracavity second harmonic generation [88] and for passively mode-locked VECSELs [20].

Open multimirror cavities allow convenient insertion of intracavity elements. As shown in Figure 1.6c, for second harmonic generation [57], an intracavity birefringent filter is used for longitudinal mode, or wavelength, control and an intracavity nonlinear optical crystal is used for second harmonic generation. Several types of intracavity frequency-selective filters have been used for VECSEL laser frequency control, such as etalons [146], birefringent filters [104], volume Bragg gratings (Chapter 6) [59], and high-reflectivity gratings [60]. Such frequency-controlled VECSELs have demonstrated single-frequency laser operation with linewidth below 5 kHz [104].

Scaling VECSELs to very high output power levels can be accomplished by overcoming some of their thermal limitations with the use of multiple gain elements in series inside a more complex laser cavity, as first proposed in Ref. [18]. Two optically pumped semiconductor gain elements have been connected in series in the five-mirror VECSEL cavity illustrated in Figure 1.6e [50], producing 19 W CW at 970 nm. Two gain chips have been used in a four-mirror cavity for high-power generation in Refs [49, 147]. Scalable optically pumped two-, three-, and four-chip cavities of the type shown in Figure 1.6f have been used to generate as much as 66 W CW at the fundamental wavelength of 1064 nm and as much as 30 W CW at the second harmonic 532 nm green wavelength [135]. In order for such multielement VECSEL optical cavities to be practical, they must be stable to component and alignment perturbations as well as to long-term aging and drift; design of such dynamically stable cavities has been described in Ref. [52].

Power scaling of VECSELs typically requires large mode diameters on the gain chip to reduce device thermal impedance. A simple two-mirror cavity would need to be very long to achieve such large mode diameters, while a compact laser cavity is desirable. The design of such compact miniaturized cavities using Keplerian or Galilean telescopes to expand laser mode size has been described in Ref. [52]. A large 480 μm diameter mode size has been demonstrated in a small 8 cm footprint Keplerian telescope folded cavity, with 531 nm green laser TEM_{00} output power in the 10–20 W CW range [52]. A Galilean telescope folded cavity also demonstrated a large 480 μm diameter spot size in a miniature footprint of only ∼1.5 cm, with 486 nm blue laser TEM_{00} output power of 7.3 W CW [52]. Such compact cavities are possible because of the thin disk nature of the optically pumped semiconductor medium with very short pump absorption depths and gain lengths. As a result, high power VECSEL lasers can be made with a footprint an order of magnitude smaller than conventional diode-pumped solid-state lasers [52].

Electrically pumped VECSELs use a modified version of the two-mirror cavity of Figure 1.6a; here a third "intracavity" partially transmitting mirror is added on the semiconductor chip such that the quantum well gain region is sandwiched between two planar on-chip mirrors (Chapter 7). This additional mirror helps reduce laser intensity and optical absorption in the lossy intracavity doped semiconductor substrate that serves as the current spreading layer and lies outside the "sandwiched" gain layer (Chapter 7). In this scenario, the low-loss laser cavity is predominantly defined by the high-reflectivity planar on-chip mirrors; relatively weak external spherical reflector serves only to stabilize the fundamental spatial transverse mode.

Another version of electrically pumped VECSELs with intracavity second harmonic generation uses a similar cavity configuration but with all planar mirrors and a volume Bragg grating serving as the output coupling mirror at the second harmonic (Chapter 7). Here, a thermal lens is used to stabilize the fundamental transverse mode of the laser.

VECSEL optical cavities acquire different dimensions when they are used for intracavity laser absorption spectroscopy [73, 75, 148, 149]. For ICLAS applications, one arm of the three-mirror resonator in Figure 1.6b, the one without the OPS gain chip, is made about 1 m long and an ∼50 cm gas absorption cell is inserted inside the laser cavity. Equivalent intracavity absorption path lengths of greater than 100 km have been obtained with this technique, making ICLAS-VECSELs an extremely sensitive tool for absorption spectroscopy.

Note that OPS chip is sometimes used at normal reflection as an end mirror of the laser cavity, such as in cavity configurations shown in Figure 1.6a, b, and d. In these cases, laser optical wave on a single round-trip in the cavity passes the OPS gain medium twice. On the other hand, at other times, OPS chip is used as a cavity folding mirror reflecting laser optical beam at an angle, such as in cavity configurations shown in Figure 1.6c, e, and f, the last two being multichip cavity configurations. In these cases, laser optical wave on a single round-trip in the cavity passes the reflection folding OPS gain medium four times. Thus, the folding-mirror OPS chip provides twice the gain per round-trip in the laser, as compared with the end-mirror OPS chip configurations; this has to be taken properly into account when designing the laser.

As we have illustrated here, VECSELs use a rich variety of optical cavities [139] that have been originally developed for other types of lasers, such as diode-pumped solid-state DPSS lasers [13, 14]. Many of these cavities have been specialized for VECSEL lasers to allow compact size, when needed, and, using intracavity elements, a wide range of optical functions, such as single-frequency operation, short pulse generation, second harmonic generation, and ultrasensitive intracavity absorption spectroscopy. Such functional richness simply would not be possible in an integrated semiconductor device without external optical cavity.

1.3.3
Optical and Electrical Pumping

VECSEL lasers have been made with two types of excitation: optical [18, 22] and electrical (Chapter 7) [92, 93, 141, 150–153]. Electrical excitation of the laser by a diode current injection across a p–n junction is very appealing, as it requires only a simple low-voltage current source to drive the laser, rather than separate pump lasers with their pump optics and power supplies. For VECSEL lasers, however, electrical pumping has significant limitations. First, intracavity laser absorption in doped semiconductor regions required for current injection degrades laser threshold and efficiency. The second major problem is the difficulty of uniform current injection across the several hundred microns wide emission areas required for high-power operation. Pump current is injected from the perimeter of the light-emitting laser aperture; thick current spreading doped semiconductor layers attempt to provide a more uniform current distribution across this aperture; however, intracavity free-carrier absorption in these doped layers is detrimental to the laser. This problem is somewhat mitigated by the three-mirror laser cavity that places current spreading absorbing regions in the lower light intensity section of the cavity (Chapter 7). Carrier transport across the multiple quantum wells and nonuniform electron and hole distributions across the wells also have to be considered. Significantly, as compared with optically pumped OPS wafers, electrically pumped VECSELs require a much more complex semiconductor wafer growth process with complex layer doping profiles, as well as post-growth lithographic processing (Chapter 7).

Optical pumping approach divides the functions of laser pumping and laser light emission between separate devices. While the final device requires multiple components and is more complex than the integrated electrically pumped approach, the individual components of an optically pumped VECSEL can be independently optimized, avoiding painful, or impossible, compromises inherent in an integrated device. Thus, a pump laser can be optimized separately for efficient high-power light generation without regard to beam spatial quality. OPS VECSEL structure is optimized for efficient pump power conversion to a high spatial quality beam, with wide output wavelength access and rich functionality, such as short pulse generation. As a result of such separate optimizations, optically pumped VECSELs have demonstrated more than an order of magnitude higher output power levels than their electrically pumped counterparts. In a hybrid approach, separate edge-emitting pump lasers have been integrated on the same substrate for optical pumping of a VECSEL structure [154].

What are the main advantages of the optical pumping of VECSELs? Optical pumping allows simple uniform transverse carrier excitation across very wide range of VECSEL emission apertures from 50 to 1000 µm in diameter. Also, no carrier transport from the surfaces through the multiple quantum wells across the device thickness is required, as pump light propagates throughout the device thickness to deliver the excitation. OPS wafer structures are undoped, which is easier to grow and produces no free-carrier absorption; also, no lithographic wafer processing is required. Multiple pump beams incident on a single pump spot from different directions can be used to excite a VECSEL. In an analogy with diode-pumped solid-state lasers [13, 14], an optical end-pumping scheme has also been used with VECSELs [155–158]. In this configuration, pump light enters through a transparent heat sink on one side of the OPS chip, and laser output is taken from the other side of the chip. In addition to single-stripe multimode semiconductor pump diode lasers, high-power multiple stripe diode arrays can also be used for direct VECSEL pumping [52]. The use of such poor beam quality pumps is made possible by the short absorption depth of semiconductors and is not possible with diode-pumped solid-state lasers that typically have absorption lengths on the scale of millimeters.

Pump diodes can be directly coupled to the OPS chip by means of relatively simple pump optics; alternatively, fiber-coupled pumps can be used. Pump optics should deliver a pump spot on the OPS chip that is approximately matched in diameter and serves as the gain aperture to the laser fundamental TEM_{00} transverse spatial mode. Note that when pump beam is incident at an angle to the OPS chip surface, say between 30° and 60°, one incident pump beam dimension is elongated upon projection onto the OPS chip. Fiber-coupled pumps require simple spherical lenses for coupling to the OPS chip. Directly coupled pump diodes can use a combination of spherical or cylindrical lenses [18] and shape the highly elongated high aspect ratio pump beam to a square-shaped pump spot on the chip with aspect ratio close to unity. Figure 1.8 shows examples of pump optics arrangement [18] for VECSEL directly pumped by a pump diode chip. It is important to note that VECSEL pump optics is not an imaging arrangement; pump spot dimensions have to be right in a very thin plane of the OPS chip where pump light is absorbed. Pump beam does not have to be in focus in this plane; it does not matter how pump beam diverges before or beyond this plane. Pump beams are typically highly spatially multimoded, only an approximately uniform pump light distribution is required across the VECSEL laser mode aperture. If the pump spot becomes too large in diameter, in-plane amplified spontaneous emission (ASE) can potentially deplete laser gain [159]. Such in-plane ASE, if not controlled, will limit the lateral size of the laser pump spot and thus limit scaling of the output power of the laser. Photonic crystal structures can help in this regard. Connecting multiple OPS gain elements in series inside VECSEL laser cavity avoids such limitation to VECSEL power scaling.

As we have already mentioned, pump wavelength flexibility is a key feature of optically pumped VECSELs as compared to diode-pumped solid-state DPSS lasers. Semiconductors absorb light for all wavelengths shorter than the material bandgap

Figure 1.8 Examples of pump optics arrangement for direct pump diode pumping of VECSEL lasers. (a) Crossed cylindrical pump optics. (b) Cylindrical graded index (GRIN) lens followed by a graded index (GRIN) lens.

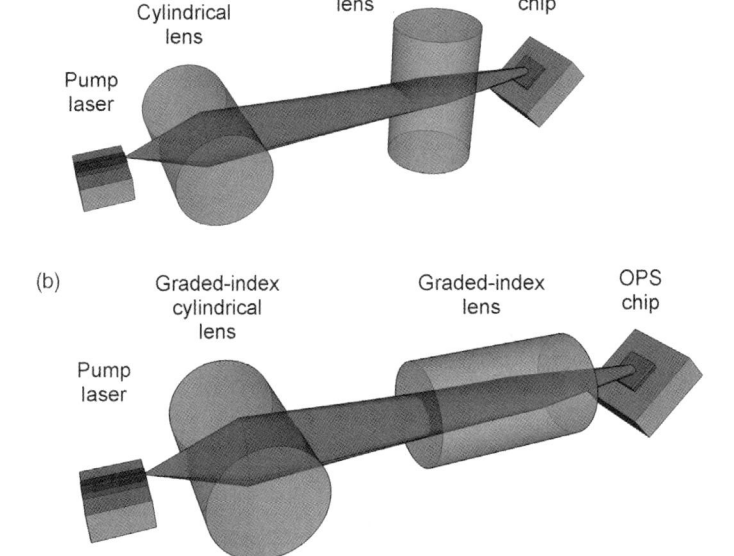

wavelength. Thus, pump diode lasers, which have a strong temperature dependence of their output wavelength, do not need to be temperature stabilized for VECSEL pumping applications. This significantly simplifies overall VECSEL laser system and avoids large power consumption of the temperature stabilization devices. Pump wavelength selection is also not critical, leading to a much higher yield, and hence lower cost, of the pump laser chips. As such, standard available wavelength pump lasers can be used for VECSELs, unlike solid-state lasers, where each laser type requires its own custom pump wavelength, for example, 808 nm for pumping ubiquitous Nd:YAG lasers or 941 nm for pumping Yb:YAG disk lasers. Pump wavelength can be very far from the laser emission wavelength; for example, 790 nm wavelength pump lasers have been used to pump 2.0 μm emission wavelength Sb-based VECSELs [94].

To prevent gain carriers from thermally escaping quantum wells, and thus depleting gain, energy difference between the confined well states of electrons and holes and the corresponding conduction and valence band edges of barriers between wells has to be at least 4–5 $k_B T$. Here, k_B is the Boltzmann constant and T is the absolute temperature; at room temperature, $k_B T$ is ~25 meV. If VECSEL pump light is to be absorbed in the barrier layers, this implies wavelength difference between pump and laser wavelengths has to be greater than ~130 nm for laser emission near 1000 nm. This energy difference between pump and laser photons, the quantum defect, reduces laser-operating efficiency and also leads to excess heat generation. To improve VECSEL efficiency, it has been proposed that instead of barrier pumping, direct in-well pumping can be used with VECSELs, without changing barrier bandgap, thus reducing pump–laser quantum defect [44, 46]. One difficulty with this approach is that quantum wells are very thin and single-pass absorption in just a few wells is weak. Placing quantum wells at the antinodes of the pump wavelength resonant sub-cavity resonantly enhances such weak in-well pump absorption [47, 48]. One disadvantage of such resonant in-well pumping scheme is that tight, ~4 nm, pump wavelength control is required [47], negating the broad acceptable pump wavelengths with barrier-pumped VECSELs. Another approach to alleviate weak in-well absorption problem is to use multiple pump passes [47], much as it is done with solid-state disk lasers [40, 41], using an on-chip pump mirror and pump-recycling optics outside the OPS chip.

An on-chip pump light mirror has been used to produce a more efficient double-pass pump absorption with barrier [64, 107] or in-well [47] pumping. Also, pump light mirror has been used to block pump light from bleaching saturable absorber in mode-locked VECSELs with integrated saturable absorber, MIXSEL [160, 161]. Such pump mirrors, however, can introduce additional undesired thermal impedance between the gain layer and the heat sink. Pump intensity decays exponentially with depth into semiconductor as pump photons are absorbed, while relatively uniform excitation of quantum wells is desired. Such a more uniform quantum well excitation can be achieved by varying the number of quantum wells at different antinodes, possibly skipping some antinodes, and also by adjusting with depth bandgap of the pump absorbing layers. Quantum well structures with graded gap barriers have been used to improve pump absorption by reducing absorption saturation in the barriers through more efficient carrier collection in the wells [162].

Because of quantum defect between pump and laser light, as well as nonunity laser emission efficiency, excess heat is generated that has to be dissipated efficiently to the heat sink. This is done with back or front, transparent, surface heat spreaders, as discussed in detail in Chapter 2. Thermal impedance between the chip active region and the heat spreaders needs to be minimized. Proper heat sinking limits temperature rise of the gain region, which is critically important for laser operation since semiconductor gain can degrade significantly with temperature as carriers spread in the band due to thermal broadening of their energy distribution. Such temperature rise is typically the limiting mechanism for laser output power at higher pump powers. Another effect of temperature rise is wavelength red shift of the semiconductor bandgap and gain peak. In parallel, temperature rise also causes increase in semiconductor refractive indices with the resulting red shifts of the Bragg resonance wavelengths of the on-chip mirror and of the resonant periodic gain spectral peak, as well as of resonance wavelengths of the on-chip subcavities. Gain peak shifts with temperature much faster than the refractive-index-related wavelengths. Optimal VECSEL operation requires design of the gain and resonance spectral positions such that they overlap at the elevated laser active region operating temperature, which depends on the heat sink temperature and the chip active region dissipated power (Chapter 2) [18].

1.3.4
VECSEL Laser Characterization

A number of measurements can be performed on VECSEL laser components, for example, chip mirror and quantum well gain region, to characterize their operation and compare them to the design parameters. Some of these measurements can be performed on the wafer level; others are done on the thinned and AR-coated OPS chip already mounted on a heat sink. Another set of measurements characterizes the complete VECSEL laser operation.

VECSEL semiconductor wafer structure with its Bragg mirror and gain region layers can be characterized very effectively by using a spectrophotometer, which measures reflectivity spectra of these structures [18]. Both front and back side spectra can be measured, giving complementary information. Figure 1.9 shows such reflectivity spectra for the window-on-substrate and mirror-on-substrate wafer structures. The window-on-substrate reflectivity in Figure 1.9a, corresponding to the wafer structure in Figure 1.5, shows the broad, ~90 nm wide, mirror reflectivity band in the front side measurement; back side measurement shows the onset of quantum well absorption for wavelengths below 976 nm in the middle of the mirror band. The absorption dip is weak due to thinness of the wells. Subcavity etalon effects are not visible here since the weakly reflecting window–substrate interface in the wafer replaces the strongly reflecting window–air interface. Upon chip mounting and substrate removal, the window–air interface is exposed and wafer reflectivities look

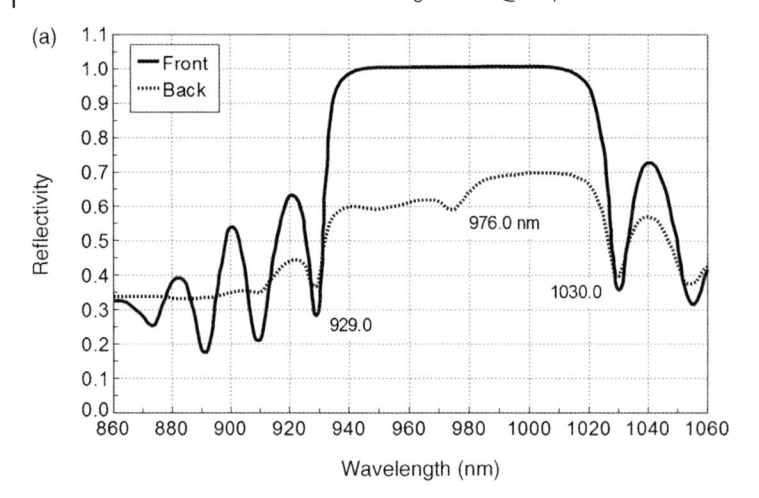

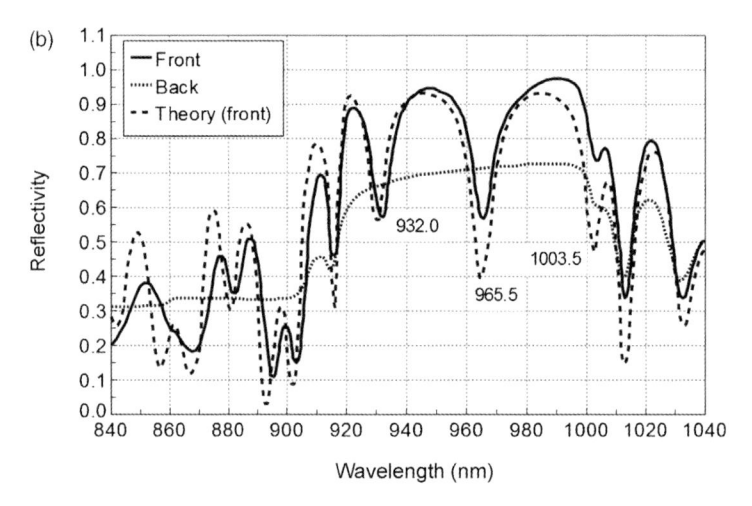

Figure 1.9 Reflectivity spectra of the OPS semiconductor wafers. (a) Window-on-substrate structure: front (epi) and back surface reflectivities; subcavity etalon effects are not visible. (b) Mirror-on-substrate wafer structure: front (epi) surface, back surface, and front surface theoretical reflectivities. Reflectivity dips at $\lambda = 932$, 966, and 1004 nm resonances of the mirror to chip surface subcavity etalon.

like those in Figure 1.9b. Figure 1.9b shows reflectivity spectra for mirror-on-substrate VECSEL structure similar to the one shown in Figure 1.5, except that the substrate appears on the mirror, rather than window, side of the structure. The back side reflectivity here is indicative of the intrinsic reflectivity spectrum of the multilayer mirror. The front side reflectivity spectrum shows strong effects of the subcavity etalon formed between the mirror and the semiconductor–air interface. In addition, weak intrinsic absorption of quantum wells located inside the subcavity etalon is strongly enhanced here by the etalon resonances. Thus, the front reflectivity of the wafer shows strong dips near λ = 932.0, 965.5, and 1003.5 nm, corresponding to enhanced quantum well absorption at the longitudinal modes of this etalon. Such strong resonant enhancement of quantum well absorption is used effectively for in-well optical pumping of VECSELs [47, 48]. Reflectivity spectrum of the structure can be calculated from the layer thicknesses, refractive indices, and absorptions, and shows good agreement with the measured data. Semiconductor layer composition and thickness growth errors are immediately apparent from such measured wafer reflectivity spectra. Applying AR coating to the front surface of the OPS chip reduces and somewhat shifts the residual etalon dips, but does not eliminate them completely [18]. Upon optical pumping, VECSEL laser lases at the wavelength of one of these residual etalon resonances that corresponds to the highest available quantum well gain [18].

Quantum well gain medium is characterized at the wafer level by measuring its photoluminescence PL spectrum [18, 23, 163, 164]. Quantum wells emit light from inside the residual subcavity etalon; their photoluminescence spectrum is strongly modified by this etalon when emitting normal to the wafer surface and largely unaffected when light is emitted from the edge of the wafer [18]. Figure 1.10 shows measured photoluminescence spectra of a VECSEL wafer: broad edge-emitted spectrum and strongly narrowed surface-emitted spectrum with PL peaks corresponding to the etalon resonances [18]. Also shown is the surface-emitted spectrum from an AR-coated wafer that shows that etalon resonances are weakened and broadened but not completely eliminated. It is important to use edge-emitted PL spectrum for determining true spectral peak location of the quantum well material gain. Combination of the wafer reflectivity spectra and photoluminescence spectra shows spectral locations of the laser material gain and resonances peaks that localize laser emission. These spectral locations have to match at the operating temperature of the laser active region.

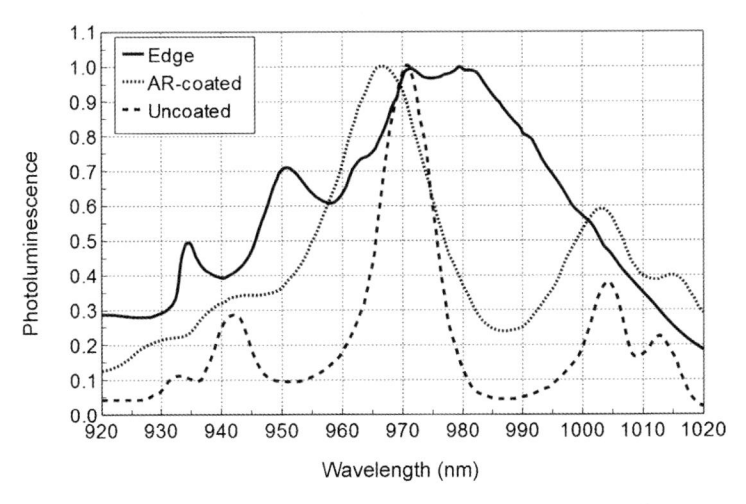

Figure 1.10 Normalized photoluminescence spectra of the OPS chips: surface-emission spectra of the AR-coated and uncoated OPS chips, and the edge-emission spectrum.

For strained quantum well OPS wafers, photoluminescence can be a strong indicator for the presence of dark line defects when total strained well thickness exceeds critical thickness [4, 8]. For example, for InGaAs/GaAs VECSELs emitting near 1 µm, when illuminating broad area of the wafer or a full OPS chip with pump light and observing photoluminescence with a camera, one can see in the PL image a pattern of crossed dark lines oriented along crystallographic [001] axes forming over time, especially about points on the chip that had been exposed to focused pump light. Such dark line defects form quickly over time and severely degrade laser performance [64]; using strain-balanced VECSEL structures [18] completely eliminates such dark line defects and produces fully reliable VECSEL lasers.

Once laser cavity and pump optics are aligned, there are a number of measurements that characterize the full VECSEL laser performance. The most revealing measurement is the output power versus pump power dependence, similar to theoretical calculation in Figure 1.4c; this measurement also produces the important values of laser threshold, output slope efficiency, total output efficiency, and maximum output power, whether limited by pump power or thermal rollover. Temperature dependence of these laser characteristics is another important measurement; here two temperatures are significant: the heat sink temperature, typically set by a thermoelectric cooler and measured by a thermistor, and the active region temperature, which can sometimes be estimated from the laser emission wavelength and its temperature dependence or from pump-dissipated power and OPS chip thermal impedance. The above laser output characteristics also depend strongly on the output coupling mirror reflectivity/transmission, as illustrated in Figure 1.4c; measuring this dependence gives an estimate of intracavity losses and enables the selection of the optimal output coupler. It is important to note that several pump powers are of relevance here: power emitted by the pump laser, pump power coupled to the OPS region by the pump optics, and finally the pump power absorbed in the OPS pump absorbing region. Spatial overlap, in size and shape, of the pump spot and laser mode should also be taken into account.

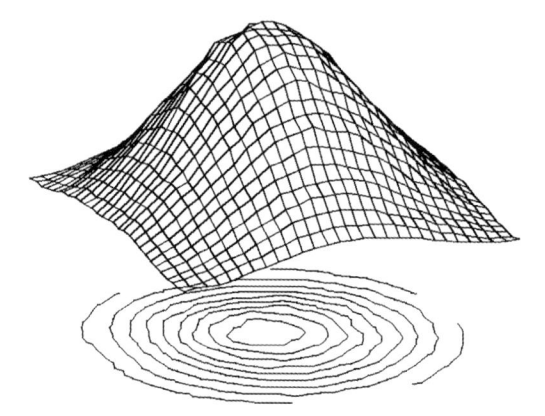

Figure 1.11 Measured output beam profile of the OPS-VECSEL laser.

For most applications, VECSEL output spatial beam quality is of utmost importance. Figure 1.11 shows an example of a measured OPS-VECSEL output beam profile with the desired circular fundamental transverse TEM_{00} mode operation. Beam quality can be characterized quantitatively by measuring its M^2 parameter [9, 50, 107, 142], which describes how much faster laser beam diverges in two transverse dimensions as compared with a diffraction limited $M^2 = 1$ Gaussian beam. The measured beam quality of $M^2 \sim 1.0$–1.5 for many VECSELs is considered to be very good. Such high-quality beam is required, for example, for confocal laser fluorescence imaging with free-space delivered [165] and single-mode fiber delivered beam [69]. $M^2 \sim 3$–4 beams have stronger divergence due to several transverse modes but are adequate for many other VECSEL applications [167], such as Coherent Inc. TracER™ laser illuminator for fingerprint detection [22]. Typically, somewhat higher powers can be extracted from VECSEL lasers in a multimode regime.

Determining spectral characteristics of VECSELs involves measurement of laser emission wavelength and spectral lineshape, as well as the dependence of these on the operating temperature and pump power [18]. For example, keeping pump power constant and adjusting chip temperature via the heat sink thermoelectric cooler, we can measure laser wavelength, typically defined by subcavity resonance, shift with temperature. This gives us a very useful "thermometer" of the active region. Later measurement of the laser wavelength shift with pump power above threshold, and using the above "thermometer" information, gives us the temperature rise of the active region with pump power and thus also an estimate of the active region thermal impedance and active region temperature at the laser-operating point. Additional measurements of the temperature and pump power dependences of the semiconductor photoluminescence spectra, both edge and surface emission, together with the above laser wavelength measurements, are important for ensuring proper spectral alignment of the laser emission and the gain and PL peaks at the laser-operating conditions (Chapter 2) [18, 163].

With intracavity tunable filter, tunable power dependence on emission wavelength becomes important [51]. For single-frequency VECSEL operation, laser linewidth and noise measurements define the relevant laser characteristics [120, 168]. Characterization of optical harmonic generation (Chapter 3) and mode-locked picosecond pulse generation (Chapter 6) in VECSELs is covered in detail in the other chapters in this book.

This section has outlined basic elements in designing, making, and characterizing VECSEL lasers. As is typical of semiconductor lasers, the many functional degrees of freedom offered by VECSELs require good understanding and precise control of the corresponding design and fabrication degrees of freedom. Extensive device characterization is critical as well, both for determining VECSEL operating parameters and for closing the design-fabrication-characterization loop and ensuring the above-mentioned laser design understanding and fabrication control. In the next section, we will describe the demonstrated performance ranges of VECSEL lasers as well as their applications and future scientific directions.

1.4
Demonstrated Performance of VECSELs and Future Directions

1.4.1
Demonstrated Power Scaling and Wavelength Coverage

Since the first demonstration of high-power optically pumped VECSELs in 1997 [17, 18], which emitted at wavelengths near 1000 nm and had output powers of about 0.5 W, many different VECSELs have been demonstrated with power levels from milliwatts to tens of watts and wavelengths from 244 nm in the UV to 5 μm in the mid-IR. Here we shall overview the power and wavelength scaling demonstrated with VECSEL lasers and the semiconductor material systems that have been used. Other chapters in this book describe in detail a variety of other aspects of VECSEL lasers and their demonstrated performance, such as mode locking and high repetition rate short pulse generation, thermal management of high-power VECSELs, visible light generation via second harmonic generation, quantum dot gain media VECSELs, long-wavelength NIR VECSELs, and electrically pumped VECSELs.

VECSEL lasers with a wide range of powers and wavelengths have been demonstrated both in university and government research laboratories across the world – United States, United Kingdom, France, Switzerland, Germany, Sweden, Finland, Korea, and Ireland – and in several commercial companies – Coherent, Novalux/Arasor, OSRAM, Samsung, Solus Technologies. VECSEL-related work has been published by scientists in Russia, Poland, Spain, Denmark, and China. There are also a number of commercial products based on the VECSEL technology. In a measure of the scientific, technological and commercial activities in VECSEL/OPSL/SDL lasers, more than 250 papers have been published in this field and more than 100 U.S. and international WIPO patents have been issued on the subject of these lasers. Utilizing flexibility of the VECSEL approach and with appropriate designs, efficient VECSEL operation has been demonstrated across a wide range of laser-operating parameters: from low to high output powers; for fixed and tunable wavelength operation; for fundamental and multiple transverse mode operation; for single-frequency and multiple longitudinal mode operation; for fundamental, second, third, and fourth harmonic wavelength operation; and for single and multiple gain chip laser operation.

Figure 1.12 summarizes in graphical form the demonstrated power and wavelength performance of VECSEL lasers, also shown are the major semiconductor material systems used in these demonstrations. Tables 1.2 and 1.3 list these results in tabular form and also include some other relevant VECSEL-operating parameters. Table 1.2 lists VECSELs with the fundamental wavelength output, while Table 1.3 has the frequency-doubled second harmonic output VECSELs.

The strain-compensated InGaAs/GaAsP/GaAs material system with fundamental output wavelengths in ∼920–1180 nm NIR range is most widely used for VECSELs. In this material system, output power greater than 20W has been demonstrated in a single-transverse mode with excellent beam quality $M^2 \sim 1.1$ [107]. Still higher 30–40 W power levels were demonstrated in a slightly multimode regime with $M^2 \sim 3$ [22, 167]. This remarkable material system from its fundamental wavelength

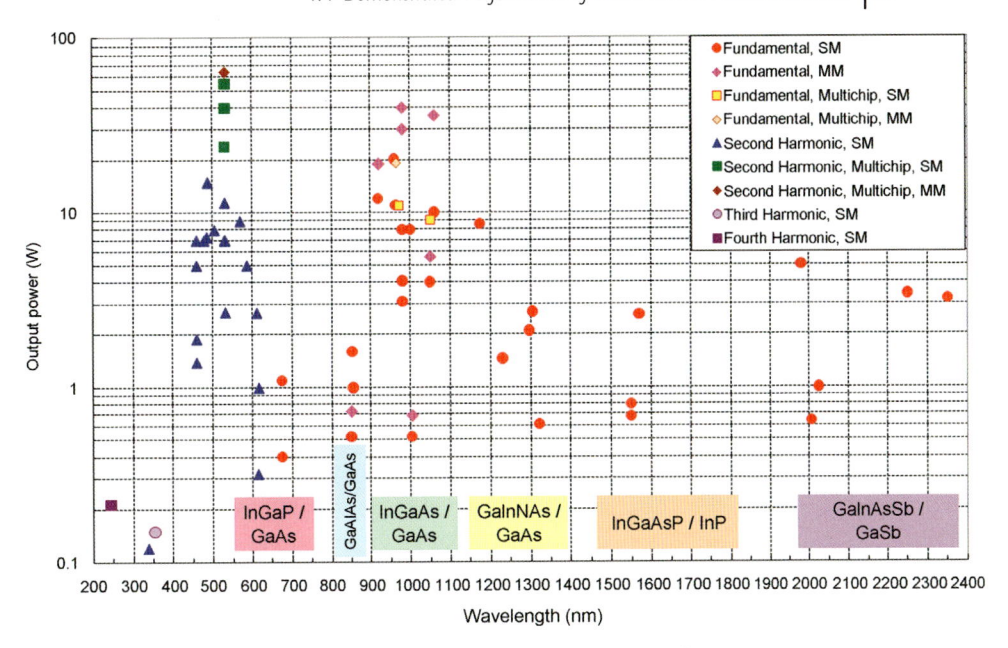

Figure 1.12 Demonstrated power and wavelength performance of VECSEL lasers.

range allows convenient and efficient intracavity frequency doubling to the visible blue–green–yellow range with the blue 460–490 nm, green 505–535 nm, and yellow 570–590 nm wavelength regimes. Power levels of 15 W in the blue [22] and 12 W in the green [52] have been demonstrated in a single-transverse mode with a single OPS gain chip. Using multiple gain chips, doubled green output powers of 40 and 55 W have been demonstrated with two and three gain chips, respectively [22], still in the fundamental transverse mode. Powers as high as 64 W in the doubled green have been demonstrated with three gain chips by going to a slightly multimode operation with $M^2 \sim 4$ [22]. Electrically pumped VECSELs in the NIR and doubled into the blue have also been made in the InGaAs/GaAs material system [177]. Using highly strained quantum wells in this material system, direct emission wavelengths can be pushed to the 1140–1180 nm range [87]. Here, intracavity-doubled output is in the yellow wavelength range, demonstrating with optical pumping output powers of 9 W at 570 nm [22] and 5 W at 587 nm [58]. Furthermore, in the same material system, using intracavity frequency tripling of the fundamental 1065 nm emission produces 355 nm ultraviolet output [108]. A commercial Genesis™ laser, from the OPSL laser family, from Coherent Inc., produces up to 150 mW at the tripled 355 nm UV wavelength [178]. In a different approach for short-wavelength UV generation, fundamental VECSEL emission at 976 nm in the NIR has been laser intracavity doubled to 488 nm blue, with this doubled VECSEL output further frequency doubled in an external cavity to the 244 nm deep UV wavelength, the fourth harmonic of the fundamental VECSEL NIR emission; 215 mW of deep UV radiation has been produced by this technique [77]. Thus, VECSELs with the InGaAs/GaAsP/GaAs material system, using fundamental, doubled, tripled, and quadrupled frequency

Table 1.2 Demonstrated wavelengths and powers of VECSEL lasers for the fundamental wavelength operation.

Operating wavelength (nm)	Pump wavelength (nm)	Output power, single mode (W)	Output power, multimode (W)	Pump power (W)	Material system	Transverse mode, M^2	Mode diameter (μm)	Transparent heat spreader or inverted structure	Output Coupling, %	No. of QWs	Tuning range, nm	References
674	532	0.4		3.3	GaInP/AlGaInP/GaAs		72	Tr.HS	2	20	10	[79]
675	532	1.1		7	GaInP/AlInGaP/GaAs		75	Tr.HS	3	20		[78]
850	660	0.523		2.4	AlGaAs/GaAs		100	Tr.HS	2.4	15	30	[66]
850	670		0.73	2.4	InAlGaAs/GaAs	5	70	Tr.HS	4	17		[86]
853	822	1.6		3.2	AlGaAs/GaAs	1.x	100	Tr.HS	3.8	14		[47]
855	806	1		30	AlGaAs/GaAs		110	Tr.HS	2	17		[45]
920	808		19	70	InGaAs/GaAsP/GaAs	3	500–900	Inv.				[167]
920	808		19	24	InGaAs/GaAsP/GaAs			Inv.				[22]
920	808	12		55	InGaAs/GaAsP/GaAs	2	250	Tr.HS	6	15		[68]
960	808	20.2		60	InGaAs/GaAsP/GaAs	1.1	430	Inv.	0.7	7		[107]
964	808	11		60	InGaAs/GaAsP/GaAs	1.75	355	Inv.	9	14/10		[50]
966	808		19.2	60	InGaAs/GaAsP/GaAs	2.14	355	Inv.	9	14/10		[50]
972	808	11		46	InGaAs/GaAsP/GaAs	1.7	355/410 + 420/480	Inv.	8	14/10	33	[51]
980	808		30	80	InGaAs/GaAsP/GaAs	3	500–900	Inv.				[167]
980	808		40	11	InGaAs/GaAsP/GaAs			Inv.				[22]
980	808	3.1		11	InGaAs/GaAsP/GaAs	1.15	220	Inv.		14		[169]
980	808	4.05		14	InGaAs/GaAsP/GaAs	1.6	230	Inv.	6	18		[130]
980	808	8		30	InGaAs/GaAsP/GaAs	TEM_{00}	430/500	Inv.	8	18	20	[61]
1000	808	8		20	InGaAs/GaAsP/GaAs	1.8	367	Inv.	6	14		[43]
1004	808	0.52	0.69	1.5	InGaAs/GaAsP/GaAs	TEM_{00}	100	Inv.	2-4	14		[17, 18]

1.4 Demonstrated Performance of VECSELs and Future Directions

1050	4		20	InGaAs/GaAs	808	1.15	165	Tr.HS	2	13	11.6	[170]
1050	9	5.6	42	InGaAs/GaAs	800	1.45	180	Tr.HS		13		[49]
1060		36	80	InGaAs/GaAsP/GaAs	808			Inv.				[22]
1060	10		24	InGaAs/GaAsP/GaAs	808	1.7	200	Tr.HS	6	15		[125]
1175	8.6		35	InGaAs/GaAsP/GaAs	808	1.5	500	Inv.	4	10	30	[58]
1220	3.5			GaInNAs/GaAs								[171]
1220	5			GaInNAs/GaAs								[172]
1230	1.46		18.2	GaInNAs/GaAs	808		180	Tr.HS	1	12	4.5	[88]
1297	2.1		23	AlGaInAs/InP	808		180	Tr.HS	1	10		[98]
1305	2.7		26	AlGaInAs/InP	980		180	Tr.HS	2.5	10		[98]
1322	0.612		8.62	GaInNAs/GaAs	810	1.2	75	Tr.HS	2	10		[67]
1550	0.68		4.6	InGaAsP/InP	1250	1.5/1.2	250	Tr.HS	1	20		[121]
1550	0.8			InGaAsP/InP	1250			Tr.HS				[80]
1570	2.6		25	AlGaInAs/InP	980		180	Tr.HS	2			[97]
1980	5		24	GaInSb/GaSb	980	1.4	160/175	Tr.HS	9	10	80	[103]
2005	0.64		16	GaInSb/GaSb	808		180	Tr.HS		15	35	[172]
2025	1		18	GaInSb/GaSb	790	1.45	180	Tr.HS	2	15		[94]
2250	3.4		21	GaInAsSb/GaSb	980	1.5	375/425	Tr.HS	5	10		[81]
2350	3.2		13	GaInAsSb/GaSb	1960		300	Tr.HS		15		[48]
5000	0.003			PbTe/PbEuTe/BaF$_2$	1550		200					[83]

Table 1.3 Demonstrated wavelengths and powers of VECSEL lasers for the frequency-doubled second harmonic output operation.

Operating wavelength (nm)	Pump wavelength (nm)	Output power, single mode (W)	Pump power (W)	Material system	Transverse mode (M^2)	Mode diameter (μm)	Transparent heat spreader or inverted structure	No. of QW	Tuning range (nm)	References
338	532	0.12	5.8	GaInP/AlInGaP/GaAs		75	Tr.HS	20	4	[78]
460	808	1.4		InGaAs/GaAsP/GaAs						[157]
460	808	1.9	20	InGaAs/GaAsP/GaAs		140	Tr.HS	30		[173]
460	808	5		InGaAs/GaAsP/GaAs	1.1	500–900	Inv.			[167]
460	808	7	52	InGaAs/GaAsP/GaAs			Inv.			[22]
479	808	7		InGaAs/GaAsP/GaAs			Inv.			[22]
486	808	7.3	38	InGaAs/GaAsP/GaAs	1.2	700	Inv.			[52]
488	808	15		InGaAs/GaAsP/GaAs	1.x	500–900	Inv.		—	[167]
488	808	15	55	InGaAs/GaAsP/GaAs			Inv.			[22]
505	808	8		InGaAs/GaAsP/GaAs			Inv.			[22]
530	808	7	33	InGaAs/GaAsP/GaAs			Inv.			[22]
531	808	11.5	40	InGaAs/GaAsP/GaAs	1.04	700	Inv.			[52]
531	808	24	110	InGaAs/GaAsP/GaAs	—	700	Inv.		—	[52]
532	808	2.7		InGaAs/GaAsP/GaAs						[157]
532	808	7	26	InGaAs/GaAsP/GaAs						[156]
532	808	40	140	InGaAs/GaAsP/GaAs	1.3	900	Inv.	Two chips		[22]
532	808	55	200	InGaAs/GaAsP/GaAs	1.3	900	Inv.	Three chips		[22]
532	808	64	250	InGaAs/GaAsP/GaAs	~4	900	Inv.	Three chips		[22]
570	808	9		InGaAs/GaAsP/GaAs			Inv.			[22]
587	808	5		InGaAs/GaAsP/GaAs		500	Inv.	10	15	[58]
610	805	0.03	2.2	GaAsSb/GaAs		80	Inv.	6		[174]
612	788	2.68	36	GaInNAs/GaAs		290	Tr.HS	10	8	[53]
615	808	0.32	18.2	GaInNAs/GaAs		180	Tr.HS	12	4.5	[88]
617	790	1	45	GaInNAs/GaAs		290	Tr.HS	12		[175]
639		1		GaInNAs/GaAs	1.5		Inv.			[176]

output, have produced emission from 244 nm in the deep UV to the 1180 nm in the NIR, a span of more than two octaves, with output powers from a few milliwatts to 64 W.

As a side note, conventional intracavity frequency-doubled solid-state lasers are susceptible to the so-called "green problem" [179], where lasers exhibit strong intensity noise due to nonlinear interactions between multiple modes of the laser. Intracavity frequency-doubled OPSL lasers do not suffer from the green problem [54] because of the short carrier lifetime in semiconductors and the resonant periodic gain structure of OPSLs that does not allow spatial hole burning. Thus, intracavity frequency-doubled OPSLs have stable, low-noise output, so important for the majority of applications.

Still using GaAs substrates, the unstrained GaAlAs/GaAs material system accesses directly the 850 nm wavelength region [55, 66, 120, 146], with powers as high as 1.6 W [45, 47] demonstrated with in-well pumping. The same wavelength window is accessed with barrier pumping by the strained InAlGaAs/InAlGaAsP/GaAs material system using quaternary quantum wells and quinary strain compensating layers [86]. Still shorter wavelengths near 675 nm in the red are accessed directly using the InGaP/AlGaInP/GaAs material system [78, 79], demonstrating 1.1 W output power. Such direct red-emitting VECSELs have also been pumped by GaN diode lasers [180]. Red-emitting VECSELs allow intracavity frequency doubling into the UV wavelength regime, demonstrating 120 mW output at 338 nm wavelength [78].

The shortest wavelength direct emission VECSEL demonstrated so far is the InGaN/AlGaN/GaN device that operated near 390 nm violet wavelength [95, 96]. One difficulty with such short-wavelength devices is finding a suitable simple pump source; in this case, the device was pumped by a frequency-tripled Q-switched pulsed output of a Nd:YAG laser [95].

Extending VECSEL wavelengths to the 1200–1350 nm NIR range has been accomplished using the GaInNAs/GaAs dilute nitride material system on GaAs substrates [127]. Output powers as high as 1.4 W at 1230 nm [88] and 0.6 W at 1322 nm [67] have been demonstrated; microchip mode of operation has also been reported in this material system [181]. Watt-level emission has been demonstrated in this material system at wavelengths as short as 1160–1210 nm [182]. Recently, using AlGaInAs/InP material system and wafer fusion with GaAs-based mirror, output power of 2.7 W has been demonstrated at 1300 nm wavelengths [98]. This wavelength region is also important since it allows frequency doubling to the visible orange, 590–620 nm, and red, 625–700 nm, wavelengths. Powers as high as 2.7 W have been demonstrated by this approach in the 612–617 nm orange–red wavelength region [53, 88, 175]. Such doubled orange–red emission VECSEL has also been demonstrated using the GaAsSb/GaAs material system [174]. In fact, the NIR range of such GaAs substrate VECSELs should be extendable all the way to the 1550 nm wavelength region using the GaInNAsSb/GaAs quinary material system [183].

The important 1500–1600 nm wavelength window for optical fiber communication applications is covered by the InGaAsP/InP or InGaAlAs/InP material systems on InP substrates. Low refractive index contrast in this material system leads to thick, high thermal impedance laser mirrors; nevertheless, VECSEL output powers as high as 0.8 W have been demonstrated [80, 121]. Using wafer fusion with GaAs-substrate-based mirrors, the InGaAlAs/InP active region has demonstrated still higher VECSEL powers of 2.6 W [97, 98]. Electrically pumped VECSELs have also been fabricated with 1550 nm emission wavelengths using InP-based material system [92, 93, 152].

The 2000–2400 nm emission wavelength mid-IR region is covered by the GaInAsSb/GaSb material system. Output power as high as 5 W has been demonstrated at 1980 nm using standard 980 nm diode pump [103]. More than 3 W output has been achieved at 2250 nm using 980 nm diode pump with a single-chip [81] and dual-chip [184] configurations, and at 2350 nm using 1960 nm thulium-doped fiber laser for in-well pumping [48, 185]. Even longer wavelength VECSELs with emission wavelength near 4.5 and 5 µm have been demonstrated using IV–VI lead–chalcogenide semiconductor material systems of PbTe (PbSe)/PbEuTe on both BaF_2 and Si substrates [82–85]. The long-wavelength VECSELs are discussed in more detail in Chapter 4.

Another approach to extend VECSEL emission to the mid-IR and THz wavelength region is to use difference frequency generation [186], for example, with a dual-wavelength VECSEL [109–111]. Such nonlinear optical difference frequency generation has been analyzed in Ref. [187], where dual emission VECSEL wavelengths at 984 and 1042 nm are considered for 17.7 µm wavelength generation.

VECSEL lasers have also been used for generation of even longer wavelength broadband terahertz radiation in the 0.1–0.8 THz frequency range [115], which corresponds to 0.4–3.0 mm wavelength range. In this terahertz spectrometer arrangement, 480 fs pulses from a passively mode-locked VECSEL laser at 1044 nm were used to illuminate both the emitter and the receiver photoconductive terahertz antennas [115]. The accessible terahertz spectral range in this technique is given by the broad spectral bandwidth of the short pulses from a passively mode-locked VECSEL.

Broad gain bandwidth of semiconductor materials allows appreciable amount of wavelength tuning from VECSELs, which has indeed been demonstrated in many of the above material systems, as indicated in Tables 1.2 and 1.3. For example, 80 nm tuning range has been demonstrated near 1980 nm emission wavelength in the GaSb material system [103]. Even higher tuning range of 156 nm from 1924 to 2080 nm is reported in Ref. [188]. A tuning range of 33 nm has been demonstrated near 972 nm emission wavelength in a two-gain-chip VECSEL laser in the InGaAs/GaAs material system [51]. Intracavity frequency-doubled VECSELs typically exhibit half the tuning range at the second harmonic wavelength as the laser is capable of at the fundamental wavelength. Thus, for example, frequency-doubled yellow laser at 587 nm wavelength exhibits 15 nm of tuning, as compared to the 30 nm tuning range at the fundamental 1174 nm wavelength [58].

To summarize, in the short time since VECSELs were first developed, their demonstrated wavelengths span from 244, 338, and 355 nm in the UV; to the violet, blue, green, yellow, orange, and red wavelengths in the visible 390–675 nm range; to the 850–1570 nm NIR wavelength range; to the 2.0–2.35 µm mid-IR range; to the

5 μm further out in the mid-IR, and even out to terahertz radiation with 0.4–3.0 mm wavelength range. What is important is that these wavelengths were targeted by design, often as required by specific application wavelength requirements. At some of these wavelengths, the materials for which had been subject of extensive development, power levels from watts to tens of watts had been demonstrated with excellent beam quality. At other wavelengths with less-developed materials, power levels of tens and hundreds of milliwatts are more typical. This wavelength and power flexibility of VECSELs, together with their beam quality and other advantageous properties, are key to the growing list of VECSEL applications, which we will discuss next.

1.4.2 Commercial Applications

In their initial application, VECSEL lasers were developed as high-power single-mode fiber-coupled sources at 980 nm for pumping Er-doped fiber and glass-waveguide amplifiers for optical fiber telecommunications systems [169, 189, 190]. Commercial optically pumped OPSL lasers from Coherent Inc. and electrically pumped NECSEL lasers from Novalux Inc. were targeting these potentially large telecom markets. Eventually, however, edge-emitting pump diode lasers came to dominate these applications. VECSEL lasers have been used in research for pumping other solid-state lasers, such as 2 μm emitting Tm- and Ho-doped lasers pumped by a 1 W GaInNAs/GaAs semiconductor disk laser at 1213 nm [182, 191, 192].

With the development of the intracavity frequency-doubled VECSEL lasers with emission wavelengths in the visible, the wavelength versatility of VECSEL, in combination with power scalability and beam quality, became key to their many successful commercial applications. Coherent Inc. pioneered the development and application of commercial OPSL lasers. Doubled into blue OPSL lasers with output at 460 and 488 nm became very successful solid-state replacements for Ar-ion lasers in a wide range of fields. As compared to the big, power inefficient, and limited lifetime Ar-ion lasers, blue OPSL lasers offer compact size, efficiency, low power consumption, beam pointing stability, excellent solid-state reliability with greater than 50 000 hours lifetime, and so on, thus bringing semiconductor laser advantages to visible wavelengths and power levels not previously addressable by these lasers. The option of fiber-coupled delivery of the OPSL laser output is also important for many applications. Relatively fast, 50 kHz–30 MHz and higher, direct modulation capability of OPSLs via pump current modulation is key to many applications, as it eliminates the need for separate acousto-optic modulators, required for the conventional diode-pumped solid-state DPSS lasers. One of the early applications of the OPSL 460 nm blue laser from Coherent Inc. was in laser-based digital to film recorders for transferring digitally produced or edited movies onto conventional 35 mm celluloid master film for cinema projection. ARRILASER film recorder from the ARRI Group uses a blue OPSL laser, together with red and green laser sources, and has been awarded a Scientific and Engineering Award, Oscar, in 2002 by the US Academy of Motion Pictures Arts and Sciences.

With doubled into blue OPSL output powers scalable in the 10–200 mW, these lasers found wide application in bioinstrumentation for confocal fluorescence microscopy [165, 166, 193–197], flow cytometry [198], optical trapping, and manipulation of cells [199], cell sorting, DNA sequencing, proteomics, drug discovery, and so on [16]. In most of these bioapplications, fluorescent marker dyes are used to label functional molecules in the cells, and laser-excited fluorescence from these marker dyes is then detected or imaged. Alternatively, green fluorescent protein gene, and its other fruit color variants [200], can be directly expressed in the cells of many different organisms, and the GFP fluorescence is then used to monitor gene expression in the cells [16]. Specific wavelengths are required for optimal excitation of this imaging fluorescence [198, 200]; OPSL lasers have been designed with emission targeting these wavelengths. The initial blue wavelengths of commercial OPSL lasers have later been expanded to the green: 532 nm, yellow: 561 and 577 nm, red: 639 nm, and ultraviolet: 355 nm emission. The availability of true CW UV emission from OPSL lasers is important for biological applications, where pulsed UV light from the more conventional solid-state lasers can damage biological samples because of the high peak powers of the pulsed lasers.

Coherent Inc. has also developed commercial high-power versions of OPSL lasers with CW output powers in the 0.5–8 W range, with single- and multitransverse mode emission, and, remarkably, with a laser head size as small as $12 \times 5 \times 7\,cm^3$. This combination of UV through visible wavelengths and a range output powers addresses a variety of new applications. Multiwatt green output OPSLs have been used to pump femtosecond mode-locked Ti:sapphire lasers [201]. Multiwatt green, blue, and yellow wavelength OPSL lasers have been used in a portable, battery-powered laser source for crime-scene fingerprint and trace evidence detection [22, 202]. For these applications, efficient battery-powered operation became possible because of the efficiency of the OPSL optical conversion of pump light and that the OPSL pump diodes do not require power-consuming tight wavelength/temperature stabilization of the kind required by the conventional diode-pumped solid-state lasers. Another application area for OPSL lasers is in entertainment, where multiwatt bright color blue, green, and yellow lasers with good beam quality are very effective for laser light shows.

Multiwatt 532 nm green and 577 nm yellow OPSL lasers have been used in ophthalmology for photocoagulation treatment of eye diseases, such as wet-form macular degeneration and diabetic retinopathy [203]. The yellow 577 nm wavelength is especially effective since it matches the absorption peak of the oxygenated hemoglobin in blood. As a result of this wavelength matching, the OPSL laser delivers effective treatment while less light is absorbed in the surrounding tissue.

Red, green, and blue lasers can be combined for RGB color projection displays [204], ranging in size from full cinema projection [205] to laser television [206], to mobile micro projectors [207]. Laser projection displays can be made in three different types [207]. The first type is a laser scanning or "flying spot" projector [208], where three primary color lasers are combined into a single beam and a mirror system is used to scan this single-pixel beam in 2D over the imaging screen surface. The second type is a line scanning projector, where primary color lasers illuminate a linear array of pixels in a 1D spatial light modulator, such as grating light valve (GLV) [209], the resulting line image is then projected onto a screen and scanned by a mirror in 1D to produce a full image. The third type is an imaging projector, where primary color lasers illuminate a 2D spatial light modulator pixel array, such as Texas Instruments' digital micromirror digital light processor (DLP) chip, the resulting full image is then projected onto a screen. For example, Microvision's mobile pico projector uses the first "flying spot" approach [207]. Sony has been developing the second line scan approach for laser TV applications using GLV technology [207]. The same GLV line scan approach is used by Evans & Sutherland in its ESLP®-8K laser projector [205]. Mitsubishi's LaserVue® TV uses the third 2D imaging approach. Such laser projection displays with saturated primary RGB colors can access a much wider color gamut [204] than is available from other display technologies, such as liquid crystal and plasma displays. Different levels of laser power, from milliwatts to many watts, are required for different size displays. Visible OPSL lasers can be used with all three types of laser displays and will play an important role here because of their power scaling, efficiency, beam quality, fast direct modulation capability, and access to the desired emission wavelengths.

A number of companies have been working to develop OPSL/SDL/VECSEL lasers for these display applications, such as Coherent, OSRAM, Samsung, and Novalux/Arasor. For mobile micro/pico projectors, such as the one developed by Microvision Inc. using the "flying spot" laser scanning approach, direct emission blue and red diode lasers are available; however, the green lasers are not available because of the semiconductor material limitations. Doubled from NIR into green, several semiconductor laser approaches are being developed, including an externally doubled diode laser from Corning and an intracavity frequency-doubled OPSL laser from OSRAM Opto Semiconductors [207, 208]. OSRAM's green OPSL laser for mobile projectors emits 70 mW at 530 nm, measures only $13 \times 6.5 \times 4.8$ mm^3, and is capable of direct modulation at rates greater than 30 MHz. Novalux/Arasor has been developing NECSEL™ electrically pumped VECSELs for display applications [210], such as DLP-based LaserVue TV from Mitsubishi [206]; these lasers are among the lasers used in the Evans & Sutherland ESLP-8K high-resolution $8\,K \times 4\,K$ digital theater laser video projector [205].

Accessing a mid-IR 1.9–2.5 μm wavelength range, Solus Technologies has developed a 100 mW narrow linewidth laser source for a range of applications from gas sensing and molecular spectroscopy [106] to medical. Such lasers are potentially capable of watt-level operating powers, 100 nm wide tuning range, and single-frequency operation.

It is remarkable for this young technology that such a diversity of commercial OPSL lasers is available, with wavelengths from UV to mid-IR and output powers from milliwatts to watts, and that such a wide variety of applications is addressed, from bio-medical to forensics and to displays. This testifies to the flexibility and versatility of OPSL lasers that can successfully address all these varied application requirements and where there is simply no other laser that can do the job as effectively.

1.4.3
Current and Future Research Directions

Future research directions and applications of VECSELs will surely reflect the already demonstrated versatility of the VECSEL laser platform. Output power scaling of VECSELs will continue to be a topic of interest. Power scaling can utilize a number of approaches, such as increasing mode area and pump spot size while controlling and limiting the in-plane amplified spontaneous emission that can deplete the gain [159]. Further improvements in efficient heat removal (Chapter 2) will help, such as using the highest thermal conductivity diamond heat spreaders and heat sinks [107], front and back side heat removal from the OPS chip, design and fabrication of low thermal impedance semiconductor structures, and so on. Further power scaling will be achieved by using multiple gain chips [22, 49, 50, 52, 184], with larger mode area per chip, from the current maximum $\sim$0.9 mm diameter, and a larger number of OPS gain chips, from the presently demonstrated maximum of three chips. Still further power scaling is possible by spectrally combining multiple laser beams at different wavelengths, as demonstrated by combining two VECSEL lasers preserving excellent beam quality using a volume Bragg grating [211]. VECSEL output powers of several hundred watts, at the fundamental or the second harmonic, should be achievable with excellent beam quality using some combination of the above techniques. At the same time, demonstrating compact and efficient, low-cost, manufacturable VECSELs with modest milliwatt- to watt-level output powers will also be important for many applications, such as mobile projection displays [207].

Wavelength scaling of VECSELs will likely proceed along two paths: one will be application-driven demonstrations of new wavelengths using the existing semiconductor material systems and techniques; the other will explore new wavelengths by using novel material systems or novel nonlinear optical techniques to expand the versatility of the VECSEL approach. In addition to novel semiconductor material systems, finding suitable pump lasers and efficient pumping schemes for new VECSEL wavelength regions will also be important. For example, direct blue emission in the GaN material system [95, 96] with high powers and good beams would be very useful; however, suitable pump diode lasers are not readily available at the required wavelengths. Perhaps a multistage cascade of intracavity frequency-doubled VECSELs can be demonstrated, where output of one stage of such cascade is used to pump the next stage of the cascade. A variety of nonlinear optical frequency conversion schemes should expand the range of accessible wavelengths: intracavity frequency doubling, tripling, and quadrupling; sum frequency generation; difference frequency generation from dual wavelength lasers for accessing new mid-IR wavelengths [109, 111], such as 17.7 µm [187]; a combination of laser intracavity and extracavity nonlinear conversions [77]; and using single and multiple lasers for nonlinear optical conversions. VECSELs can become an important compact source of broadband terahertz radiation by using ultrashort pulse mode-locked VECSELs in combination with photoconductive antennas [115]. To summarize, VECSELs wavelength coverage range will extend deeper into the UV and further into mid-IR, as well as start filling in the gaps in the existing wavelength coverage.

Tunable wavelength operation will prove important for many VECSEL applications. The target here will be to extend the relative wavelength tuning ranges from the currently demonstrated 3–8%, 33 nm tuning near 972 nm [51], 80 nm tuning near 1980 nm [103], and 156 nm tuning near 2000 nm [188], to the 10–15% level. The ways to achieve this will include using broadband gain semiconductor materials, such as quantum dots (Chapter 5); broadband OPS chip gain structures with quantum well, or dot, layers displaced somewhat from their periodic RPG positions [18, 130]; multiple, nonidentical gain chips in the cavity [51]; and, also important, low-loss tunable filters. Single-frequency tuning of VECSELs [55, 140, 164] is also important and will see further development for some applications, such as laser sources for atomic clocks [120, 146] and spectroscopy [106].

Development of new semiconductor materials both for gain and on-chip mirror structures will enable better performance of VECSELs and expand their operating range in terms of power, wavelength, tuning range, and so on. New semiconductor material systems with binary, ternary, quaternary, and quinary alloys can enable easier materials growth with fewer defects for the gain region and better index contrast and lower thermal impedance for the VECSEL mirrors. Unstrained or strain-compensated quantum well structures will be required in these material systems. Further development of the quantum dot active region (Chapter 5) [99–102] will enable VECSELs with reduced thresholds, wider gain bandwidth, and reduced temperature dependence [101]. Bonding wafers of different semiconductor materials using wafer fussion [97, 98], such as for bonding separate VECSEL mirror and gain wafers, eliminates some of the material limitations in VECSELs. Expanding the use of this technique will enable a range of novel better performing devices, for example, with higher powers and at new wavelengths. Wafer fusion of other VECSEL functional blocks might also prove useful, such as bonding of gain regions at different wavelengths, bonding of diode pump lasers to the gain region, and bonding gain and saturable absorbers for mode locking. New materials and epitaxial growth techniques have been driving the development of semiconductor lasers for more than 40 years, and they will also be critical for the future VECSEL development. Complex multilayer semiconductor VECSEL structures require state-of-the-art semiconductor materials and growth techniques. In addition, finding optimal nonlinear optical materials, including periodically poled crystals, will boost performance of the frequency-doubled visible emitting VECSELs.

Further development of electrically pumped VECSELs is an important topic of future research. While optically pumped VECSELs will reach comparatively higher output powers, electrical pumping capability will enable more compact, efficient, and simple devices for many applications of VECSELs in a variety of operating regimes:

cw and pulsed, tunable, intracavity doubled, and so on. Electrically pumped VECSELs are discussed in detail in Chapter 7.

Another important future direction for VECSEL research is the integration of multiple functional blocks on a common semiconductor substrate. For example, integration of pump lasers with the gain-mirror VECSEL structure has already been demonstrated [154]. Further development of such pump-integrated VECSELs can produce highly manufacturable high-power lasers that require significantly reduced component assembly, thus reducing the device cost and expanding the potential markets for these lasers. Another example of functional component integration is the MIXSEL laser (Chapter 6) [160, 161], where gain and saturable absorber regions of short-pulse mode-locked VECSELs are integrated on a common substrate. An important direction of functional integration is the development of various forms of VECSEL-integrated optical cavities; such integrated cavities are more compact and better manufacturable and might prove very valuable for some applications. This includes plane–plane microchip cavities with thermally stabilized transverse modes [69–71, 181]; compact plane–curved cavities where the curved surface is produced as a microlens on an optical substrate [141–144, 212, 213]; and single-transverse mode optical resonators, for example, in the pillbox configuration [145]. Such functional integration, including combinations of the above schemes, can produce simpler, more compact, easier manufacturable, and cheaper devices, as well as enable better device performance and novel functions. Future development of functional integration can help make VECSELs even more widely used commercially, especially in low-cost and high-volume applications, such as mobile projection displays.

A variety of optical cavity configurations have been used with VECSELs; development of optical cavities optimized for different applications will play an important role in expanding VECSEL function and application. While integrated optical cavities mentioned above are important, compact bulk optical cavities with large mode area for high-power operation [52] will also be valuable. Optical cavity design is an important factor in VECSELs incorporating multiple OPS chips [22, 49–52, 135, 184]. VECSELs using intracavity nonlinear optical crystals for harmonic generation or saturable absorbers for mode locking require optical cavities that control the absolute and relative laser mode diameters in the gain and the nonlinear element. In contrast with most VECSELs that use a linear laser cavity, even if with multiple cavity elements, ring laser cavity has been used in a VECSEL-based laser gyro [137] and a passively mode-locked VECSEL [136]. A diffractive unstable optical resonator has been used with VECSELs where Gaussian beam output was extracted from a hard-edged outcoupling aperture [138]. VECSELs have also been operated in unstable resonator regime with plane–spherical resonator cavities [18]. Optimized and novel optical cavities will enable VECSELs with enhanced performance, while making them more compact, stable, and manufacturable.

Array operation of VECSELs has been demonstrated both with electrical (Chapter 7) and optical [70, 143] pumping. This is a promising research direction for producing higher output powers from multiple beam incoherent arrays (Chapter 7) or phase-locked arrays [214]. Individually addressable VECSEL arrays can be used for optical micromanipulation applications, such as optical trapping and transport of biological cells [215]. Incoherent arrays of VECSELs also produce reduced laser speckle for illumination applications, such as in imaging projection displays [210], where laser speckle reduces as the square root of the number of array elements.

VECSEL laser power modulation is important for many applications, such as for laser projection displays. Laser power can be modulated directly by current for electrically pumped VECSELs [49] or via pump diode current modulation for optically pumped VECSELs [207, 208]. Exploring modulation properties of VECSELs of both types and optimizing their high-speed modulation performance is an important future research direction. For example, pumping a VECSEL with multiple pump diodes connected in series can yield a high slope differential modulation capability useful in microwave photonics applications. Fast laser wavelength tuning might also prove important; such fast tuning is possible with MEMS-tunable optically pumped VECSELs that were developed for optical fiber communication applications [216–218].

A range of VECSEL applications require single-frequency laser operation, both tunable and nontunable, for example, molecular spectroscopy [106, 219], pumping of atomic clocks [120, 146], ring laser gyro [137], and microwave photonics applications [56, 220]. A number of studies have explored properties of VECSELs in their single-mode operating regime, such as laser linewidth, which can be very narrow in the kilohertz range [55, 104, 106, 119], and relative intensity noise (RIN) and its spectrum [56, 140]. Such single-mode operation has been achieved both with intracavity filters [55, 56, 59, 221], such as etalons, birefringent filters, and volume Bragg gratings, or alternatively free running without intracavity filters [106, 140]. Single-mode operation has been demonstrated both for the fundamental and intracavity frequency-doubled VECSELs [222]. Further investigation of the single-mode VECSEL regime and understanding of its properties can yield devices with unique characteristics for the above-mentioned applications, such as hertz-level quantum-limited linewidth [119] and shot-noise limited RIN [56], while at the same time with watt-level output powers and potentially tunable over several tens of nanometers wavelength range.

Short pulses at gigahertz repetition rates have been generated by saturable absorber passive mode locking of VECSELs [20]; this topic is discussed in detail in Chapter 6. Such passive mode locking has been demonstrated with both optically [62, 134, 223–226] and electrically [227–230] pumped VECSELs. Some of the remarkable characteristics demonstrated with passively mode-locked VECSELs are transform-limited pulses as short as 220 fs [231], repetition rates as high as 50 GHz [62], timing jitter as low as 160 fs with actively stabilized laser [232], and average output powers as high as 2.1 W [133], all with excellent circular output beams. Saturable absorbers have also been integrated with the gain medium on the same semiconductor substrate for a very simple and compact implementation of a passively mode-locked VECSEL or MIXSEL [160, 161]. Alternatively to saturable absorber passive mode locking, VECSELs have also been actively mode-locked by synchronous pumping with picosecond and femtosecond pulses from a Ti:sapphire laser [233]. The resulting chirped picosecond pulses were compressed down to 185 fs by a grating pair compressor. Semiconductor gain medium, unlike traditional solid-state laser materials, for example, Yb:YAG, does not have long storage times; therefore, high pulse energy regime is outside the scope of VECSEL characteristics. VECSEL sources of high-power ultrashort pulses at gigahertz repetition rates will find applications in such areas as optical clock distribution in high-speed computing systems and metrology. Future research will expand the short pulse generation performance of VECSELs in the direction of shorter femtosecond pulses, higher repetition rates beyond 100 GHz, new operating wavelengths from UV to mid-IR with different material systems and nonlinearly converted optical frequencies, compact integrated implementations, and further development of electrically pumped devices. Such compact high-power ultrashort pulse sources will find many applications and will displace more conventional solid-state pulsed sources in many application areas.

Microwave photonics applications transmit analogue radio frequency (RF) or microwave signals modulated on top of an optical carrier. VECSELs are exceptionally suited as light sources for such microwave links because of their high power and shot-noise-limited RIN capability [56]. With a high-Q external cavity, VECSELs operate in the flat-noise oscillation-relaxation-free class-A laser regime, achieved because photon lifetime in the VECSEL laser cavity is much larger than the carrier lifetime in the gain chip [119, 220]. Application of VECSELs to microwave photonics links is a promising direction of future development.

With their high-power, tunable, single-mode, and narrow-linewidth operation capability, VECSELs are a good laser light source for pumping atomic clocks, such as Cs atomic clock near 852 nm [120, 146]. Here, VECSELs can be used for trapping, cooling, manipulation, and optical detection of atoms. The use of VECSELs for optical tweezers for trapping and manipulation of biological samples [199] is also an area of future research interest; output power, wavelength, and beam quality, all in compact size, are key VECSEL qualities for these applications.

Another interesting application of VECSELs is for rotation sensing using VECSEL-based ring laser gyro [137]; the initial demonstration of such laser gyros is certain to be followed by many others. High power, excellent beam quality, and wavelength selection of VECSELs have a natural application for a sodium laser guide star [58, 234]. Here a multiwatt 589 nm yellow laser beam is used to excite the D_2 sodium line in the mesospheric layer of sodium atoms, creating a glowing artificial "star" in the regions of the sky where no bright stars are available. This sodium guide star is then used for adaptive optics compensation of atmospheric distortion in ground-based astronomical telescopes [234, 235] and in space–ground optical links.

Important for the future development and broad application of VECSELs is detailed physical modeling, including analytical and numerical models, of various aspects of their behavior. One example that is already bearing fruit is the modeling of the thermal heat dissipation issues in VECSELs [20, 72, 236, 237]; this subject is discussed in detail in Chapter 2. Modeling and understanding of the short pulse generation mechanisms in VECSELs, including cavity group delay dispersion (GDD), nonlinear self-phase modulation (SPM), and gain and loss saturation dynamics, has enabled decreasing mode-locked pulse widths from the picosecond to the femtosecond regime [20, 134, 238], with further pulse width decreases expected in the future. Modeling of the VECSEL single-mode regime nonlinear dynamics and noise [119, 239] will produce higher performing devices for microwave photonics and atomic clock applications. Further application of full numerical microscopic models of semiconductor lasers, including gain properties of semiconductor materials [117, 118], as well as modeling of gain and carrier behavior in VECSELs [240, 241], will lead to better understanding of the laser behavior and design of improved performance lasers. Detailed modeling is also required for improving the performance of electrically pumped VECSELs (Chapter 7).

An important application area of VECSELs of much research is in molecular spectroscopy. Tunable single-frequency VECSELs have been used directly to acquire spectra in the NIR by scanning the laser line across absorption features of interest, for example, methane absorption near 2.3 μm [106, 219, 242]. Detection sensitivity of VECSEL spectroscopy can be magnified with cavity-enhanced absorption spectroscopy [243], where a mode-locked VECSEL frequency comb is matched to the comb of resonances of a high-finesse cavity containing gas under study. Using this technique, acetylene absorption near 1 μm was measured with a 25 m effective absorption path length for a 13 cm long cell [243].

Even greater enhancement of detection sensitivity in molecular spectroscopy is obtained with an ultrasensitive ICLAS technique, first demonstrated with VECSELs in Refs [73, 74]. In this technique, the absorption cell is placed inside optical cavity of a homogeneously broadened multimode laser, the laser is turned on, and the lasing spectrum is measured by a spectrometer a time t_g later within a narrow time slot Δt_g. This approach gives effective absorption path length $L_p = \varrho c t_g$, where ϱ is the length fraction of the laser occupied by the absorption cell and c is the speed of light. For the typical generation times of 200–400 μs and a 50 cm long absorption cell, the effective absorption path lengths are 30–60 km; effective path lengths as long as 130 km have been demonstrated for a 1 m long absorbing path [244]. Due to the long effective path lengths, ICLAS-VECSEL allows measurement of absorption coefficients as low as 10^{-9}–10^{-10} cm^{-1} (sensitivity $\sim 10^{-11}$ cm^{-1} Hz$^{-1/2}$). Acousto-optic modulator is used to switch the laser beam to the measuring spectrometer; both grating and Fourier transform spectrometers have been used for spectrum acquisition in ICLAS. Several properties of VECSELs enable their application to ICLAS: wide range of available wavelengths, broad homogeneous gain bandwidth, and very low loss laser cavity; because of their high loss, edge-emitting semiconductor lasers are not very suitable for ICLAS. ICLAS-VECSEL technique has been applied for molecular spectroscopy measurements at wavelengths near 1.0 μm (8900–10 100 cm^{-1}) [74, 244, 245], 1.55 μm (6500 cm^{-1}) [148], and 2.3 μm (4300 cm^{-1}) [246]. With its high sensitivity, ICLAS-VECSEL can be used for trace gas detection or for measurement of weak absorption lines of atmospheric gases. A variety of molecular gases have been measured by ICLAS-VECSEL, detecting thousands of spectral lines, including new ones. Rovibrational analysis has been used to assign the detected lines [247]. Some of the measured molecular species are nitrous oxide (N_2O) [245], water (H_2O) [148, 244, 246–248], carbon dioxide (CO_2) [148, 246], hydrogen sulfide (H_2S) [249], acetylene (HCCH) [248], and carbonyl sulfide (OCS) [250], as well as a variety of isotopic variants such as D_2O [251]. Prior to the development of semiconductor VECSELs, ICLAS relied primarily on Ti:sapphire laser in the 700–900 nm range and dye lasers in the visible range. Semiconductor VECSELs opened new wavelengths for ICLAS in the NIR and mid-IR 0.9–2.5 µm wavelength, 4000–11 000 cm^{-1} wave number, including the spectrally rich molecular overtone region. VECSELs also offer a simple and compact laser source for spectroscopy. Using VECSELs for ICLAS revealed an unexpected spectral condensation effect, where spectral absorption dips are replaced by blue-shifted emission peaks at longer generation times [248]. ICLAS-VECSEL promises to be an active research area in the future with more molecular species observed, improved detection sensitivities, and expanded spectral detection regions in the NIR and mid-IR. Highly sensitive ICLAS-VECSEL gas detection can have applications in understanding of atmospheric gases and in medicine.

To summarize this section, current research work has explored an amazing variety of applications that exploit the unique properties of VECSELs. Further research into expanding, improving, and understanding of the VECSEL properties will lead to even greater growth in the future of the research and commercial applications of these remarkable lasers.

1.4.4
Future of VECSEL Lasers: Scalable Power with Beam Quality from UV to IR

This chapter has outlined a new semiconductor laser technology that comes with several alternative names: vertical-external-cavity surface-emitting laser, semiconductor disk laser, or optically pumped semiconductor laser. This is one technology that, uniquely among various other laser technologies, addresses *simultaneously* multiple laser application requirements: power scaling, wavelength versatility, beam quality, compact size, efficiency, reliability, and so on. This technology allows laser *design* with characteristics to match application requirements optimally, for example, output power from milliwatts to tens of watts, wavelength from UV to mid-IR, and an excellent beam quality. It should be emphasized that VECSEL technology achieves the desired laser characteristics by *design*, rather than by hoping for a *serendipitous coincidence*, as with solid-state and gas laser wavelength targeting.

What is the source of these remarkable properties of VECSEL technology? The answer is, as this chapter has described, that VECSELs combine some of the best qualities of semiconductor lasers and solid-state lasers, while avoiding many of the weaknesses of these two laser classes.

What are VECSEL strengths acquired from semiconductor lasers? VECSELs take from semiconductor lasers the use of semiconductor materials with their various binary compounds and multicomponent alloys that can target a very wide range of emission wavelengths. VECSELs have also incorporated bandgap-engineered semiconductor heterostructures that are used to engineer by design the desired laser emission and pump absorption properties. Due to the semiconductor laser materials, optically pumped VECSELs have strong pump absorption in a very thin layer, just a few microns thick, and over extremely wide range of pump wavelengths, tens and hundreds of nanometers. VECSELs have also inherited the excellent circular output beam of a VCSEL surface-emitting semiconductor laser.

What are VECSEL strengths acquired from the solid-state lasers? The key contribution of solid-state lasers to VECSELs is the flexible optical laser cavity, which enables many of the key VECSEL properties. Optical cavity provides transverse mode control to VECSELs, which enables their excellent beam quality. VECSEL optical cavity also enables mode size scaling for laser output power scaling. More important, flexible open laser cavity makes possible for VECSELs the insertion of various intracavity functional optical elements, such as nonlinear optical crystals for harmonic generation, saturable absorbers for short pulse passive mode locking, multiple gain chips, optical filters for spectral control, and so on. This gives VECSELs many different regimes of operation and underlies their tremendous functional flexibility. Significantly, both solid-state laser and VECSEL optical cavity has low intracavity loss. This has important consequences by allowing efficient use of low transmission output coupling, which in turn results in high intracavity power compared to output laser power and thus enables very high nonlinear optical conversion efficiency. Another feature acquired by VECSELs from solid-state lasers is optical pumping. While electrical pumping of VECSELs is preferred in some cases, optical pumping produces much higher VECSEL output powers and allows much simpler VECSEL laser wafer structure.

Many VECSEL properties are enabled by avoiding some of the key weaknesses of semiconductor lasers. Edge-emitting semiconductor lasers have their light locked in a waveguide; good beam quality requires this to be a single-mode waveguide, which results in a nonscalable small laser mode size that limits laser output power with good beam quality. Surface-emitting VCSEL lasers have no adequate transverse mode control and good beam quality single-transverse mode output power is again limited. In contrast, VECSELs avoid these limitations; they have open cavities, allowing both transverse mode control for fundamental circular mode operation and mode size scaling for higher output powers. Another disadvantage of edge-emitting semiconductor lasers is their lossy optical cavity, both due to lossy waveguides, which give loss of several tens of percent per cavity pass, and due to lossy coupling to external optical cavity. Lossy cavities of edge-emitting lasers require for output power efficiency the use of high-transmission output couplers, hence intracavity power becomes comparable to output power, and thus no efficient intracavity-enhanced nonlinear optical frequency doubling is possible. Lossy cavity also produces strong spontaneous emission noise. In contrast, VECSELs with their low-loss laser cavities do not have these limitations.

What are the weaknesses of solid-state lasers avoided by VECSELs? The key weakness of solid-state lasers is very inflexible emission wavelengths, which are constrained to the few available transitions of active ions. The second weakness, for similar reasons, is the very inflexible narrow range of pump wavelengths; pump diode lasers at such wavelengths might not be readily available, when available, they require costly wavelength stabilization, for example, by temperature stabilization. Another limitation of solid-state lasers is relatively weak, compared to semiconductors, pump absorption, thus requiring thicker crystals for efficient pump absorption. Thicker absorbing crystals require complex pump optics and relatively good quality pump beams; thicker crystals also produce thermal lensing, which distorts beam and limits output power. VECSELs avoid these limitations due to strong broadband pump absorption of semiconductors.

The solid-state lasers typically have fairly long excited-state lifetimes, which can be as long as a millisecond, allowing energy storage and high-energy pulse generation in these lasers. On the other hand, the long carrier lifetime of solid-state lasers severely limits their output modulation rate and typically requires external modulation. In contrast, semiconductor materials of VECSEL have very short carrier lifetimes of the order of a nanosecond. As a result, VECSELs lose the energy storage and the high pulse energy capabilities of solid-state lasers. However, due to the short carrier lifetime, VECSELs acquire fast modulation capability, either directly via current modulation for electrically pumped VECSELs or via pump diode power modulation for optically pumped VECSELs.

This combination of advantageous laser properties in VECSELs produces a very flexible and versatile laser technology, as illustrated by the multitude of research directions described above: power scaling from milliwatts to tens of watts with single or multiple gain chips, UV and visible output wavelengths via harmonic generation, wavelength tunable operation, femtosecond short pulse generation at gigahertz high rates, single-frequency operation, intracavity laser absorption spectroscopy, laser gyros, microwave photonics, terahertz spectroscopy, and so on. Versatile VECSEL laser technology leads to an impressive and surprising variety of commercial applications already demonstrated for this new technology: from biomedical fluorescence imaging to ophthalmic photocoagulation treatment, to projection displays, to forensic evidence analysis, to pumping Ti:sapphire lasers. VECSEL applications so far have been mostly in the UV and visible wavelengths, where compact and efficient sources with other laser technologies have been lacking. The accomplishments until now for the young VECSEL laser technology only illustrate its rich potentials; future research and development are going to expand even wider the scope of VECSEL performance, functional versatility, and the range of scientific and commercial applications.

What is the future of VECSEL lasers? VECSELs will expand the range of their capabilities both through technology push and a market pull. VECSEL technology will expand the range of VECSEL-operating parameters: output powers will increase to the hundreds of watts levels maintaining excellent beam quality; wavelength coverage will expand deeper into UV below 200 nm and further into mid-IR beyond 5 µm with more complete wavelength coverage inside this range; shorter pulses in the tens of femtoseconds will become possible, with increased average power levels of multiple watts; single-frequency operation with subkilohertz linewidth will be demonstrated;

and VECSELs will become efficient sources of terahertz radiation. At the same time, various VECSEL performance parameters will be optimized: expect higher power efficiencies; smaller laser packages; further integration of various functional components, such as pump lasers, saturable absorbers, and optical cavities; and electrical pumping will be used more widely when needed and possible. VECSELs will be used more widely in novel scientific applications, such as molecular spectroscopy with ICLAS, laser gyros, microwave photonics, and atomic clocks. Market pull from existing and developing applications will stimulate appearance of new commercial devices, for example, multiwatt red, green, and blue (RGB) VECSEL lasers for cinema projection applications. The availability of compact, efficient, and high-performance lasers with targeted custom-designed wavelength and other parameters will expand their use in existing commercial applications and will enable new applications that could not be addressed previously with other laser technologies. VECSELs will be displacing other laser technologies in some applications areas, for example, Ar-ion gas lasers with the blue and green wavelength outputs have been used in the past as the only suitable source for confocal fluorescence microscopy and for pumping Ti: sapphire lasers. These gas lasers have largely been displaced by diode-pumped solid-state lasers at these wavelengths; solid-state lasers in turn are in the process of being themselves displaced by VECSELs in these application areas. For these applications, VECSELs offer wavelength flexibility, power scalability, beam quality, compact size, good efficiency, and high reliability.

We are entering a new era of VECSEL/SDL/OPSL lasers that are poised to become ubiquitous laser sources in scientific and commercial applications. Perhaps some of the more interesting future VECSEL developments are the unexpected ones; future of VECSELs will be full of bright surprises.

Acknowledgments

I would like to acknowledge my former colleagues at Micracor Inc. who greatly contributed to the development of the first VECSEL lasers: Aram Mooradian, Farhad Hakimi, and Robert Sprague. I am also grateful to Oleg Okhotnikov for putting this book collection together.

References

1 Svelto, O. (1998) *Principles of Lasers*, 4th edn, Springer.
2 Agarwal, G.P. (2002) *Fiber-Optic Communication Systems*, 3rd edn, Wiley–Interscience.
3 Agrawal, G.P. and Dutta, N.K. (1986) *Long-Wavelength Semiconductor Lasers*, Van Nostrand Reinhold Co., New York.
4 Zory, P.S. (ed.) (1993) *Quantum Well Lasers*, Academic Press, San Diego.
5 Chuang, S.L. (1995) *Physics of Optoelectronic Devices*, John Wiley & Sons, Inc., New York.
6 Coldren, L.A. and Corzine, S.W. (1995) *Diode Lasers and Photonic Integrated Circuits*, John Wiley & Sons, Inc., New York.
7 Chow, W. and Koch, S.W. (1999) *Semiconductor Laser Fundamentals*, Springer, Berlin.

8. Kapon, E. (ed.) (1999) *Semiconductor Lasers I: Fundamentals*, Academic Press.
9. Kapon, E. (ed.) (1999) *Semiconductor Lasers II: Materials and Structures*, Academic Press.
10. Wilmsen, C.W., Temkin, H., and Coldren, L.A. (eds) (1999) *Vertical-Cavity Surface-Emitting Lasers: Design, Fabrication, Characterization, and Applications*, Cambridge University Press.
11. Li, H. and Iga, K. (eds) (2002) *Vertical-Cavity Surface-Emitting Laser Devices*, Springer.
12. Diehl, R. (ed.) (2000) *High Power Diode Lasers*, Springer.
13. Koechner, W. (2006) *Solid-State Laser Engineering*, 6th edn, Springer.
14. Koechner, W. and Bass, M. (2003) *Solid-State Lasers: A Graduate Text*, Springer.
15. Digonnet, M.J.F. (ed.) (2001) *Rare-Earth-Doped Fiber Lasers and Amplifiers*, 2nd edn, CRC Press.
16. Prasad, P.N. (2003) *Introduction to Biophotonics*, Wiley-Interscience.
17. Kuznetsov, M., Hakimi, F., Sprague, R., and Mooradian, A. (1997) High-power (>0.5-W CW) diode-pumped vertical-external-cavity surface-emitting semiconductor lasers with circular TEM_{00} beams. *IEEE Photon. Technol. Lett.*, **9**, 1063–1065.
18. Kuznetsov, M., Hakimi, F., Sprague, R., and Mooradian, A. (1999) Design and characteristics of high-power (>0.5-W CW) diode-pumped vertical-external-cavity surface-emitting semiconductor lasers with circular TEM_{00} beams. *IEEE J. Sel. Top. Quantum Electron.*, **5**, 561–573.
19. Tropper, A.C., Foreman, H.D., Garnache, A., Wilcox, K.G., and Hoogland, S.H. (2004) Vertical-external-cavity semiconductor lasers. *J. Phys. D: Appl. Phys.*, **37**, R75–R85.
20. Keller, U. and Tropper, A.C. (2006) Passively modelocked surface-emitting semiconductor lasers. *Phys. Rep.*, **429**, 67–120.
21. Tropper, A.C. and Hoogland, S. (2006) Extended cavity surface-emitting semiconductor lasers. *Prog. Quantum Electron.*, **30**, 1–43.
22. Chilla, J., Shu, Q.-Z., Zhou, H., Weiss, E., Reed, M., and Spinelli, L. (2007) Recent advances in optically pumped semiconductor lasers. *Proc. SPIE*, **6451**, 645109.
23. Schulz, N., Hopkins, J.M., Rattunde, M., Burns, D., and Wagner, J. (2008) High-brightness long-wavelength semiconductor disk lasers. *Laser Photon. Rev.*, **2**, 160–181.
24. Calvez, S., Hastie, J.E., Guina, M., Okhotnikov, O.G., and Dawson, M.D. (2009) Semiconductor disk lasers for the generation of visible and ultraviolet radiation. *Laser Photon. Rev.*, **3** (5), 407–434.
25. Boyd, R.W. (2008) *Nonlinear Optics*, 3rd edn, Academic Press.
26. Seurin, J.-F., Xu, G., Khalfin, V., Miglo, A., Wynn, J.D., Pradhan, P., Ghosh, C.L., and D'Asaro, L.A. (2009) Progress in high-power high-efficiency VCSEL arrays. *Proc. SPIE*, **7229**, 722903–722911.
27. Delfyett, P.J., Shi, H., Gee, S., Barty, C.P.J., Alphones, G., and Connolly, J. (1998) Intracavity spectral shaping in external cavity mode-locked semiconductor diode lasers. *IEEE J. Sel. Top. Quantum Electron.*, **4**, 216–223.
28. Hadley, M.A., Wilson, G.C., Lau, K.Y., and Smith, J.S. (1993) High single-transverse-mode output from external-cavity surface-emitting laser diodes. *Appl. Phys. Lett.*, **63**, 1607–1609.
29. Chinn, S.R., Rossi, J.A., Wolfe, C.M., and Mooradian, A. (1973) Optically pumped room-temperature GaAs lasers. *IEEE J. Quantum Electron.*, **QE-9**, 294–300.
30. Stone, J., Wiesenfeld, J.M., Dentai, A.G., Damen, T.C., Duguay, M.A., Chang, T.Y., and Caridi, E.A. (1981) Optically pumped ultrashort cavity $In_{1-x}Ga_xAs_yP_{1-y}$ lasers: picosecond operation between 0.83 and 1.59 μm. *Opt. Lett.*, **6**, 534–536.
31. Roxlo, C.B. and Salour, M.M. (1981) Synchronously pumped mode-locked CdS platelet laser. *Appl. Phys. Lett.*, **38**, 738–740.
32. Le, H.Q., Di Cecca, S., and Mooradian, A. (1991) Scalable high-power optically

pumped GaAs laser. *Appl. Phys. Lett.*, **58**, 1967–1969.

33 McDaniel, D.L., Jr., McInerney, J.G., Raja, M.Y.A., Schaus, C.F., and Brueck, S.R.J. (1990) Vertical cavity surface-emitting semiconductor laser with cw injection laser pumping. *IEEE Photon. Technol. Lett.*, **2**, 156–158.

34 Sandusky, J.V. and Brueck, S.R.J. (1996) A CW external-cavity surface-emitting laser. *IEEE Photon. Technol. Lett.*, **8**, 313–315.

35 Hanaizumi, O., Jeong, K.T., Kashiwada, S.-Y., Syuaib, I., Kawase, K., and Kawakami, S. (1996) Observation of gain in an optically pumped surface-normal mutiple-quantum-well optical amplifier. *Opt. Lett.*, **21**, 269–271.

36 Jiang, W.B., Friberg, S.R., Iwamura, H., and Yamamoto, Y. (1991) High powers and subpicosecond pulses from an external-cavity surface-emitting InGaAs/InP multiple quantum well laser. *Appl. Phys. Lett.*, **58**, 807–809.

37 Sun, D.C., Friberg, S.R., Watanabe, K., Machida, S., Horikoshi, Y., and Yamamoto, Y. (1992) High power and high efficiency vertical cavity surface emitting GaAs laser. *Appl. Phys. Lett.*, **61**, 1502–1503.

38 Le, H.Q., Goodhue, W.D., and Di Cecca, S. (1992) High-brightness diode-laser-pumped semiconductor heterostructure lasers. *Appl. Phys. Lett.*, **60**, 1280–1282.

39 Le, H.Q., Goodhue, W.D., Maki, P.A., and Di Cecca, S. (1993) Diode-laser-pumped InGaAs/GaAs/AlGaAs heterostructure lasers with low internal loss and 4-W average power. *Appl. Phys. Lett.*, **63**, 1465–1467.

40 Stewen, C., Contag, K., Larionov, M., Giesen, A., and Hugel, H. (2000) A 1-kW CW thin disc laser. *IEEE J. Sel. Top. Quantum Electron.*, **6**, 650–657.

41 Giesen, A. and Speiser, J. (2007) Fifteen years of work on thin-disk lasers: results and scaling laws. *IEEE J. Sel. Top. Quantum Electron.*, **13**, 598–609.

42 Corzine, S.W., Geels, R.S., Scott, J.W., Yan, R.-H., and Coldren, L.A. (1989) Design of Fabry–Perot surface-emitting lasers with a periodic gain structure. *IEEE J. Quantum Electron.*, **QE-25**, 1513–1524.

43 Lutgen, S., Albrecht, T., Brick, P., Reill, W., Luft, J., and Spath, W. (2003) 8-W high-efficiency continuous-wave semiconductor disk laser at 1000 nm. *Appl. Phys. Lett.*, **82**, 3620–3622.

44 Schmid, M., Benchabane, S., Torabi-Goudarzi, F., Abram, R., Ferguson, A.I., and Riis, E. (2004) Optical in-well pumping of a vertical-external-cavity surface-emitting laser. *Appl. Phys. Lett.*, **84**, 4860–4862.

45 Zhang, W., Ackemann, T., McGinily, S., Schmid, M., Riis, E., and Ferguson, A.I. (2006) Operation of an optical in-well-pumped vertical-external-cavity surface-emitting laser. *Appl. Opt.*, **45**, 7729–7735.

46 Beyertt, S.S., Zorn, M., Kubler, T., Wenzel, H., Weyers, M., Giesen, A., Trankle, G., and Brauch, U. (2005) Optical in-well pumping of a semiconductor disk laser with high optical efficiency. *IEEE J. Quantum Electron.*, **41**, 1439–1449.

47 Beyertt, S.S., Brauch, U., Demaria, F., Dhidah, N., Giesen, A., Kubler, T., Lorch, S., Rinaldi, F., and Unger, P. (2007) Efficient gallium–arsenide disk laser. *IEEE J. Quantum Electron.*, **43**, 869–875.

48 Schulz, N., Rattunde, M., Ritzenthaler, C., Rosener, B., Manz, C., Kohler, K., and Wagner, J. (2007) Resonant optical in-well pumping of an (AlGaIn)(AsSb)-based vertical-external-cavity surface-emitting laser emitting at 2.35 µm. *Appl. Phys. Lett.*, **91**, 091113.

49 Saarinen, E.J., Harkonen, A., Suomalainen, S., and Okhotnikov, O.G. (2006) Power scalable semiconductor disk laser using multiple gain cavity. *Opt. Express*, **14**, 12868–12871.

50 Fan, L., Fallahi, M., Hader, J., Zakharian, A.R., Moloney, J.V., Murray, J.T., Bedford, R., Stolz, W., and Koch, S.W. (2006) Multichip vertical-external-cavity surface-emitting lasers: a coherent power scaling scheme. *Opt. Lett.*, **31**, 3612–3614.

51 Fan, L., Fallahi, M., Zakharian, A., Hader, J., Moloney, J.V., Bedford, R., Murray, J.T., Stolz, W., and Koch, S.W. (2007) Extended tunability in a two-chip VECSEL. *IEEE Photon. Technol. Lett.*, **19**, 544–546.

52 Hunziker, L.E., Ihli, C., and Steingrube, D.S. (2007) Miniaturization and power

scaling of fundamental mode optically pumped semiconductor lasers. *IEEE J. Sel. Top. Quantum Electron.*, **13**, 610–618.

53 Rautiainen, J., Harkanen, A., Korpijarvi, V.M., Tuomisto, P., Guina, M., and Okhotnikov, O.G. (2007) 2.7 W tunable orange–red GaInNAs semiconductor disk laser. *Opt. Express*, **15**, 18345–18350.

54 Seelert, W., Butterworth, S., Rosperich, J., Walter, C., Elm, R.v., Ostroumov, V., Chilla, J., Zhou, H., Weiss, E., and Caprara, A. (2005) Optically pumped semiconductor lasers: a new reliable technique for realizing highly efficient visible lasers. *Proc. SPIE*, **5707**, 33–37.

55 Holm, M.A., Ferguson, D., and Dawson, M.D. (1999) Actively stabilized single-frequency vertical-external-cavity AlGaAs laser. *IEEE Photon. Technol. Lett.*, **11**, 1551–1553.

56 Baili, G., Alouini, M., Dolfi, D., Bretenaker, F., Sagnes, I., and Garnache, A. (2007) Shot-noise-limited operation of a monomode high-cavity-finesse semiconductor laser for microwave photonics applications. *Opt. Lett.*, **32**, 650–652.

57 Hilbich, S., Seelert, W., Ostroumov, V., Kannengiesser, C., Elm, R.v., Mueller, J., Weiss, E., Zhou, H., and Chilla, J. (2007) New wavelengths in the yellow orange range between 545 nm to 580 nm generated by an intracavity frequency-doubled optically pumped semiconductor laser. *Proc. SPIE*, **6451**, 64510C.

58 Fallahi, M., Fan, L., Kaneda, Y., Hessenius, C., Hader, J., Li, H., Moloney, J.V., Kunert, B., Stolz, W., Koch, S.W., Murray, J., and Bedford, R. (2008) 5-W yellow laser by intracavity frequency doubling of high-power vertical-external-cavity surface-emitting laser. *IEEE Photon. Technol. Lett.*, **20**, 1700–1702.

59 Giet, S., Sun, H.D., Calvez, S., Dawson, M.D., Suomalainen, S., Harkonen, A., Guina, M., Okhotnikov, O., and Pesa, M. (2006) Spectral narrowing and locking of a vertical-external-cavity surface-emitting laser using an intracavity volume Bragg grating. *IEEE Photon. Technol. Lett.*, **18**, 1786–1788.

60 Giet, S., Lee, C.-L., Calvez, S., Dawson, M.D., Destouches, N., Pommier, J.-C., and Parriaux, O. (2007) Stabilization of a semiconductor disk laser using an intra-cavity high reflectivity grating. *Opt. Express*, **15**, 16520–16526.

61 Fan, L., Fallahi, M., Murray, J.T., Bedford, R., Kaneda, Y., Zakharian, A.R., Hader, J., Moloney, J.V., Stolz, W., and Koch, S.W. (2006) Tunable high-power high-brightness linearly polarized vertical-external-cavity surface-emitting lasers. *Appl. Phys. Lett.*, **88**, 021105.

62 Lorenser, D., Maas, D., Unold, H.J., Bellancourt, A.R., Rudin, B., Gini, E., Ebling, D., and Keller, U. (2006) 50-GHz passively mode-locked surface-emitting semiconductor laser with 100-mW average output power. *IEEE J. Quantum Electron.*, **42**, 838–847.

63 Haring, R., Paschotta, R., Gini, E., Morier-Genoud, F., Martin, D., Melchior, H., and Keller, U. (2001) Picosecond surface-emitting semiconductor laser with >200 mW average power. *Electron. Lett.*, **37**, 766–767.

64 Haring, R., Paschotta, R., Aschwanden, A., Gini, E., Morier-Genoud, F., and Keller, U. (2002) High-power passively mode-locked semiconductor lasers. *IEEE J. Quantum Electron.*, **38**, 1268–1275.

65 Alford, W.J., Raymond, T.D., and Allerman, A.A. (2002) High power and good beam quality at 980 nm from a vertical external-cavity surface-emitting laser. *J. Opt. Soc. Am. B*, **19**, 663–666.

66 Hastie, J.E., Hopkins, J.M., Calvez, S., Jeon, C.W., Burns, D., Abram, R., Riis, E., Ferguson, A.I., and Dawson, M.D. (2003) 0.5-W single transverse-mode operation of an 850-nm diode-pumped surface-emitting semiconductor laser. *IEEE Photon. Technol. Lett.*, **15**, 894–896.

67 Hopkins, J.M., Smith, S.A., Jeon, C.W., Sun, H.D., Burns, D., Calvez, S., Dawson, M.D., Jouhti, T., and Pessa, M. (2004) 0.6 W CW GaInNAs vertical external-cavity surface emitting laser operating at 1.32 μm. *Electron. Lett.*, **40**, 30–31.

68 Kim, K.S., Yoo, J., Kim, G., Lee, S., Cho, S., Kim, J., Kim, T., and Park, Y. (2007) 920-nm vertical-external-cavity surface-emitting lasers with a slope efficiency of

58% at room temperature. *IEEE Photon. Technol. Lett.*, **19**, 1655–1657.

69 Hastie, J.E., Hopkins, J.M., Jeon, C.W., Calvez, S., Burns, D., Dawson, M.D., Abram, R., Riis, E., Ferguson, A.I., Alford, W.J., Raymond, T.D., and Allerman, A.A. (2003) Microchip vertical external cavity surface emitting lasers. *Electron. Lett.*, **39**, 1324–1326.

70 Hastie, J.E., Morton, L.G., Calvez, S., Dawson, M.D., Leinonen, T., Pessa, M., Gibson, G., and Padgett, M.J. (2005) Red microchip VECSEL array. *Opt. Express*, **13**, 7209–7214.

71 Kemp, A.J., Maclean, A.J., Hastie, J.E., Smith, S.A., Hopkins, J.M., Calvez, S., Valentine, G.J., Dawson, M.D., and Burns, D. (2006) Thermal lensing, thermal management and transverse mode control in microchip VECSELs. *Appl. Phys. B*, **83**, 189–194.

72 Kemp, A.J., Maclean, A.J., Hopkins, J.M., Hastie, J.E., Calvez, S., Dawson, M.D., and Burns, D. (2007) Thermal management in disc lasers: doped-dielectric and semiconductor laser gain media in thin-disc and microchip formats. *J. Mod. Opt.*, **54**, 1669–1676.

73 Garnache, A., Kachanov, A.A., Stoeckel, F., and Planel, R. (1999) High-sensitivity intracavity laser absorption spectroscopy with vertical-external-cavity surface-emitting semiconductor lasers. *Opt. Lett.*, **24**, 826–828.

74 Garnache, A., Kachanov, A.A., Stoeckel, F., and Houdré, R. (2000) Diode-pumped broadband vertical-external-cavity surface-emitting semiconductor laser applied to high-sensitivity intracavity absorption spectroscopy. *J. Opt. Soc. Am. B*, **17**, 1589–1598.

75 Picqué, N., Guelachvili, G., and Kachanov, A.A. (2003) High-sensitivity time-resolved intracavity laser Fourier transform spectroscopy with vertical-cavity surface-emitting multiple-quantum-well lasers. *Opt. Lett.*, **28**, 313–315.

76 Raymond, T.D., Alford, W.J., Crawford, M.H., and Allerman, A.A. (1999) Intracavity frequency doubling of a diode-pumped external-cavity surface-emitting semiconductor laser. *Opt. Lett.*, **24**, 1127–1129.

77 Kaneda, Y., Yarborough, J.M., Li, L., Peyghambarian, N., Fan, L., Hessenius, C., Fallahi, M., Hader, J., Moloney, J.V., Honda, Y., Nishioka, M., Shimizu, Y., Miyazono, K., Shimatani, H., Yoshimura, M., Mori, Y., Kitaoka, Y., and Sasaki, T. (2008) Continuous-wave all-solid-state 244 nm deep-ultraviolet laser source by fourth-harmonic generation of an optically pumped semiconductor laser using $CsLiB_6O_{10}$ in an external resonator. *Opt. Lett.*, **33**, 1705–1707.

78 Hastie, J.E., Morton, L.G., Kemp, A.J., Dawson, M.D., Krysa, A.B., and Roberts, J.S. (2006) Tunable ultraviolet output from an intracavity frequency-doubled red vertical-external-cavity surface-emitting laser. *Appl. Phys. Lett.*, **89**, 061114.

79 Hastie, J., Calvez, S., Dawson, M., Leinonen, T., Laakso, A., Lyytikäinen, J., and Pessa, M. (2005) High power CW red VECSEL with linearly polarized TEM_{00} output beam. *Opt. Express*, **13**, 77–81.

80 Lindberg, H., Strassner, A., Gerster, E., and Larsson, A. (2004) 0.8 W optically pumped vertical external cavity surface emitting laser operating CW at 1550 nm. *Electron. Lett.*, **40**, 601–602.

81 Rosener, B., Schulz, N., Rattunde, M., Manz, C., Kohler, K., and Wagner, J. (2008) High-power high-brightness operation of a 2.25-µm (AlGaIn)(AsSb)-based barrier-pumped vertical-external-cavity surface-emitting laser. *IEEE Photon. Technol. Lett.*, **20**, 502–504.

82 Rahim, M., Arnold, M., Felder, F., Behfar, K., and Zogg, H. (2007) Midinfrared lead–chalcogenide vertical external cavity surface emitting laser with 5 µm wavelength. *Appl. Phys. Lett.*, **91**, 151102.

83 Rahim, M., Felder, F., Fill, M., and Zogg, H. (2008) Optically pumped 5 µm IV–VI VECSEL with Al-heat spreader. *Opt. Lett.*, **33**, 3010–3012.

84 Rahim, M., Felder, F., Fill, M., Boye, D., and Zogg, H. (2008) Lead chalcogenide VECSEL on Si emitting at 5 µm. *Electron. Lett.*, **44** (25), 1467–1469.

85 Rahim, M., Khiar, A., Felder, F., Fill, M., and Zogg, H. (2009) 4.5 µm wavelength vertical external cavity surface emitting laser operating above room temperature. *Appl. Phys. Lett.*, **94**, 201112.

86 McGinily, S.J., Abram, R.H., Gardner, K.S., Riis, E., Ferguson, A.I., and Roberts, J.S. (2007) Novel gain medium design for short-wavelength vertical-external-cavity surface-emitting laser. *IEEE J. Quantum Electron.*, **43**, 445–450.

87 Fan, L., Hessenius, C., Fallahi, M., Hader, J., Li, H., Moloney, J.V., Stolz, W., Koch, S.W., Murray, J.T., and Bedford, R. (2007) Highly strained InGaAs/GaAs multiwatt vertical-external-cavity surface-emitting laser emitting around 1170 nm. *Appl. Phys. Lett.*, **91**, 131114.

88 Harkonen, A., Rautiainen, J., Guina, M., Konttinen, J., Tuomisto, P., Orsila, L., Pessa, M., and Okhotnikov, O.G. (2007) High power frequency doubled GaInNAs semiconductor disk laser emitting at 615 nm. *Opt. Express*, **15**, 3224–3229.

89 Lindberg, H., Larsson, A., and Strassner, M. (2005) Single-frequency operation of a high-power, long-wavelength semiconductor disk laser. *Opt. Lett.*, **30**, 2260–2262.

90 Lindberg, H., Sadeghi, M., Westlund, M., Wang, S.M., Larsson, A., Strassner, M., and Marcinkevicius, S. (2005) Mode locking a 1550 nm semiconductor disk laser by using a GaInNAs saturable absorber. *Opt. Lett.*, **30**, 2793–2795.

91 Lindberg, H., Strassner, M., Gerster, E., Bengtsson, J., and Larsson, A. (2005) Thermal management of optically pumped long-wavelength InP-based semiconductor disk lasers. *IEEE J. Sel. Top. Quantum Electron.*, **11**, 1126–1134.

92 Kurdi, M.E., Bouchoule, S., Bousseksou, A., Sagnes, I., Plais, A., Strassner, M., Symonds, C., Garnache, A., and Jacquet, J. (2004) Room-temperature continuous-wave laser operation of electrically-pumped 1.55 µm VECSEL. *Electron. Lett.*, **40**, 671–672.

93 Bousseksou, A., Kurdi, M.E., Salik, M.D., Sagnes, I., and Bouchoule, S. (2004) Wavelength tunable InP-based EP-VECSEL operating at room temperature and in CW at 1.55 µm. *Electron. Lett.*, **40**, 1490–1491.

94 Harkonen, A., Guina, M., Okhotnikov, O., Rossner, K., Hummer, M., Lehnhardt, T., Muller, M., Forchel, A., and Fischer, M. (2006) 1-W antimonide-based vertical external cavity surface emitting laser operating at 2-µm. *Opt. Express*, **14**, 6479–6484.

95 Park, S.-H., Kim, J., Jeon, H., Sakong, T., Lee, S.-N., Chae, S., Park, Y., Jeong, C.-H., Yeom, G.-Y., and Cho, Y.-H. (2003) Room-temperature GaN vertical-cavity surface-emitting laser operation in an extended cavity scheme. *Appl. Phys. Lett.*, **83**, 2121–2123.

96 Park, S.-H. and Jeon, H. (2006) Microchip-type InGaN vertical external-cavity surface-emitting laser. *Opt. Rev.*, **13**, 20–23.

97 Rautiainen, J., Lyytikäinen, J., Sirbu, A., Mereuta, A., Caliman, A., Kapon, E., and Okhotnikov, O.G. (2008) 2.6 W optically-pumped semiconductor disk laser operating at 1.57-µm using wafer fusion. *Opt. Express*, **16**, 21881–21886.

98 Lyytikäinen, J., Rautiainen, J., Toikkanen, L., Sirbu, A., Mereuta, A., Caliman, A., Kapon, E., and Okhotnikov, O.G. (2009) 1.3-µm optically-pumped semiconductor disk laser by wafer fusion. *Opt. Express*, **17**, 9047–9052.

99 Strittmatter, A., Germann, T.D., Pohl, J., Pohl, U.W., Bimberg, D., Rautiainen, J., Guina, M., and Okhotnikov, O.G. (2008) 1040 nm vertical external cavity surface emitting laser based on InGaAs quantum dots grown in Stranski–Krastanow regime. *Electron. Lett.*, **44**, 290–291.

100 Germann, T.D., Strittmatter, A., Pohl, J., Pohl, U.W., Bimberg, D., Rautiainen, J., Guina, M., and Okhotnikov, O.G. (2008) High-power semiconductor disk laser based on InAs/GaAs submonolayer quantum dots. *Appl. Phys. Lett.*, **92**, 101123.

101 Germann, T.D., Strittmatter, A., Pohl, J., Pohl, U.W., Bimberg, D., Rautiainen, J., Guina, M., and Okhotnikov, O.G. (2008) Temperature-stable operation of a quantum dot semiconductor disk laser. *Appl. Phys. Lett.*, **93**, 051104.

102 Butkus, M., Wilcox, K.G., Rautiainen, J., Okhotnikov, O.G., Mikhrin, S.S., Krestnikov, I.L., Kovsh, A.R., Hoffmann, M., Südmeyer, T., Keller, U., and Rafailov, E.U. (2009) High-power quantum-dot-based semiconductor disk laser. *Opt. Lett.*, **34**, 1672–1674.

103 Hopkins, J.M., Hempler, N., Rosener, B., Schulz, N., Rattunde, M., Manz, C., Kohler, K., Wagner, J., and Burns, D. (2008) High-power, (AlGaIn)(AsSb) semiconductor disk laser at 2.0 μm. *Opt. Lett.*, **33**, 201–203.

104 Abram, R.H., Gardner, K.S., Riis, E., and Ferguson, A.I. (2004) Narrow linewidth operation of a tunable optically pumped semiconductor laser. *Opt. Express*, **12**, 5434–5439.

105 Harkonen, A., Guina, M., Rossner, K., Hummer, M., Lehnhardt, T., Muller, M., Forchel, A., Fischer, M., Koeth, J., and Okhotnikov, O.G. (2007) Tunable self-seeded semiconductor disk laser operating at 2 μm. *Electron. Lett.*, **43**, 457–458.

106 Ouvrard, A., Garnache, A., Cerutti, L., Genty, F., and Romanini, D. (2005) Single-frequency tunable Sb-based VCSELs emitting at 2.3 μm. *IEEE Photon. Technol. Lett.*, **17**, 2020–2022.

107 Rudin, B., Rutz, A., Hoffmann, M., Maas, D.J.H.C., Bellancourt, A.-R., Gini, E., Südmeyer, T., and Keller, U. (2008) Highly efficient optically pumped vertical-emitting semiconductor laser with more than 20 W average output power in a fundamental transverse mode. *Opt. Lett.*, **33**, 2719–2721.

108 Spinelli, L.A. and Caprara, A. (2008) Intracavity frequency-tripled cw laser. US Patent 7,463,657, issued December 9.

109 Leinonen, T., Morozov, Y.A., Harkonen, A., and Pessa, M. (2005) Vertical external-cavity surface-emitting laser for dual-wavelength generation. *IEEE Photon. Technol. Lett.*, **17**, 2508–2510.

110 Morozov, Y.A., Leinonen, T., Harkonen, A., and Pessa, M. (2006) Simultaneous dual-wavelength emission from vertical external-cavity surface-emitting laser: a numerical modeling. *IEEE J. Quantum Electron.*, **42**, 1055–1061.

111 Leinonen, T., Ranta, S., Laakso, A., Morozov, Y., Saarinen, M., and Pessa, M. (2007) Dual-wavelength generation by vertical external cavity surface-emitting laser. *Opt. Express*, **15**, 13451–13456.

112 Harkonen, A., Rautiainen, J., Leinonen, T., Morozov, Y.A., Orsila, L., Guina, M., Pessa, M., and Okhotnikov, O.G. (2007) Intracavity sum-frequency generation in dual-wavelength semiconductor disk laser. *IEEE Photon. Technol. Lett.*, **19**, 1550–1552.

113 Fan, L., Fallahi, M., Hader, J., Zakharian, A.R., Moloney, J.V., Stolz, W., Koch, S.W., Bedford, R., and Murray, J.T. (2007) Linearly polarized dual-wavelength vertical-external-cavity surface-emitting laser. *Appl. Phys. Lett.*, **90**, 181124.

114 Andersen, M.T., Schlosser, P.J., Hastie, J.E., Tidemand-Lichtenberg, P., Dawson, M.D., and Pedersen, C. (2009) Singly-resonant sum frequency generation of visible light in a semiconductor disk laser. *Opt. Express*, **17**, 6010–6017.

115 Mihoubi, Z., Wilcox, K.G., Elsmere, S., Quarterman, A., Rungsawang, R., Farrer, I., Beere, H.E., Ritchie, D.A., Tropper, A., and Apostolopoulos, V. (2008) All-semiconductor room-temperature terahertz time domain spectrometer. *Opt. Lett.*, **33**, 2125–2127.

116 Raja, M.Y.A., Brueck, S.R.J., Osinsky, M., Schaus, C.F., McInerney, J.G., Brennan, T.M., and Hammons, B.E. (1989) Resonant periodic gain surface-emitting semiconductor lasers. *IEEE J. Quantum Electron.*, **25**, 1500–1512.

117 Hader, J., Moloney, J.V., Fallahi, M., Fan, L., and Koch, S.W. (2006) Closed-loop design of a semiconductor laser. *Opt. Lett.*, **31**, 3300–3302.

118 Moloney, J.V., Hader, J., and Koch, S.W. (2007) Quantum design of semi-conductor active materials: laser and amplifier applications. *Laser Photon. Rev.*, **1**, 24–43.

119 Garnache, A., Ouvrard, A., and Romanini, D. (2007) Single-frequency operation of external-cavity VCSELs: non-linear multimode temporal dynamics and quantum limit. *Opt. Express*, **15**, 9403–9417.

120 Cocquelin, B., Holleville, D., Lucas-Leclin, G., Sagnes, I., Garnache, A., Myara, M., and Georges, P. (2009) Tunable single-frequency operation of a diode-pumped vertical external-cavity laser at the cesium D_2 line. *Appl. Phys. B*, **95** (2), 315–321.

121 Lindberg, H., Strassner, M., and Larsson, A. (2005) Improved spectral properties of an optically pumped semiconductor disk laser using a thin diamond heat spreader as an intracavity filter. *IEEE Photon. Technol. Lett.*, **17**, 1363–1365.

122 Tourrenc, J.P., Bouchoule, S., Khadour, A., Decobert, J., Miard, A., Harmand, J.C., and Oudar, J.L. (2007) High power single-longitudinal-mode OP-VECSEL at 1.55 μm with hybrid metal-metamorphic Bragg mirror. *Electron. Lett.*, **43**, 754–755.

123 Tourrenc, J.P., Bouchoule, S., Khadour, A., Harmand, J.C., Miard, A., Decobert, J., Lagay, N., Lafosse, X., Sagnes, I., Leroy, L., and Oudar, J.L. (2008) Thermal optimization of 1.55 μm OP-VECSEL with hybrid metal-metamorphic mirror for single-mode high power operation. *Opt. Quantum Electron.*, **40**, 155–165.

124 Symonds, C., Dion, J., Sagnes, I., Dainese, M., Strassner, M., Leroy, L., and Oudar, J.L. (2004) High performance 1.55 μm vertical external cavity surface emitting laser with broadband integrated dielectric-metal mirror. *Electron. Lett.*, **40**, 734–735.

125 Kim, K.S., Yoo, J.R., Cho, S.H., Lee, S.M., Lim, S.J., Kim, J.Y., Lee, J.H., Kim, T., and Park, Y.J. (2006) 1060 nm vertical-external-cavity surface-emitting lasers with an optical-to-optical efficiency of 44% at room temperature. *Appl. Phys. Lett.*, **88**, 091107.

126 Zorn, M., Klopp, P., Saas, F., Ginolas, A., Krüger, O., Griebner, U., and Weyers, M. (2008) Semiconductor components for femtosecond semiconductor disk lasers grown by MOVPE. *J. Cryst. Growth*, **310**, 5187–5190.

127 Konttinen, J., Härkönen, A., Tuomisto, P., Guina, M., Rautiainen, J., Pessa, M., and Okhotnikov, O. (2007) High-power (>1 W) dilute nitride semiconductor disk laser emitting at 1240 nm. *New J. Phys.*, **9**, 140.

128 Manz, C., Yang, Q., Rattunde, M., Schulz, N., Rösener, B., Kirste, L., Wagner, J., and Köhler, K. (2009) Quaternary GaInAsSb/AlGaAsSb vertical-external-cavity surface-emitting lasers: a challenge for MBE growth. *J. Cryst. Growth*, **311**, 1920–1922.

129 Korpijärvi, V.-M., Guina, M., Puustinen, J., Tuomisto, P., Rautiainen, J., Härkönen, A., Tukiainen, A., Okhotnikov, O., and Pessa, M. (2009) MBE grown GaInNAs-based multi-Watt disk lasers. *J. Cryst. Growth*, **311**, 1868–1871.

130 Fan, L., Hader, J., Schillgalies, M., Fallahi, M., Zakharian, A.R., Moloney, J.V., Bedford, R., Murray, J.T., Koch, S.W., and Stolz, W. (2005) High-power optically pumped VECSEL using a double-well resonant periodic gain structure. *IEEE Photon. Technol. Lett.*, **17**, 1764–1766.

131 Maclean, A.J., Kemp, A.J., Calvez, S., Kim, J.Y., Kim, T., Dawson, M.D., and Burns, D. (2008) Continuous tuning and efficient intracavity second-harmonic generation in a semiconductor disk laser with an intracavity diamond heatspreader. *IEEE J. Quantum Electron.*, **44**, 216–225.

132 Seelert, W., Kubasiak, S., Negendank, J., von Elm, R., Chilla, J., Zhou, H., and Weiss, E. (2006) Optically-pumped semiconductor lasers at 505-nm in the power range above 100 mW. *Proc. SPIE*, **6100**, 610002.

133 Aschwanden, A., Lorenser, D., Unold, H.J., Paschotta, R., Gini, E., and Keller, U. (2005) 2.1-W picosecond passively mode-locked external-cavity semiconductor laser. *Opt. Lett.*, **30**, 272–274.

134 Garnache, A., Hoogland, S., Tropper, A.C., Sagnes, I., Saint-Girons, G., and Roberts, J.S. (2002) Sub-500-fs soliton-like pulse in a passively mode-locked broadband surface-emitting laser with 100 mW average power. *Appl. Phys. Lett.*, **80**, 3892–3894.

135 Hunziker, L.E., Shu, Q.-Z., Bauer, D., Ihli, C., Mahnke, G.J., Rebut, M., Chilla, J.R., Caprara, A.L., Zhou, H., Weiss, E.S., and Reed, M.K. (2007) Power-scaling of optically pumped semiconductor lasers. *Proc. SPIE*, **6451**, 64510.

136 Ochalski, T.J., de Burca, A., Huyet, G., Lyytikainen, J., Guina, M., Pessa, M., Jasik, A., Muszalski, J., and Bugajski, M. (2008) Passively modelocked bi-directional vertical external ring cavity surface emitting laser. Conference on Lasers and Electro-Optics, and Conference on Quantum Electronics and Laser Science (CLEO/QELS).

137 Mignot, A., Feugnet, G., Schwartz, S., Sagnes, I., Garnache, A., Fabre, C., and Pocholle, J.-P. (2009) Single-frequency external-cavity semiconductor ring-laser gyroscope. *Opt. Lett.*, **34**, 97–99.

138 Eckstein, H.-C. and Zeitner, U.D. (2008) Experimental realization of a diffractive unstable resonator with Gaussian outcoupled beam using a VECSEL amplifier. Conference on Lasers and Electro-Optics, and Conference on Quantum Electronics and Laser Science (CLEO/QELS).

139 Hodgson, N. and Weber, H. (2005) *Laser Resonators and Beam Propagation*, 2nd edn, Springer.

140 Laurain, A., Myara, M., Beaudoin, G., Sagnes, I., and Garnache, A. (2009) High power single-frequency continuously-tunable compact extended-cavity semiconductor laser. *Opt. Express*, **17**, 9503–9508.

141 Keeler, G.A., Serkland, D.K., Geib, K.M., Peake, G.M., and Mar, A. (2005) Single transverse mode operation of electrically pumped vertical-external-cavity surface-emitting lasers with micromirrors. *IEEE Photon. Technol. Lett.*, **17**, 522–524.

142 Laurand, N., Lee, C.L., Gu, E., Hastie, J.E., Calvez, S., and Dawson, M.D. (2007) Microlensed microchip VECSEL. *Opt. Express*, **15**, 9341–9346.

143 Laurand, N., Lee, C.-L., Gu, E., Hastie, J.E., Kemp, A.J., Calvez, S., and Dawson, M.D. (2008) Array-format microchip semiconductor disk lasers. *IEEE J. Quantum Electron.*, **44**, 1096–1103.

144 Laurand, N., Lee, C.-L., Gu, E., Calvez, S., and Dawson, M.D. (2009) Power-scaling of diamond microlensed microchip semiconductor disk lasers. *IEEE Photon. Technol. Lett.*, **21**, 152–154.

145 Kuznetsov, M., Stern, M., and Coppeta, J. (2005) Single transverse mode optical resonators. *Opt. Express*, **13**, 171–181.

146 Cocquelin, B., Lucas-Leclin, G., Georges, P., Sagnes, I., and Garnache, A. (2008) Design of a low-threshold VECSEL emitting at 852 nm for Cesium atomic clocks. *Opt. Quantum Electron.*, **40**, 167–173.

147 Okhotnikov, O.G. (2008) Power scalable semiconductor disk lasers for frequency conversion and mode-locking. *Quantum Electron.*, **38**, 1083–1096.

148 Jacquemet, M., Picqu,é, N., Guelachvili, G., Garnache, A., Sagnes, I., Strassner, M., and Symonds, C. (2007) Continuous-wave 1.55 µm diode-pumped surface emitting semiconductor laser for broadband multiplex spectroscopy. *Opt. Lett.*, **32**, 1387–1389.

149 Campargue, A., Wang, L., Cermak, P., and Hu, S.-M. (2005) ICLAS-VeCSEL and FTS spectroscopies of C_2H_2 between 9000 and 9500 cm^{-1}. *Chem. Phys. Lett.*, **403**, 287–292.

150 Mooradian, A. (2001) High brightness cavity-controlled surface emitting GaInAs lasers operating at 980 nm. Optical Fiber Communication Conference (OFC 2001), vol. 4, pp. PD17-1–PD17-3.

151 McInerney, J.G., Mooradian, A., Lewis, A., Shchegrov, A.V., Strzelecka, E.M., Lee, D., Watson, J.P., Liebman, M.K., Carey, G.P., Umbrasas, A., Amsden, C.A., Cantos, B.D., Hitchens, W.R., Heald, D.L., and Doan, V. (2003) Novel 980-nm and 490-nm light sources using vertical-cavity lasers with extended coupled cavities. *Proc. SPIE*, **4994**, 21.

152 Bousseksou, A., Bouchoule, S., El Kurdi, M., Strassner, M., Sagnes, I., Crozat, P., and Jacquet, J. (2006) Fabrication and characterization of 1.55 µm single transverse mode large diameter electrically pumped VECSEL. *Opt. Quantum Electron.*, **38**, 1269–1278.

153 Kardosh, I., Demaria, F., Rinaldi, F., Riedl, M.C., and Michalzik, R. (2008) Electrically pumped frequency-doubled surface emitting lasers operating at

485 nm emission wavelength. *Electron. Lett.*, **44**, 524–525.

154 Illek, S., Albrecht, T., Brick, P., Lutgen, S., Pietzonka, I., Furitsch, M., Diehl, W., Luft, J., and Streubel, K. (2007) Vertical-external-cavity surface-emitting laser with monolithically integrated pump lasers. *IEEE Photon. Technol. Lett.*, **19**, 1952–1954.

155 Lee, J.H., Kim, J.Y., Lee, S.M., Yoo, J.R., Kim, K.S., Cho, S.H., Lim, S.J., Kim, G.B., Hwang, S.M., Kim, T., and Park, Y.J. (2006) 9.1-W high-efficient continuous-wave end-pumped vertical-external-cavity surface-emitting semiconductor laser. *IEEE Photon. Technol. Lett.*, **18**, 2117–2119.

156 Lee, J.H., Lee, S.M., Kim, T., and Park, Y.J. (2006) 7 W high-efficiency continuous-wave green light generation by intracavity frequency doubling of an end-pumped vertical external-cavity surface emitting semiconductor laser. *Appl. Phys. Lett.*, **89**, 241107.

157 Kim, G.B., Kim, J.-Y., Lee, J., Yoo, J., Kim, K.-S., Lee, S.-M., Cho, S., Lim, S.-J., Kim, T., and Park, Y.J. (2006) End-pumped green and blue vertical external cavity surface emitting laser devices. *Appl. Phys. Lett.*, **89**, 181106.

158 Cho, S., Kim, G.B., Kim, J.-Y., Kim, K.-S., Lee, S.-M., Yoo, J., Kim, T., and Park, Y. (2007) Compact and efficient green VECSEL based on novel optical end-pumping scheme. *IEEE Photon. Technol. Lett.*, **19**, 1325–1327.

159 Bedford, R.G., Kolesik, M., Chilla, J.L.A., Reed, M.K., Nelson, T.R., and Moloney, J.V. (2005) Power-limiting mechanisms in VECSELs. *Proc. SPIE*, **5814**, 199–208.

160 Maas, D., Bellancourt, A.R., Rudin, B., Golling, M., Unold, H.J., Sudmeyer, T., and Keller, U. (2007) Vertical integration of ultrafast semiconductor lasers. *Appl. Phys. B*, **88**, 493–497.

161 Bellancourt, A.-R., Maas, D.J.H.C., Rudin, B., Golling, M., Südmeyer, T., and Keller, U. (2009) Modelocked integrated external-cavity surface emitting laser. *IET Optoelectron.*, **3**, 61–72.

162 Saas, F., Talalaev, V., Griebner, U., Tomm, J.W., Zorn, M., Knigge, A., and Weyers, M. (2006) Optically pumped semiconductor disk laser with graded and step indices. *Appl. Phys. Lett.*, **89**, 151120.

163 Schulz, N., Rattunde, A., Manz, C., Kohler, K., Wild, C., Wagner, J., Beyertt, S.S., Brauch, U., Kubler, T., and Giesen, A. (2006) Optically pumped GaSb-based VECSEL emitting 0.6 W at 2.3 μm. *IEEE Photon. Technol. Lett.*, **18**, 1070–1072.

164 Hopkins, J.M., Maclean, A.J., Burns, D., Riis, E., Schulz, N., Rattunde, M., Manz, C., Kohler, K., and Wagner, J. (2007) Tunable, single-frequency, diode-pumped 2.3 μm VECSEL. *Opt. Express*, **15**, 8212–8217.

165 Esposito, E., Keatings, S., Gardner, K., Harris, J., Riis, E., and McConnell, G. (2008) Confocal laser scanning microscopy using a frequency doubled vertical external cavity surface emitting laser. *Rev. Sci. Instrum.*, **79**, 083702.

166 Ra, H., Piyawattanametha, W., Mandella, M.J., Hsiung, P.-L., Hardy, J., Wang, T.D., Contag, C.H., Kino, G.S., and Solgaard, O. (2008) Three-dimensional *in vivo* imaging by a handheld dual-axes confocal microscope. *Opt. Express*, **16**, 7224–7232.

167 Chilla, J.L.A., Butterworth, S.D., Zeitschel, A., Charles, J.P., Caprara, A.L., Reed, M.K., and Spinelli, L. (2004) High-power optically pumped semiconductor lasers. *Proc. SPIE*, **5332**, 143.

168 Garnache, A., Myara, M., Bouchier, A., Perez, J.-P., Signoret, P., Sagnes, I., and Romanini, D. (2008) Single frequency free-running low noise compact external-cavity VCSELs at high power level (50 mW). 34th European Conference on Optical Communication (ECOC).

169 Fan, L., Fallahi, M., Hader, J., Zakharian, A.R., Kolesik, M., Moloney, J.V., Qiu, T., Schulzgen, A., Peyghambarian, N., Stolz, W., Koch, S.W., and Murray, J.T. (2005) Over 3 W high-efficiency vertical-external-cavity surface-emitting lasers and application as efficient fiber laser pump sources. *Appl. Phys. Lett.*, **86**, 211116.

170 Harkonen, A., Suomalainen, S., Saarinen, E., Orsila, L., Koskinen, R., Okhotnikov, O., Calvez, S., and Dawson, M. (2006) 4 W single-transverse

mode VECSEL utilising intra-cavity diamond heat spreader. *Electron. Lett.*, **42**, 693–694.

171 Guina, M., Korpijarvi, V.-M., Rautiainen, J., Tuomisto, P., Puustinen, J., Harkonen, A., and Okhotnikov, O. (2008) 3.5 W GaInNAs disk laser operating at 1220 nm. *Proc. SPIE*, **6997**, 69970.

172 Harkonen, A., Guina, M., Rossner, K., Hummer, M., Lehnhardt, T., Muller, M., Forchel, A., Fischer, M., Koeth, J., and Okhotnikov, O.G. (2007) Tunable self-seeded semiconductor disk laser operating at 2 μm. *Electron. Lett.*, **43**, 457–458.

173 Kim, J.-Y., Cho, S., Lim, S.-J., Yoo, J., Kim, G.B., Kim, K.-S., Lee, J., Lee, S.-M., Kim, T., and Park, Y. (2007) Efficient blue lasers based on gain structure optimizing of vertical-external-cavity surface-emitting laser with second harmonic generation. *J. Appl. Phys.*, **101**, 033103.

174 Gerster, E., Ecker, I., Lorch, S., Hahn, C., Menzel, S., and Unger, P. (2003) Orange-emitting frequency-doubled GaAsSb/GaAs semiconductor disk laser. *J. Appl. Phys.*, **94**, 7397–7401.

175 Rautiainen, J., Harkonen, A., Tuornisto, P., Konttinen, J., Orsila, L., Guina, M., and Okhotnikov, O.G. (2007) 1 W at 617 nm generation by intracavity frequency conversion in semiconductor disk laser. *Electron. Lett.*, **43**, 980–981.

176 Genesis™ Taipan™ 639M OPSL laser from Coherent Inc., www.coherent.com.

177 Rafailov, E.U., Sibbett, W., Mooradian, A., McInerney, J.G., Karlsson, H., Wang, S., and Laurell, F. (2003) Efficient frequency doubling of a vertical-extended-cavity surface-emitting laser diode by use of a periodically poled KTP crystal. *Opt. Lett.*, **28**, 2091–2093.

178 Genesis™ 355 OPSL laser from Coherent Inc., www.coherent.com.

179 Risk, W.P., Gosnell, T.R., and Nurmikko, A.V. (2003) *Compact Blue–Green Lasers*, Cambridge University Press.

180 Smith, A., Hastie, J.E., Foreman, H.D., Leinonen, T., Guina, M., and Dawson, M.D. (2008) GaN diode-pumping of a red semiconductor disk laser. *Electron. Lett.*, **44**, 1195–1196.

181 Smith, S.A., Hopkins, J.-M., Hastie, J.E., Burns, D., Calvez, S., Dawson, M.D., Jouhti, T., Kontinnen, J., and Pessa, M. (2004) Diamond-microchip GaInNAs vertical external-cavity surface-emitting laser operating CW at 1315 nm. *Electron. Lett.*, **40**, 935–936.

182 Vetter, S.L., Hastie, J.E., Korpijarvi, V.-M., Puustinen, J., Guina, M., Okhotnikov, O., Calvez, S., and Dawson, M.D. (2008) Short-wavelength GaInNAs/GaAs semiconductor disk lasers. *Electron. Lett.*, **44**, 1069–1070.

183 Laurand, N., Calvez, S., Sun, H.D., Dawson, M.D., Gupta, J.A., and Aers, G.C. (2006) C-band emission from GaInNAsSb VCSEL on GaAs. *Electron. Lett.*, **42**, 28–30.

184 Rosener, B., Schulz, N., Rattunde, M., Moser, R., Manz, C., Kohler, K., and Wagner, J. (2008) Optically pumped (AlGaIn)(AsSb) semiconductor disk laser employing a dual-chip cavity. IEEE Lasers and Electro-Optics Society Annual Meeting (LEOS), pp. 854–855.

185 Schulz, N., Rosener, B., Moser, R., Rattunde, M., Manz, C., Kohler, K., and Wagner, J. (2008) An improved active region concept for highly efficient GaSb-based optically in-well pumped vertical-external-cavity surface-emitting lasers. *Appl. Phys. Lett.*, **93**, 181113.

186 Hoffmann, S. and Hofmann, M.R. (2007) Generation of Terahertz radiation with two color semiconductor lasers. *Laser Photon. Rev.*, **1**, 44–56.

187 Morozov, Y.A., Nefedov, I.S., Leinonen, T., and Morozov, M.Y. (2008) Nonlinear-optical frequency conversion in a dual-wavelength vertical-external-cavity surface-emitting laser. *Semiconductors*, **42**, 463–469.

188 Harkonen, A. (2009) Antimonide disk lasers achieve multiwatt power and a wide tuning range. *SPIE Newsroom*. doi:

189 Giudice, G.E., Guy, S.C., Crigler, S.G., Zenteno, L.A., and Hallock, B.S. (2002) Effect of pump laser noise on an erbium-doped fiber-amplified signal. *IEEE Photon. Technol. Lett.*, **14**, 1403–1405.

190 Schlager, J.B., Callicoatt, B.E., Mirin, R.P., Sanford, N.A., Jones, D.J., and Ye, J.

(2003) Passively mode-locked glass waveguide laser with 14-fs timing jitter. *Opt. Lett.*, **28**, 2411–2413.

191 Vetter, S., Calvez, S., Dawson, M.D., Fusari, F., Lagatsky, A.A., Sibbett, W., Brown, C.T.A., Korpijarvi, V.-M., Guina, M., Richards, B., Jose, G., and Jha, A. (2008) 1213 nm semiconductor disk laser pumping of a Tm^{3+}-doped tellurite glass laser. IEEE Lasers and Electro-Optics Society Annual Meeting, pp. 840–841.

192 Vetter, S.L., McKnight, L.J., Calvez, S., Dawson, M.D., Fusari, F., Lagatsky, A.A., Sibbett, W., Brown, C.T.A., Korpijärvi, V.–M., Guina, M.D., Richards, B., Jose, G., and Jha, A. (2009) GaInNAs semiconductor disk lasers as pump sources for Tm^{3+} ($,Ho^{3+}$)-doped glass, crystal and fibre lasers. *Proc. SPIE*, **7193**, 719317.

193 Wang, T.D., Mandella, M.J., Contag, C.H., and Kino, G.S. (2003) Dual-axis confocal microscope for high-resolution *in vivo* imaging. *Opt. Lett.*, **28**, 414–416.

194 Wang, T.D., Contag, C.H., Mandella, M.J., Chan, N.Y., and Kino, G.S. (2004) Confocal fluorescence microscope with dual-axis architecture and biaxial postobjective scanning. *J. Biomed. Opt.*, **9**, 735–742.

195 Simbuerger, E., Pflanz, T., and Masters, A. (2008) Confocal microscopy: new lasers enhance live cell imaging. *Physik Journal, (Physics' Best)*, 10.

196 Schulze, M. (2008) Laser microscopy continues to diversify, *Photonik International Online*, February, p. 36.

197 Pawley, J.B. (ed.) (2006) *Handbook of Biological Confocal Microscopy*, 3rd edn, Springer Science + Business Media, LLC.

198 Schiemann, M. and Busch, D.H. (2009) Selection and combination of fluorescent dyes, in *Cellular Diagnostics. Basics, Methods and Clinical Applications of Flow Cytometry* (eds U. Sack, A. Tárnok, and G. Rothe), Karger, Basel, pp. 107–140.

199 Lake, T., Carruthers, A., Taylor, M., Paterson, L., Gunn-Moore, F., Allen, J., Sibbett, W., and Dholakia, K. (2004) Optical trapping and fluorescence excitation with violet diode lasers and extended cavity surface emitting lasers. *Opt. Express*, **12**, 670–678.

200 Su, W.W. (2005) Fluorescent proteins as tools to aid protein production. *Microb. Cell Fact.*, **4**, 12.

201 Resan, B., Coadou, E., Petersen, S., Thomas, A., Walther, P., Viselga, R., Heritier, J.-M., Chilla, J., Tulloch, W., and Fry, A. (2008) Ultrashort pulse Ti:sapphire oscillators pumped by optically pumped semiconductor (OPS) pump lasers. *Proc. SPIE*, **6871**, 687116.

202 Harris, R. (2009) OPSL technology provides new wavelengths that benefit forensics, *Photonik International Online*, March, p. 2.

203 Masters, A. (2007) Yellow lasers target macular degeneration, *Biophotonics International*, June.

204 Masters, A. and Seaton, C. (2006) Laser-based displays will deliver superior images, *Laser Focus World*, November.

205 Evans & Sutherland ESLP® (2009) 8 K laser video projector, www.es.com.

206 Mitsubishi LaserVue® (2009) TV, www.laservuetv.com.

207 Freeman, M., Champion, M., and Madhavan, S. (2009) Scanned laser pico-projetors, *Optics and Photonics News*, May.

208 Schmitt, M. and Steegmüller, U. (2008) Green laser meets mobile projection requirements. *Opt. Laser Europe*, 17–19.

209 Trisnadi, J.I., Carlisle, C.B., and Monteverde, R., (2004) Overview and applications of grating-light-valve-based optical write engines for high-speed digital imaging. *Proc. SPIE*, **5348**, 52–64.

210 Niven, G. and Mooradian, A. (2006) Trends in laser light sources for projection display. 13th International Display Workshop (IDW 2006), Otsu, Japan, December 2006.

211 Kaneda, Y., Fan, L., Hsu, T.-C., Peyghambarian, N., Fallahi, M., Zakharian, A.R., Hader, J., Moloney, J.V., Stoltz, W., Koch, S., Bedford, R., Sevian, A., and Glebov, L. (2006) High brightness spectral beam combination of high-power vertical-external-cavity surface-emitting lasers. *IEEE Photon. Technol. Lett.*, **18**, 1795–1797.

212 Aldaz, R.I., Wiemer, M.W., Miller, D.A.B., and Harris, J.S. (2004) Monolithically-

integrated long vertical cavity surface emitting laser incorporating a concave micromirror on a glass substrate. *Opt. Express*, **12**, 3967–3971.

213 Wiemer, M.W., Aldaz, R.I., Miller, D.A.B., and Harris, J.S. (2005) A single transverse-mode monolithically integrated long vertical-cavity surface-emitting laser. *IEEE Photon. Technol. Lett.*, **17**, 1366–1368.

214 Khurgin, J.B., Vurgaftman, I., and Meyer, J.R. (2005) Analysis of phase locking in diffraction-coupled arrays of semiconductor lasers with gain/index coupling. *IEEE J. Quantum Electron.*, **41**, 1065–1074.

215 Flynn, R.A., Birkbeck, A.L., Gross, M., Ozkan, M., Shao, B., and Esener, S.C. (2003) Simultaneous transport of multiple biological cells by VCSEL array optical traps, in *Optics in Computing* (ed. A. Sawchuk), OSA Trends in Optics and Photonics, vol. 90, Paper OThB2, Optical Society of America.

216 Vakhshoori, D., Tayebati, P., Lu, C.C., Azimi, M., Wang, P., Zhou, J.H., and Canoglu, E. (1999) 2 mW CW single mode operation of a tunable 1550 nm vertical cavity surface emitting laser with 50 nm tuning range. *Electron. Lett.*, **35**, 900–901.

217 Matsui, Y., Vakhshoori, D., Wang, P., Chen, P., Lu, C.-C., Jiang, M., Knopp, K., Burroughs, S., and Tayebati, P. (2003) Complete polarization mode control of long-wavelength tunable vertical-cavity surface-emitting lasers over 65-nm tuning, up to 14-mW output power. *IEEE J. Quantum Electron.*, **39**, 1037–1048.

218 Knopp, K.J., Vakhshoori, D., Wang, P.D., Azimi, M., Jiang, M., Chen, P., Matsui, Y., McCallion, K., Baliga, A., Sakhitab, F., Letsch, M., Johnson, B., Huang, R., Jean, A., DeLargy, B., Pinzone, C., Fan, F., Liu, J., Lu, C., Zhou, J., Zhu, H., Gurjar, R., Tayebati, P., MacDaniel, D., Baorui, R., Waterson, R., and VanderRhodes, G. (2001) High power MEMs-tunable vertical-cavity surface-emitting lasers. Advanced Semiconductor Lasers and Applications/Ultraviolet and Blue Lasers and Their Applications/Ultralong Haul DWDM Transmission and Networking/WDM Components, IEEE LEOS Summer Topical Meetings.

219 Triki, M., Cermak, P., Cerutti, L., Garnache, A., and Romanini, D. (2008) Extended continuous tuning of a single-frequency diode-pumped vertical-external-cavity surface-emitting laser at 2.3 μm. *IEEE Photon. Technol. Lett.*, **20**, 1947–1949.

220 Baili, G., Alouini, M., Malherbe, T., Dolfi, D., Huignard, J.-P., Merlet, T., Chazelas, J., Sagnes, I., and Bretenaker, F. (2008) Evidence of ultra low microwave additive phase noise for an optical RF link based on a class-A semiconductor laser. *Opt. Express*, **16**, 10091–10097.

221 Gardner, K., Abram, R., and Riis, E. (2004) A birefringent etalon as single-mode selector in a laser cavity. *Opt. Express*, **12**, 2365–2370.

222 Jacquemet, M., Domenech, M., Lucas-Leclin, G., Georges, P., Dion, J., Strassner, M., Sagnes, I., and Garnache, A. (2007) Single-frequency cw vertical external cavity surface emitting semiconductor laser at 1003 nm and 501 nm by intracavity frequency doubling. *Appl. Phys. B*, **86**, 503–510.

223 Hoogland, S., Dhanjal, S., Tropper, A.C., Roberts, J.S., Haring, R., Paschotta, R., Morier-Genoud, F., and Keller, U. (2000) Passively mode-locked diode-pumped surface-emitting semiconductor laser. *IEEE Photon. Technol. Lett.*, **12**, 1135–1137.

224 Hoffmann, M., Barbarin, Y., Maas, D.J.H.C., Golling, M., Krestnikov, I.L., Mikhrin, S.S., Kovsh, A.R., Südmeyer, T., and Keller, U. (2008) Modelocked quantum dot vertical external cavity surface emitting laser. *Appl. Phys. B*, **93**, 733–736.

225 Klopp, P., Saas, F., Zorn, M., Weyers, M., and Griebner, U. (2008) 290-fs pulses from a semiconductor disk laser. *Opt. Express*, **16**, 5770–5775.

226 Wilcox, K.G., Mihoubi, Z., Daniell, G.J., Elsmere, S., Quarterman, A., Farrer, I., Ritchie, D.A., and Tropper, A. (2008) Ultrafast optical Stark mode-locked semiconductor laser. *Opt. Lett.*, **33**, 2797–2799.

227 Jasim, K., Zhang, Q., Nurmikko, A.V., Mooradian, A., Carey, G., Ha, W., and Ippen, E. (2003) Passively modelocked vertical extended cavity surface emitting diode laser. *Electron. Lett.*, **39**, 373–375.

228 Jasim, K., Zhang, Q., Nurmikko, A.V., Ippen, E., Mooradian, A., Carey, G., and Ha, W. (2004) Picosecond pulse generation from passively modelocked vertical cavity diode laser at up to 15 GHz pulse repetition rate. *Electron. Lett.*, **40**, 34–36.

229 Zhang, Q., Jasim, K., Nurmikko, A.V., Mooradian, A., Carey, G., Ha, W., and Ippen, E. (2004) Operation of a passively mode-locked extended-cavity surface-emitting diode laser in multi-GHz regime. *IEEE Photon. Technol. Lett.*, **16**, 885–887.

230 Zhang, Q., Jasim, K., Nurmikko, A.V., Ippen, E., Mooradian, A., Carey, G., and Ha, W. (2005) Characteristics of a high-speed passively mode-locked surface-emitting semiconductor InGaAs laser diode. *IEEE Photon. Technol. Lett.*, **17**, 525–527.

231 Klopp, P., Griebner, U., Zorn, M., Klehr, A., Liero, A., Erbert, G., and Weyers, M. (2009) InGaAs–AlGaAs disk laser generating sub-220-fs pulses and tapered diode amplifier with ultrafast pulse picking, in *Advanced Solid-State Photonics*, Optical Society of America.

232 Wilcox, K.G., Foreman, H.D., Roberts, J.S., and Tropper, A.C. (2006) Timing jitter of 897 MHz optical pulse train from actively stabilised passively modelocked surface-emitting semiconductor laser. *Electron. Lett.*, **42**, 159–160.

233 Zhang, W., Ackemann, T., Schmid, M., Langford, N., and Ferguson, A.I. (2006) Femtosecond synchronously mode-locked vertical-external cavity surface-emitting laser. *Opt. Express*, **14**, 1810–1821.

234 Max, C.E. *et al.* (1997) Image improvement from a sodium-layer laser guide star adaptive optics system. *Science*, **277**, 1649–1652.

235 Rutten, T.P., Veitch, P.J., and Munch, J. (2007) Development of a sodium laser guide star for astronomical adaptive optics systems. Conference on Lasers and Electro-Optics (CLEO).

236 Kemp, A.J., Valentine, G.J., Hopkins, J.M., Hastie, J.E., Smith, S.A., Calvez, S., Dawson, M.D., and Burns, D. (2005) Thermal management in vertical-external-cavity surface-emitting lasers: finite-element analysis of a heatspreader approach. *IEEE J. Quantum Electron.*, **41**, 148–155.

237 Kemp, A.J., Hopkins, J.M., Maclean, A.J., Schulz, N., Rattunde, M., Wagner, J., and Burns, D. (2008) Thermal management in 2.3-μm semiconductor disk lasers: a finite element analysis. *IEEE J. Quantum Electron.*, **44**, 125–135.

238 Love, D., Kolesik, M., and Moloney, J.V. (2009) Optimization of ultrashort pulse generation in passively mode-locked vertical external-cavity semiconductor lasers. *IEEE J. Quantum Electron.*, **45**, 439–445.

239 Baili, G., Bretenaker, F., Alouini, M., Morvan, L., Dolfi, D., and Sagnes, I. (2008) Experimental investigation and analytical modeling of excess intensity noise in semiconductor class-A lasers. *J. Lightwave Technol.*, **26**, 952–961.

240 Zakharian, A.R., Hader, J., Moloney, J.V., Koch, S.W., Brick, P., and Lutgen, S. (2003) Experimental and theoretical analysis of optically pumped semiconductor disk lasers. *Appl. Phys. Lett.*, **83**, 1313–1315.

241 Zakharian, A.R., Hader, J., Moloney, J.V., and Koch, S.W. (2005) VECSEL threshold and output power-shutoff dependence on the carrier recombination rates. *IEEE Photon. Technol. Lett.*, **17**, 2511–2513.

242 Ouvrard, A., Cerutti, L., and Garnache, A. (2009) Broad continuous tunable range with single frequency Sb-based external-cavity VCSEL emitting in MIR. *Electron. Lett.*, **45**, 629–631.

243 Gherman, T., Romanini, D., Sagnes, I., Garnache, A., and Zhang, Z. (2004) Cavity-enhanced absorption spectroscopy with a mode-locked diode-pumped vertical external-cavity surface-emitting laser. *Chem. Phys. Lett.*, **390**, 290–295.

244 Picqué, N., Guelachvili, G., and Kachanov, A.A. (2003) High-sensitivity time-resolved intracavity laser Fourier

transform spectroscopy with vertical-cavity surface-emitting multiple-quantum-well lasers. *Opt. Lett.*, **28**, 313–315.

245 Bertseva, E., Kachanov, A.A., and Campargue, A. (2002) Intracavity laser absorption spectroscopy of N_2O with a vertical external cavity surface emitting laser. *Chem. Phys. Lett.*, **351**, 18–26.

246 Garnache, A., Liu, A., Cerutti, L., and Campargue, A. (2005) Intracavity laser absorption spectroscopy with a vertical external cavity surface emitting laser at 2.3 μm: application to water and carbon dioxide. *Chem. Phys. Lett.*, **416**, 22–27.

247 Naumenko, O. and Campargue, A. (2003) Rovibrational analysis of the absorption spectrum of H_2O around 1.02 μm by ICLAS-VECSEL. *J. Mol. Spectrosc.*, **221**, 221–226.

248 Bertseva, E. and Campargue, A. (2004) Spectral condensation near molecular transitions in intracavity laser spectroscopy with vertical external cavity surface emitting lasers. *Opt. Commun.*, **232**, 251–261.

249 Ding, Y., Naumenko, O., Hu, S.-M., Zhu, Q., Bertseva, E., and Campargue, A. (2003) The absorption spectrum of H_2S between 9540 and 10 000 cm^{-1} by intracavity laser absorption spectroscopy with a vertical external cavity surface emitting laser. *J. Mol. Spectrosc.*, **217**, 222–238.

250 Bertseva, E., Campargue, A., Ding, Y., Fayt, A., Garnache, A., Roberts, J.S., and Romanini, D. (2003) The overtone spectrum of carbonyl sulfide in the region of the v1 + 4v3 and 5v3 bands by ICLAS-VECSEL. *J. Mol. Spectrosc.*, **219**, 81–87.

251 Naumenko, O.V., Leshchishina, O., Shirin, S., Jenouvrier, A., Fally, S., Vandaele, A.C., Bertseva, E., and Campargue, A. (2006) Combined analysis of the high sensitivity Fourier transform and ICLAS-VeCSEL absorption spectra of D_2O between 8800 and 9520 cm^{-1}. *J. Mol. Spectrosc.*, **238**, 79–90.

2
Thermal Management, Structure Design, and Integration Considerations for VECSELs

Stephane Calvez, Jennifer E. Hastie, Alan J. Kemp,
Nicolas Laurand, and Martin D. Dawson

2.1
Introduction

In this chapter, we introduce the design criteria, structural and optical characteristics, cavity format, and laser performance considerations for VECSEL.

As described earlier, the VECSEL concept is that of an optically pumped inorganic semiconductor platelet giving gain in a surface normal direction and operated in an external optical cavity. The gain chip design is largely based on principles already established for vertical-cavity surface-emitting lasers and the production of these multilayer thin-film structures relies similarly on the advanced capabilities of modern epitaxial growth techniques. However, there are simplifications (and indeed some subtleties) arising from the use of photopumping and the requirement to accommodate only one mirror within the gain chip. An important consequence of this simplification is the emergence of common design principles for a wide range of III–V quantum well and quantum dot gain structures, underpinning the flexibility of the technology for wide wavelength coverage. These design principles are covered in this chapter. It is important to emphasize, however, that optimization of any structure for any particular wavelength and desired performance usually proceeds in a "semiempirical" manner, where the detailed effects of, for example, number and distribution of quantum wells, barrier and well compositions, resonant or antiresonant structures, and so on, are compared experimentally for a systematic series of growths, allowing iteration to a final design.

Thermal issues are critical in VECSELs not only because of the temperature dependence of gain spectra, carrier recombination, and thermalization but also because of the different temperature rate of shift of gain and cavity resonance features. These issues and their consequences for a range of VECSEL device formats are also considered in detail in this chapter.

Semiconductor Disk Lasers. Physics and Technology. Edited by Oleg G. Okhotnikov
Copyright © 2010 WILEY-VCH Verlag GmbH & Co. KGaA, Weinheim
ISBN: 978-3-527-40933-4

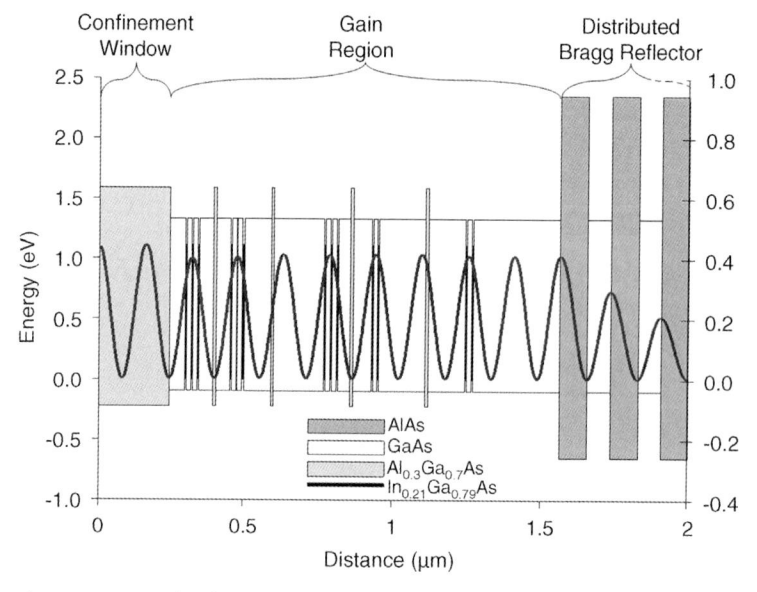

Figure 2.1 Example of the active region design for a 1060 nm emitting VECSEL.

2.2
VECSEL Structure Design

Typically, a VECSEL semiconductor structure includes four subsections (see Figure 2.1): a mirror, a gain section, a carrier-confinement window, and a cap layer to protect the previous layer and those underneath from oxidation. To offer the most wavelength-versatile design, the gain section commonly exploits quantum wells (QWs) as active elements although quantum dots (QDs – see Chapter 5) and indeed bulk semiconductors [1] have also been used. The remainder of this chapter will be dedicated to QW-based VECSELs since they are the most widespread type of device. As noted above, however, the design and operation of VECSELs with bulk or QD-active regions follow very similar principles and considerations.

2.2.1
Material System Selection

The design of the semiconductor structure starts with the use of a lattice constant versus bandgap energy diagram, as exemplified in Figure 2.2, to select an appropriate substrate and alloy material system to obtain, with a direct energy gap, the desired signal emission wavelength λ_s. The chosen semiconductor alloys should present the same lattice constant as the substrate, a rule known as lattice matching, as this leads to an epitaxial growth with low density of defects and provides conditions most appropriate for strong luminescence efficiency. Strained but mechanically stable layers with relative deviations in lattice parameter up to ∼3.5% can be grown under

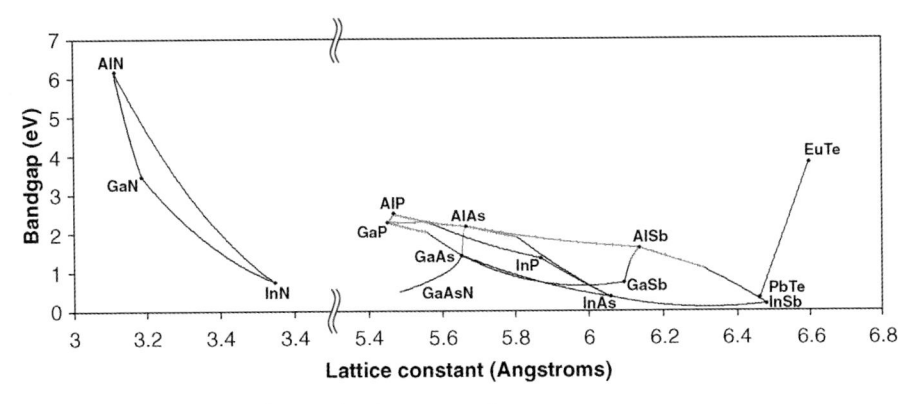

Figure 2.2 Semiconductor lattice – bandgap energy diagram.

compressive ($a > a_{subs}$) or tensile ($a < a_{subs}$) strain, provided their thickness is lower than the so-called "critical thickness" [2].

Table 2.1 lists the respective materials and substrates selected for reported VECSEL active regions in accordance with the above-mentioned principles.

It should also be pointed out that the growth of high-quality wafers with a thickness of strained material exceeding the critical layer thickness but with relatively higher interfacial strain is possible as long as alternating materials with larger and smaller lattice constants than the substrate are used. This technique is generally used to provide mechanical stability of multi-QW structures where the thickness of a layer with strain of opposite sign is adjusted to exactly cancel the in-built strain energy induced during the QW growth, a method known as "strain compensation" [3].

III–V semiconductor material systems, including the effects of strain, have been widely reviewed, and the reader is referred to such references for further detail [4, 5].

Table 2.1 Selected materials for demonstrated VECSELs.

Wavelength span (nm)	Material	Substrate	Type of gain structure	References
370–440	InGaN/GaN	Sapphire/SiC	QWs	[101]
640–690	InGaP/AlInGaP	GaAs	QWs	[56]
700–740	InP/AlInGaP	GaAs	QDs	[102]
780–900	GaAs/AlGaAs	GaAs	QWs	[30]
	InAlGaAs/GaAs		QWs	[68]
920–1350	InGaAs(N)/GaAs	GaAs	QWs	[78]
	InGaAs/GaAs		QDs	
1500–1900	InGaAsP	InP	QWs	[103]
	InGaAlAs		QWs	[12]
1900–2800	GaInAsSb	GaSb	QWs	[104, 105]
~5000	PbTe	BaF$_2$	Bulk	[65]

2.2.2
Gain

In this section, we will review the design principles of the heart of the VECSEL semiconductor chip: the gain section. As indicated earlier, this will focus on QW-based active regions as they have been the most commonly used to date. The reader can however refer to Chapter 5 for a more in-depth description of QD-based VECSELs.

QWs are thin (<10 nm) planar layers embedded within another semiconductor material of higher bandgap referred to as the barrier. Their thickness dictates that the emission transition energy is no longer set by the QW semiconductor bandgap alone rather it also depends on the barrier bandgap and the QW thickness. Optical emission occurs from the lowest level of the conduction band to the highest level of the valence band and its exact energy can be calculated by k·p analysis of the appropriate well potential. As illustrated in Figure 2.3 for InGaAs/GaAs QWs, several combinations of well width and composition can usually be found for emission at the target wavelength (here, 1060 nm). Typically, the chosen QW would offer a sufficiently large conduction band offset to limit thermal leakage of the electrons to the higher energy (barrier) levels and not too deep valence band offsets to ensure equal hole occupancy throughout the structure [6]. Narrow QWs are thus generally favored for their larger sub-band separation and reduced number of populated sub-bands for a given band-filling energy, which results in high-gain and lower threshold densities [7]. However, growth reproducibility and carrier thermalization set a lower limit to the practically used QW width (this limit would be the 5 nm $In_{0.29}Ga_{0.71}As$/GaAs QW in the chosen example). The gain provided by each QW can then be evaluated and optimized [6, 8].

As shown in the previous example, the use of compressively strained (i.e., under biaxial compressive strain in the plane of the structure) material as the QW low bandgap material is frequent as it is accompanied by a reduction in threshold because of changes in carrier effective mass, the induced energy splitting of the valence bands,

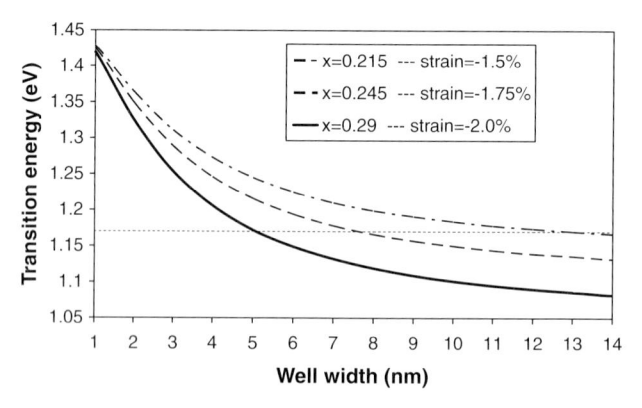

Figure 2.3 Calculated $In_xGa_{1-x}As$/GaAs quantum well transition energy dependence on well width for different indium content.

and because it promotes the required TE-polarized gain [9]. Besides, to maximize gain extraction, the QWs are usually placed at the antinodes of the standing wave of the electric field pattern established by the semiconductor mirror. This configuration is known as "resonant periodic gain" [10, 11]. It enhances the gain around the design wavelength λ_{RPG} and avoids spatial hole burning effects (see Figure 2.1).

To operate the devices, two optical pumping approaches referred to as "barrier" and "in-well" pumping are available, with the pump photon energy chosen to be respectively greater and lower than the QW barrier bandgap energy. The former and most common arrangement creates carriers in the continuum states of the barriers and as such benefits from a relative wavelength insensitivity and short absorption length (typically 1–2 μm). The exponential pump profile will however provide a graded carrier distribution, a factor that is somewhat mitigated by carrier diffusion and/or by special structural design features such as the use of nonuniform QW distribution [12] and/or the insertion of pump dividers [13]. The alternative pumping configuration creates carriers directly in the QWs where they are needed. As each QW absorbs a small fraction of the pump power (typically less than 1%), uniform pumping is readily achieved but more complex pump recirculation is needed to exhaust the pump [14, 15]. Typically, this involves the use of dual band mirrors to reflect both pump and signal and exploiting resonance effects at the pump wavelength.

Once the pumping scheme has been selected, the actual number of QWs and their distribution are usually decided on the basis of modeled performance [16–19]. Given that, the QWs have to be located within the pump absorption length so as to be pumped effectively, accommodating the optimum number of QWs regularly involves positioning several of them within a given antinode of the standing-wave electric field. Ideally, these QWs would be closely spaced to maximize gain overlap but carrier tunneling and coupling effects set a lower limit to the usable QW separation. For the 5 nm $In_{0.29}Ga_{0.71}As$/GaAs QW considered earlier, that distance would be ~8 nm as shown in Figure 2.4. Continuing with this design example, Figures 2.5 and 2.6

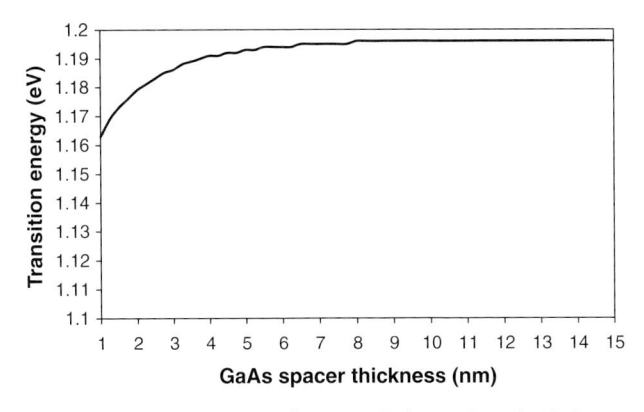

Figure 2.4 Transition energy of two 5 nm thick $In_{0.29}Ga_{0.71}As$/GaAs quantum wells as a function of GaAs separation.

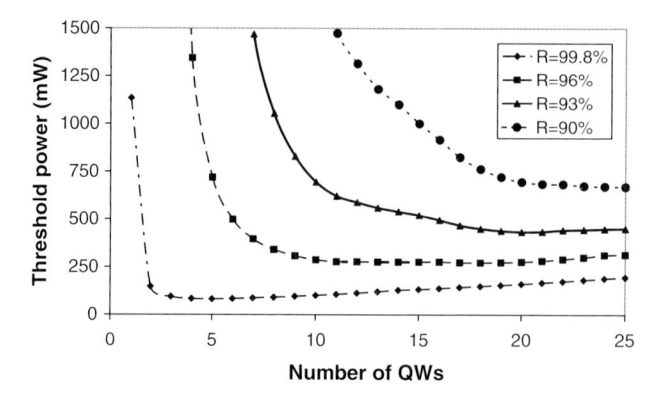

Figure 2.5 Modeled threshold performance of 1060 nm quantum well VECSEL under 808 nm pumping.

represent the performance predictions for an 808 nm pumped 1060 nm VECSEL using the model reported in Ref. [16], where the carrier distribution is assumed to be uniform, thermal effects are neglected, and an internal efficiency of 75% is considered. These plots demonstrate that structures with 10–15 QWs should allow operation with an optimum external mirror reflectivity of ~93%, with threshold powers of ~0.75 W and output powers reaching ~2 W under 10 W of excitation. The resulting 13 QW structure with its carrier dividing layers and nonuniform distribution of QWs to define areas of constant average carrier density per QW is presented in Figure 2.1.

Although heating effects were neglected in the previous model, they play a very significant role in VECSEL performance, as further discussed in Section 2.3. Indeed, as the structure heats up (through a rise of the ambient temperature or because of pump-induced heating), the gain provided by the QWs reduces as carrier nonradiative recombination or thermalization loss increases. This gain reduction will ultimately prevent laser action and therefore effectively control the operating temper-

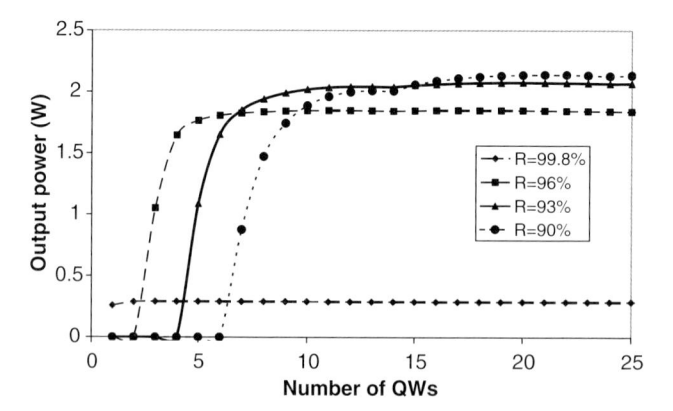

Figure 2.6 Modeled output power of 1060 nm quantum well VECSEL for 10 W of 808 nm incident pump power.

ature range of the devices. In addition, the QW peak emission wavelength is found to shift at a rate that generally differs from the cavity/resonant periodic gain wavelength shift, exacerbating the previously described thermal shutoff. A "gain offset" detuning between the QWs emission wavelength and the cold RPG and subcavity is therefore commonly implemented as a strategy to mitigate these effects [20, 21].

2.2.3
Mirrors

Given that QWs or QDs are only likely to provide a small percentage of gain per cavity round-trip and that the substrate may be absorptive at the wavelength of operation, VECSEL semiconductor structures usually contain a high-reflectivity mirror. In most circumstances, these mirrors are distributed Bragg reflectors (DBRs), that is, stacks of N periodic repeats of alternating quarter-wave thick layers of low and high refractive indices (n_{low} and n_{high}), embodied monolithically as part of the semiconductor structure.

The reflectivity and E-field patterns within a VECSEL epilayer, and in the DBRs in particular, are generally calculated using the transfer matrix method as commonly employed in thin-film software [22, 23]. It is however convenient to note that DBRs can be modeled as fixed hard mirrors having an effective reflectivity R_{DBR} and bandwidth $\Delta\lambda_{DBR}$, given by [24]

$$R_{DBR} = \left(\frac{1-qp^{2N}}{1+qp^{2N}}\right)^2, \qquad (2.1)$$

$$\frac{\Delta\lambda_{DBR}}{\lambda_c} = \frac{4}{\pi}\arcsin\left(\frac{1-p}{1+p}\right), \qquad (2.2)$$

where $p = n_{low}/n_{high}$ and $q = n_I/n_E$ are, respectively, the refractive index ratio at the DBR internal interfaces and the incident to exit refractive index ratio.

From the above equations and previously mentioned considerations, it is evident that the preferred DBR constituting alloys are not only transparent at the signal wavelength and lattice matched to the substrate but also offer a high-index contrast ratio (small p factor) to minimize the number of pairs (N) required to achieve a given reflectivity and obtain broadband reflectivity mirrors. This is especially important to reduce growth time and reduce effects of changes in growth rates.

In addition, to facilitate efficient heat extraction (see Section 2.3), these constituents are often chosen to maximize the DBR effective longitudinal (k_l) and lateral (k_r) thermal conductivities, given by Refs [25, 26]:

$$k_l = \sum_i t_i \bigg/ \sum_i t_i/k_i, \qquad (2.3)$$

$$k_r = \sum_i t_i k_i \bigg/ \sum_i t_i, \qquad (2.4)$$

where the sums are calculated over all the layers of thickness t_i and thermal conductivity k_i, constituting each part.

2 Thermal Management, Structure Design, and Integration Considerations for VECSELs

Table 2.2 Distributed Bragg reflector materials and properties.

Wavelength of operation	Mirror compositions	p	N	Δλ (nm)	Total mirror thickness (μm)	k_r	k_l	References
390	SiO_2/HfO_2	0.7318	11.5	77.2	1.263	1.5	1.5	[101]
670	$AlAs/Al_{0.45}Ga_{0.55}As$	0.8835	40	52.8	4.08	53.7	20.5	[56]
850	$AlAs/Al_{0.2}Ga_{0.8}As$	0.860	30	81.1	3.97	56.0	26.6	[30]
940	AlAs/GaAs	0.8839	25.5	108.5	3.64	69.9	61.4	[78]
1060	AlAs/GaAs	0.8840	25.5	114	4.1	69.8	61.2	[78]
1220	AlAs/GaAs	0.8505	25.5	125.6	4.83	69.7	61.2	[59]
1310	AlAs/GaAs	0.8528	25.5	132.6	5.21	69.7	61.2	[35]
1550	$In_{0.18}Ga_{0.82}As_{0.4}P_{0.6}/InP$	0.9545	48	45.9	11.47	37.9	11.8	[103]
2000	$AlAsSb_{0.08}/GaSb$	0.8021	21.5	280.2	5.95	20.1	14.3	[104]
5000	$Pb_{0.93}Eu_{0.07}Te/BaF_2$	0.2804	2.5	2287	2.80	9.2	1.4	[65]

Table 2.2 provides mirror compositions and parameters of a representative set of reported devices to illustrate this selection process.

To reduce the mirror thermal resistance and relieve some of the growth efforts, hybrid mirrors, which exploit a combination of a metallic (gold) layer with a dielectric DBR presenting a reduced pair number requirement, have also been explored [12]. Although high (reflectivity and thermal) performance can be obtained, their use is often associated with the demanding processing step of fully removing the substrate.

2.2.4
Subcavity Designs

The association of a micrometer-thickness gain section with a mirror constitutes a subcavity within the laser resonator whose resulting Fabry–Perot resonance effects significantly affect laser performance. Indeed, the effective gain of the semiconductor chip is enhanced and narrowed because of the resonance-induced recirculating field within the active section, which in turn results in a lower threshold but more temperature-sensitive device. These phenomena can easily be understood and illustrated (see Figure 2.7) by comparing the E-field pattern, reflectivity curve, and the chip gain G_{chip} for a *resonant* (antinode of the field at the semiconductor interface) and *antiresonant* (node of the field at the semiconductor interface) device. The effective chip gain is given by [27]

$$G_{chip}(\lambda) = \frac{(\sqrt{R_i}-\sqrt{R_{DBR}}g_s)^2 + 4\sqrt{R_i R_{DBR}}g_s \sin^2(2\pi nL_c/\lambda)}{(1-\sqrt{R_i R_{DBR}}g_s)^2 + 4\sqrt{R_i R_{DBR}}g_s \sin^2(2\pi nL_c/\lambda)}, \quad (2.5)$$

where g_s is the single-pass gain of the structure, R_{DBR} and R_i are the DBR mirror and air–semiconductor interface reflectivities, respectively, and L_c is the length of the cavity.

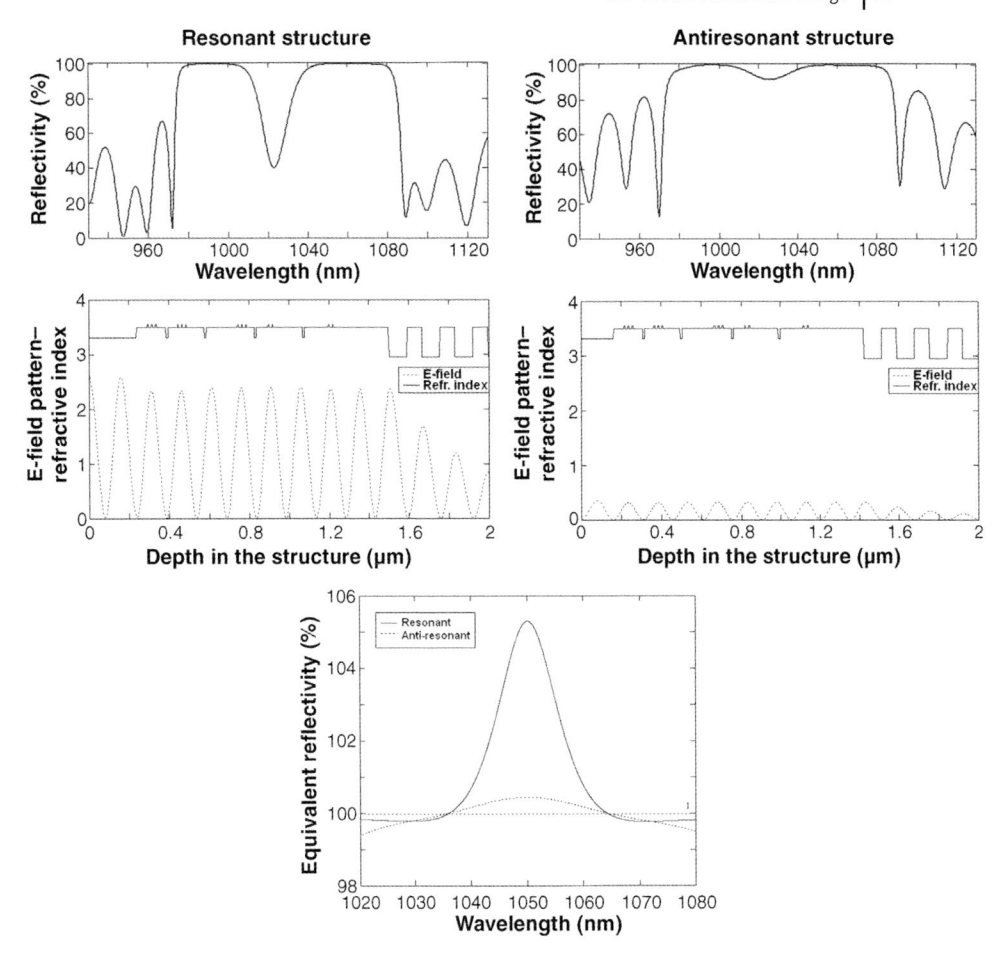

Figure 2.7 Comparison of the characteristics of resonant and antiresonant VECSEL structures.

The position of the air–semiconductor interface with respect to the E-field pattern is therefore carefully chosen at the device design stage. This is generally accomplished through the thickness tuning of the confinement window, a layer of high bandgap and preferably high thermal conductivity inserted to prevent pump-generated carriers from reaching the surface where they would be lost to nonradiative recombination. Resonant devices are commonly used in frequency-doubled VECSELs because they intrinsically benefit from an additional wavelength selection process. However, for wide laser tunability, antiresonant structures are generally preferable.

2.2.5
Growth

The growth of the designed multilayer semiconductor heterostructure is a key step. For this type of advanced structure, it is a standard practice to use either molecular beam epitaxy (MBE) or metal-organic chemical vapor deposition (MOCVD) technologies as they allow the preparation of low dimensional structures with high purity and structural perfection.

In a MBE, high-purity elemental constituents are produced using independently controlled (heated) effusion cells and the generated fluxes recombine as epitaxial layers on a substrate placed in an ultrahigh vacuum. In contrast, in an MOCVD system, the deposition occurs under near-ambient pressure conditions and is based on a surface reaction of metal organic elements or metal hybrids containing the required chemical species. In both cases, the substrate is heated to control the deposition morphology and rotated for improved uniformity. The production of the full structure can take several hours at typical growth rates of $\sim 1\,\mu m\,h^{-1}$. A detailed description of these well-established techniques is beyond the scope of this chapter, but the reader is referred to the literature for further information [28].

2.2.6
Structure Characterization

As already explained, the growth of the VECSEL epilayer is a rather long process that needs to be accurately controlled to correctly position the mirror, subcavity resonance, and QW/QD emission. To routinely assess the success of such operations, temperature-dependent photoluminescence and reflectivity measurements are usually performed on the as-grown wafers. Figure 2.8 shows a typical characterization setup

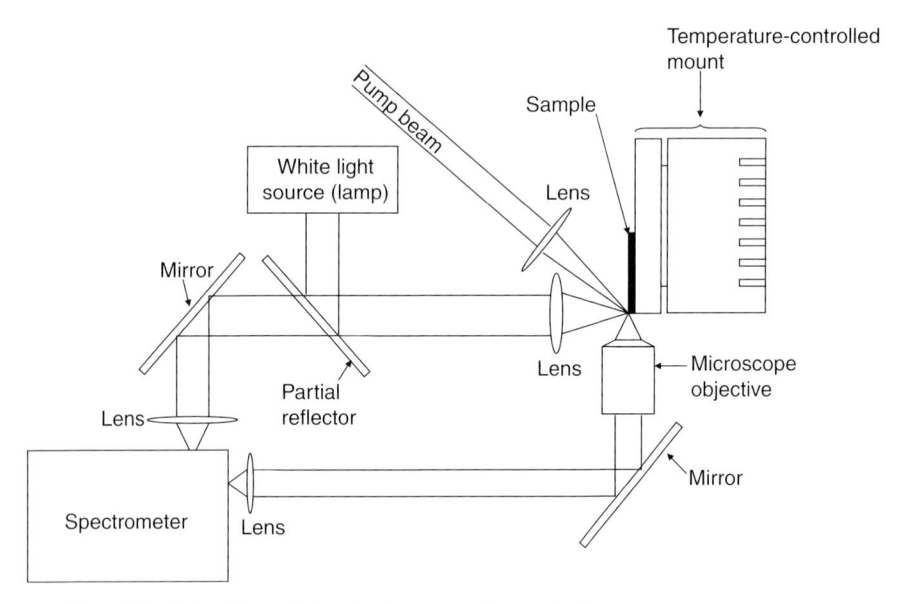

Figure 2.8 Reflectivity and photoluminescence characterization setup.

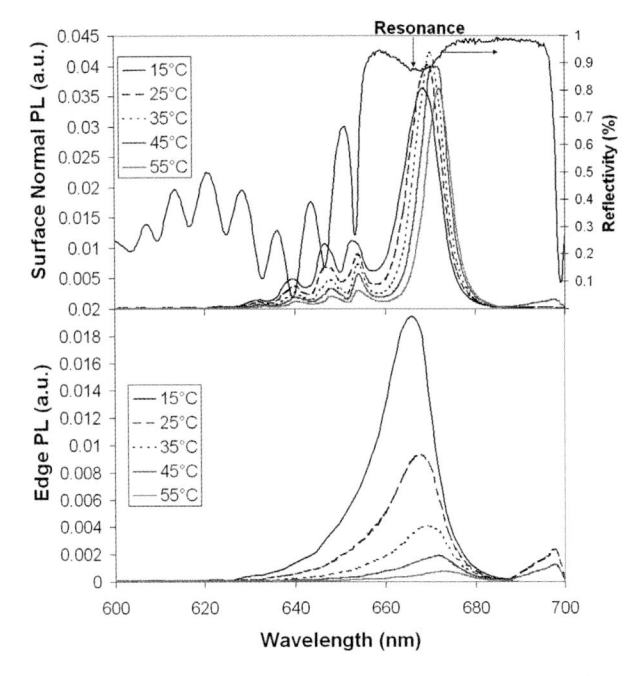

Figure 2.9 Reflectivity and photoluminescence characteristics of a 670 nm VECSEL.

where the photoluminescence is measured from both the surface and the edge of the sample to respectively position the device and the underlying QW/QD gain. Figure 2.9 features an example of the characteristics of a red-emitting AlInGaP/GaAs VECSEL structure. In this case the room-temperature reflectivity shows that the mirror stop band is 47 nm broad and centered on 676 nm and that the resonance marked by a reflectivity dip is at 668 nm at 15 °C. The QW photoluminescence at 25 °C is shown to have an offset of 3 nm from the surface peak PL, and their respective peak wavelengths match at 35 °C, considered as the optimum internal operating temperature for this device.

2.2.7
Laser Cavity

A typical VECSEL laser cavity is setup with the semiconductor chip used as an optically pumped active end mirror. In its simplest form, it exploits a single additional bulk curved mirror acting as the output coupler (OC). To obtain a high-efficiency and a high-quality beam output, the mirror radius of curvature and distance to the semiconductor chip are chosen to provide mode matching, that is, a cavity mode at the semiconductor whose size is closely matched to the gain aperture size. More complex resonators are however routinely used to produce lasers with advanced performance. In particular, fundamental wavelength-emitting VECSELs with three-mirror cavities (see Figure 2.10) are common as this is a way to set up dynamically

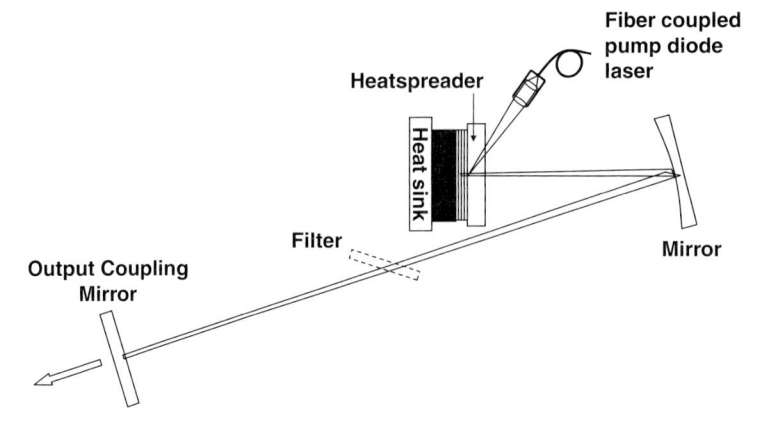

Figure 2.10 Typical three-mirror cavity for fundamental emitting VECSELs.

stable lasers, a category of sources whose resonator meets mode matching conditions irrespective of the thermal lens incurred in the semiconductor chip [29].

2.3
Thermal Management

2.3.1
Introduction: Why Is Thermal Management Important?

Efficient thermal management is vital for virtually all solid-state lasers and VECSELs are no exception – indeed, thermal management choices affect both output power and spectral coverage. However, the thermal management challenges are different from those in a conventional diode-pumped solid-state laser in two important respects. First, the power scaling limits are set by the absolute temperature and not by temperature gradients as is often the case in a conventional diode-pumped solid-state laser (e.g., thermal lensing, stress fracture). Second, the heat density is much larger than in a conventional diode-pumped solid-state laser. The pump absorption length in a semiconductor is of the order of 1 μm compared to 1 mm in 1 atm% Nd: YAG. Given that similar pump spot radii are used in both cases (tens to hundreds of μm), the heat density is typically three orders of magnitude higher in a VECSEL.

VECSELs are sensitive to the absolute temperature for two reasons. First, the quantum well gain falls as the temperature rises. Second (see Section 2.2.4), VECSELs often oscillate at the resonance of the etalon formed by the mirror and the top of the structure. This resonance and the quantum well gain peak shift spectrally with temperature but typically at different rates. As a result, the two walk out of alignment and the device switches off above a certain temperature and hence above a certain pump power. Thus, the fundamental objective of thermal management in a VECSEL is to minimize the temperature rise per unit pump power.

2.3.2
Thermal Management Strategies in VECSELs

In a basic VECSEL, a piece of as-grown material is soldered directly to a heat sink. The heat sink is typically cooled by water or a Peltier device. Optical pumping deposits heat near the top surface of the device, in the gain section; however, the heat is removed from the bottom of the substrate (see Figure 2.11). The mirror structure (or distributed Bragg reflector) and the substrate thus represent a significant thermal resistance on the heat flow path from source to sink. As a result, the output powers

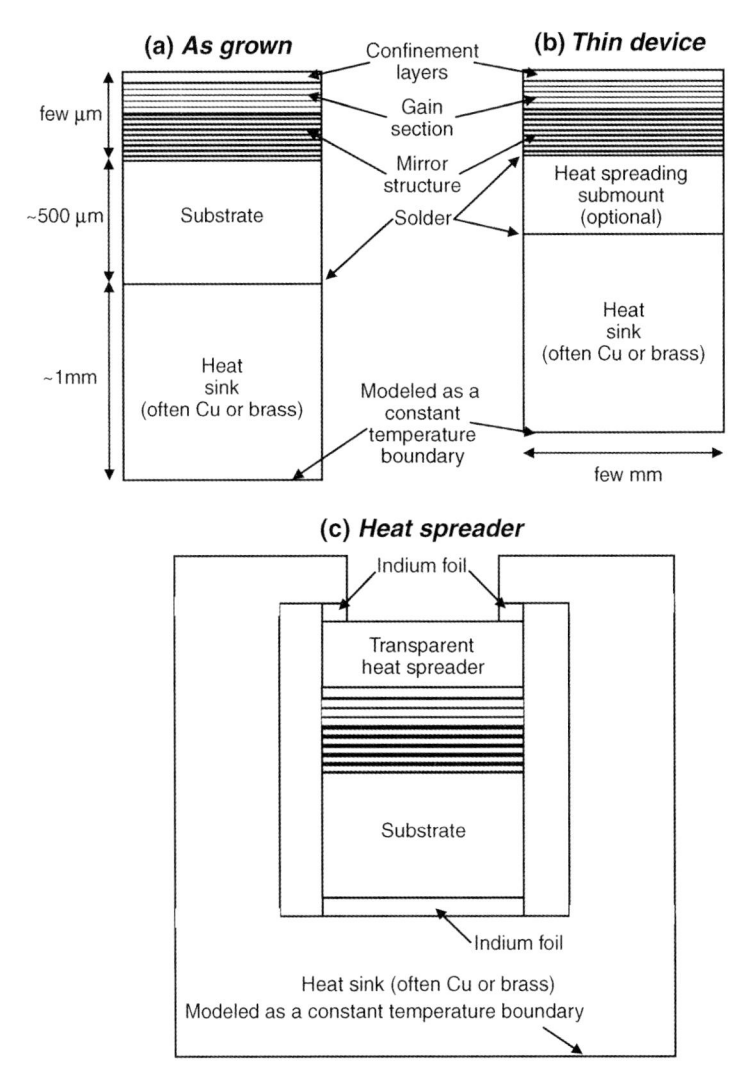

Figure 2.11 Schematic diagrams of heat sinking approaches for VECSELs: (a) as-grown, (b) thin device, and (c) heat spreader.

from such systems have typically been limited to a few tens to a few hundreds of milliwatts [30, 31].

Two approaches are widely used to permit scaling to higher powers. The first is to remove the substrate to reduce the thermal resistance (Figure 2.11b); and the second is to bypass that resistance by bonding a high thermal conductivity transparent material to the top surface of the device (Figure 2.11c). These alternatives will be discussed in turn.

Removing the substrate leaves a thin semiconductor structure of a few micrometers thickness soldered to a heat sink (Figure 2.11b). This approach to thermal management will be referred to as the thin device approach. Where the mirror structure has good thermal conductivity – as in systems emitting around 1 µm based on GaAs/AlAs mirrors (see Table 2.2), the thin device approach represents an excellent means to extract heat from the gain section. Indeed, it is by using this approach that the highest powers have been demonstrated from VECSELs (Figure 2.20). Due to the disparity between the thickness of the epitaxial semiconductor layers (a few micrometers) and the width of the pumped region (a few hundred micrometers), heat flow in these layers is largely one dimensional, down through the mirror structure into the heat sink. By analogy to the thin disk concept in doped dielectric lasers [32], this might be expected to give considerable potential for power scaling. And to a degree it does; however, the power scaling achievable is limited by three-dimensional heat flow in the submount (see Section 2.3.6).

Although the highest output powers have been demonstrated using the thin device approach, it has a number of disadvantages. The most important of these is its dependence on the thermal conductivity of the mirror structure. This makes it unsuitable for many semiconductor material systems and hence for many wavelength bands (see Section 2.3.7). An alternative approach is to bond a high thermal conductivity transparent material to the top surface of the semiconductor structure [33]; adhesive-free bonding is usually achieved using liquid-assisted optical contacting [34]. This thermal management scheme will be referred to as the heat spreader approach. If a very high thermal conductivity material such as diamond [35] or silicon carbide [36] is used as the heat spreader, it bypasses the thermal resistance of the mirror structure and enables higher power operation at wavelengths that require less thermally conductive semiconductor materials (see Section 2.3.7). In contrast to the thin device case where heat is extracted directly through the mirror structure, heat predominantly flows out of the semiconductor into the heat spreader in the vicinity of the pumped region. It is then removed both by extraction down through the semiconductor structure – over a larger area and hence with lower thermal resistance – or directly out of the heat spreader into the surrounding mount. The three-dimensional nature of this heat flow means that indefinite power scaling with pump radius would not be expected even in an idealized scenario (see Section 2.3.6). In addition, any losses introduced by intracavity use of the heat spreader have to be taken into account [37, 38]. On the other hand, the heat spreader approach is effective for a wider range of semiconductor material systems. Hence, it is important to examine the relative merits of the heat spreader and thin device approaches in more detail.

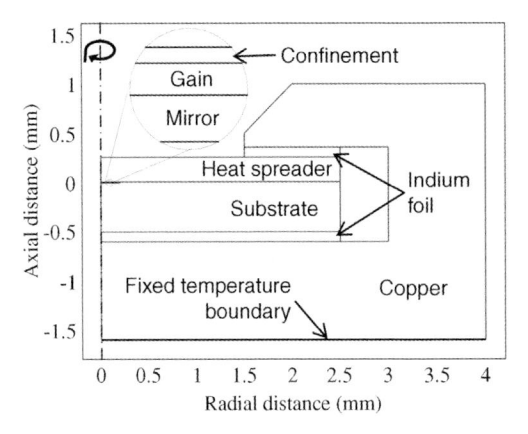

Figure 2.12 Typical setup for a finite-element simulation of a VECSEL with a heat spreader. The dash-dot line is the axis of rotational symmetry.

2.3.3
Modeling of Heat Flow in VECSELs: Guidelines

In general, the heat flow in VECSELs is relatively complicated and does not lend itself to analytical analysis (an exception is the insightful analysis of the thin device approach in Ref. [39]); hence, finite-element numerical approaches are typically used [25, 39–44]. However, the structure of a semiconductor disk lasers presents problems even for a numerical analysis. These relate to the disparity in the dimensions: typically, the submount dimensions and semiconductor chip widths are on the millimeter scale; the pump spot radius is of the order of tens of micrometers; and the thickness of the epitaxially grown semiconductor layers are in the tens to hundreds of nanometer regime. This makes it difficult to find a suitable compromise between the fine grid required to adequately capture heat flow over the smaller dimensions and the computational time required for such a large number of elements. To facilitate analysis on a standard desktop computer, a number of approximations are usually made. First, a cylindrically symmetric geometry is assumed to effectively reduce the problem to two dimensions. Second, the epitaxial structure is usually grouped into layers of similar function: the mirror structure, the gain region (quantum well and barriers), and the confinement region (window layer and any cap). Average radial and axial thermal conductivities are then used for these regions [40]. A typical model geometry is summarized in Figure 2.12.

2.3.4
The Thin Device and Heat Spreader Approaches at 1 and 2 μm

The thin device and heat spreader approaches have rather different thermal properties and hence are better adapted to different VECSELs: where the thermal conductivity of the mirror structure is high, the thin device approach excels; where the mirror represents more of a thermal resistance, the heat spreader approach is

Table 2.3 The maximum temperature rise per unit pump power in an InGaAs-based VECSEL for 0.98 μm and a GaSb-based VECSEL for 2.35 μm when different thermal management strategies are used.

Thermal management approach	Maximum temperature rise per unit pump power for a pump spot radius of 100 μm (K W^{-1})	
	InGaAs based for 0.98 μm	GaSb based for 2.35 μm
As-grown (Figure 2.11a)	21	86
Thin, diamond submount (Figure 2.11b)	2.7	26
Diamond heat spreader (Figure 2.11c)	2.1	4.6

favorable. This can be illustrated by a head-to-head comparison of the temperature rise in an InGaAs-based device designed for operation at 0.98 μm [45] and a GaSb-based device for 2.35 μm [46]. A comparison of three thermal management scenarios – the as-grown material, the thin device case (bonded to a diamond submount), and the diamond heat spreader case – is summarized in Table 2.3. The details of the structures can be found in Refs [25, 26, 42]. The fraction of the absorbed pump power that results in heating is not well characterized for VECSELs. For the purposes of the simulations discussed here, this fraction has been assumed to be equal to the quantum defect for light absorbed in the gain section and equal to that for light absorbed elsewhere. Hence, the predicted temperature rises are likely to be conservative.)

If a piece of as-grown wafer is simply bonded to a heat sink and the pump spot radius is 100 μm, the maximum temperature rise predicted per unit pump power is 21 K W^{-1} in the 0.98 μm case and 86 K W^{-1} in the 2.35 μm case. Typically, the semiconductor structural design accommodates a temperature rise of a few tens of Kelvin on pumping. Hence, the pump power before thermal rollover occurs is likely to be limited to the watt level for the 0.98 μm device; for a GaSb-based device operating at 2.35 μm, these powers are likely to be fourfold lower. Operation at output powers of a few watts requires improved heat sinking in both cases.

If the substrate is removed and the resulting thin slice of epitaxial material is soldered (mirror side down) to an extra-cavity diamond submount and then to copper [47, 48], the temperature rise per unit pump power is reduced by almost an order of magnitude for the 0.98 μm emitting InGaAs-based device to 2.7 K W^{-1}. In contrast, the temperature rise for the GaSb-based 2.35 μm emitter is still substantial at 26 K W^{-1}. If the intracavity surface of the as-grown semiconductor wafer is bonded to a high thermal conductivity transparent heat spreader, heat can be extracted directly from the gain region. This results in maximum temperature rises of 2.1 K W^{-1} and 4.6 K W^{-1} for the InGaAs-based 0.98 μm device and the GaSb-based 2.35 μm device, respectively, using diamond as the heat spreader material. The larger temperature rise in the latter case can largely be accounted for by the threefold higher quantum defect. That is to say, the heat spreader essentially bypasses the thermal resistance of the mirror structure, which in turn means the temperature rise is nearly independent of the thermal conductivity of the mirror structure (see also Figure 2.15 and Section 2.3.7).

2.3.5
Important Parameters: The Thermal Conductivity of the Mirror Structure, Submount, and Heat Spreader

The effect of mirror structure thermal conductivity on the temperature rise in the device for heat spreader and thin device approaches to thermal management is shown in Figure 2.13. In this simulation, only the mirror structure thermal conductivity has been changed; all other parameters, including the mirror structure thickness, are based on a 2.35 μm VECSEL [42]. (For all simulations, 10 W of pump power has been assumed to be deposited in a pump spot radius of 100 μm unless otherwise stated.) The difference between the heat spreader and thin device cases is marked: where an intracavity heat spreader is used, the temperature rise is essentially independent of the thermal conductivity of the mirror structure; where the thin device approach is used, maintaining a reasonable temperature rise within the device is contingent on having a high thermal conductivity mirror structure.

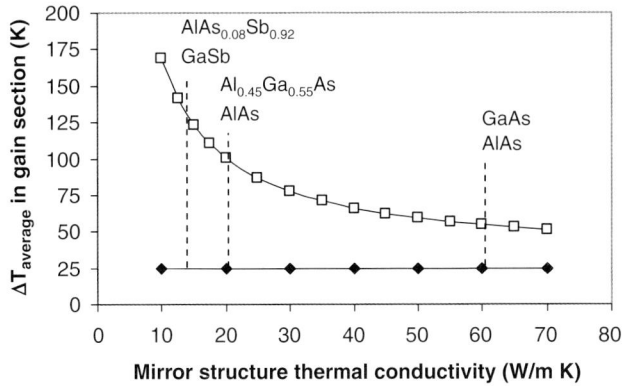

Figure 2.13 Average temperature rise in the gain section as a function of the average thermal conductivity of the mirror structure for the heat spreader and thin device approaches to thermal management of a VECSEL. The dashed vertical lines indicate the average axial thermal conductivities of various mirror structures: GaAs/AlAs mirrors used in lasers operating around 1 μm [47]; $Al_{0.45}Ga_{0.55}As$/AlAs mirrors used in lasers operating around 0.67 μm [55]; and $AlAs_{0.08}Sb_{0.92}$/GaSb mirrors used in lasers operating around 2.35 μm [42]. A weighted average of the temperature has been taken over the Gaussian pump/laser mode (both assumed to be 100 μm in radius) within the gain section [42]. The thermal conductivity of the mirror structure was assumed to be isotropic: the anisotropy typical of a distributed Bragg reflector structure only changes the result by a small percentage. The lines joining the data points are not fits and are only intended to guide the eye.

For a high thermal conductivity mirror, the relative performance of the thin device and intracavity heat spreader approaches depends on other factors, including the required thickness of the mirror structure and the pump spot radius. The large pump spot radii and thin mirror structures tend to favor the thin device approach (see Sections 2.3.6 and 2.3.7). In Figure 2.13, a relatively thick mirror structure and a medium-size pump spot lead to the heat spreader approach being favorable for all the mirror structure thermal conductivities examined.

In calculating the temperature rises shown in Figure 2.13, diamond was assumed as the submount material in the thin device case and as the intracavity heat spreader material. In both cases, the use of diamond turns out to significantly reduce the pump-induced temperature rise and hence significantly increase the potential for power scaling. However, is the use of such a high thermal conductivity material really justified?

In Figure 2.14, the average temperature rise in the gain region is plotted as a function of the submount thermal conductivity for a thin device case (based on a 0.98 μm device [25]). In these simulations, it is assumed that the semiconductor structure (with the substrate removed) is first soldered to a 0.25 mm thick submount, and this submount is then soldered to a 1 mm thick copper block. The bottom surface of this block is held at constant temperature (Figure 2.11b). The effect of the submount thermal conductivity is marked: even moving from diamond to silicon carbide (SiC) almost doubles the temperature rise; move to the widely used heat sink material copper–tungsten (CuW), and the temperature rise trebles. Thus, a high thermal conductivity submount is really required for the effective thermal management in the thin device case.

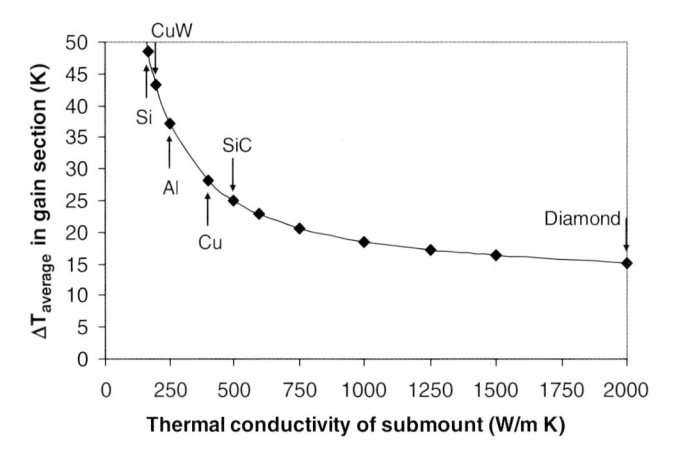

Figure 2.14 Average temperature rise in the gain section as a function of the thermal conductivity of the submount for a 0.98 μm VECSEL using a thin device approach to thermal management. A weighted average of the temperature has been taken over the Gaussian pump/laser mode (both assumed to be 100 μm in radius) within the gain section [42]. The line joining the data points is not a fit and is only intended to guide the eye.

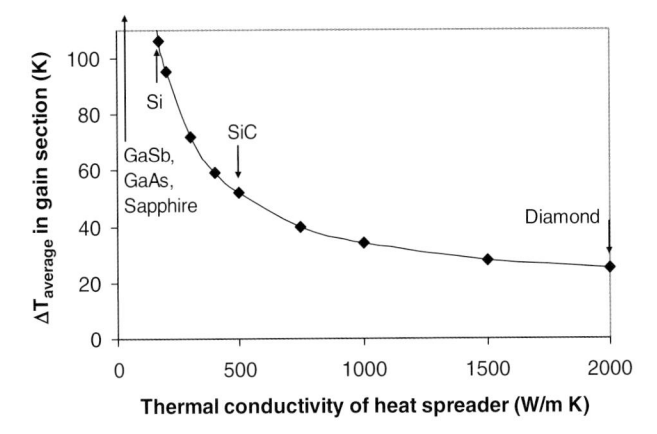

Figure 2.15 Average temperature rise in the gain section as a function of the thermal conductivity of the intracavity heat spreader. A weighted average of the temperature has been taken over the Gaussian pump/laser mode (both assumed to be 100 μm in radius) within the gain section [42]. The line joining the data points is not a fit and is only intended to guide the eye.

A similar requirement for a high thermal conductivity material is seen in the heat spreader case (see Figure 2.15). If a 2.35 μm VECSEL pumped with 10 W in a 100 μm pump spot radius is taken as an example, the effect of moving from diamond to silicon carbide (SiC) is the doubling of temperature rise. For silicon, the temperature rise is fourfold higher; for sapphire, it is an order of magnitude higher. Hence, the utility of the heat spreader approach depends on the use of a transparent material of very high thermal conductivity.

Although both the thin device and the heat spreader approaches require the use of high thermal conductivity materials, the heat flow paths in the two cases are rather different; hence, differences in the way these approaches scale to higher output powers might be expected. This will be the subject of the next section.

2.3.6
Power Scaling of VECSELs

By analogy to doped dielectric thin-disk lasers, power scaling of VECSELs might be envisaged by increasing the pump power and pump spot radius simultaneously to keep the pump intensity constant. (Although thermal effects are not the only limit to power scaling in semiconductor disk lasers [49], they are typically the dominant effect). In an idealized thin-disk laser, the temperature rise would remain constant in this scenario [32]. However, this picture assumes that heat flow in the disk laser is one dimensional: down through the disk to the heat sink. To what extent does this picture apply to a VECSEL?

Figures 2.16 and 2.17 illustrate the power scaling properties of a 0.98 and 2.35 μm VECSEL, respectively, as the pump power is increased at a constant pump intensity of 318 W mm^{-2} (the pump is assumed to have a "top-hat" transverse profile – that is to say,

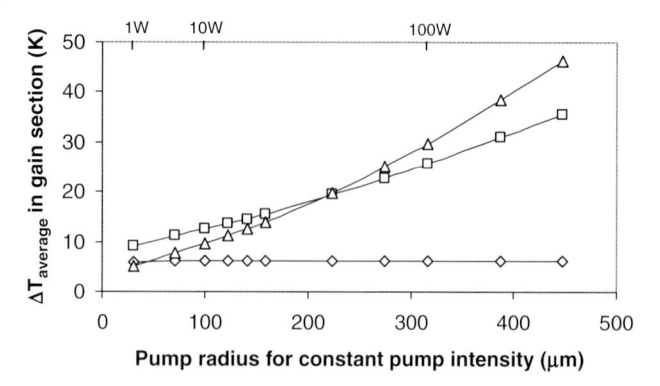

Figure 2.16 Average temperature rise in the gain section of a 0.98 μm VECSEL as a function of the pump spot radius when the pump power is increased at constant pump intensity for three thermal management scenarios. The pump powers for selected pump spot radii are indicated at the top of the graph. A weighted average of the temperature within the gain section has been taken over the laser mode (assumed to have a Gaussian transverse profile where the $1/e^2$ radius equals the pump spot radius) [42]. The lines joining the data points are not fits and are only intended to guide the eye.

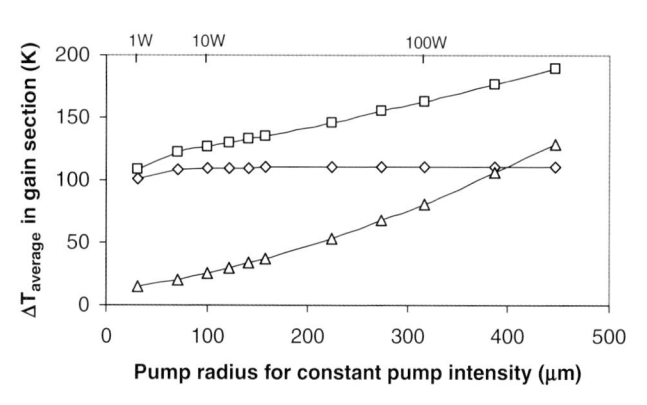

Figure 2.17 Average temperature rise in the gain section of a 2.35 μm VECSEL as a function of the pump spot radius when the pump power is increased at constant pump intensity for three thermal management scenarios. The pump powers for selected pump spot radii are indicated at the top of the graph. A weighted average of the temperature within the gain section has been taken over the laser mode (assumed to have a Gaussian transverse profile where the $1/e^2$ radius equals the pump spot radius) [42]. The lines joining the data points are not fits and are only intended to guide the eye.

318 W mm^{-2} within the pump spot radius and 0 W mm^{-2} outside). Three thermal management scenarios are examined: an idealized thin device where the substrate has been removed and a fixed temperature boundary condition imposed directly at the rear of the mirror structure; a more realistic thin device case where the semiconductor structure (minus the substrate) is soldered to a 0.25 mm thick diamond submount, which in turn is soldered to a 1 mm thick copper heat sink (Figure 2.11b); and a case where an intracavity diamond heat spreader is used (Figure 2.11c).

For the idealized thin device case, the temperature rise would be constant, as expected by analogy to the conventional thin-disk laser. For the thermal management techniques used in practice, this indefinite scalability however breaks down because the heat flow is no longer one dimensional. For a thin device (substrate removed) soldered to diamond and copper, lateral heat flow in the diamond submount and copper heat sink means that the temperature increases as the pump power is scaled at constant intensity; for the intracavity heat spreader case, lateral heat flow in the heat spreader has an even more marked effect.

The differences in the material parameters between the 0.98 and 2.35 μm cases imply that there are important differences in the scaling behavior. For the 0.98 μm device (Figure 2.16), the idealized thin device would give the lowest temperature rise for all pump spot radii above about 50 μm. Of the two realistic scenarios, the intracavity heat spreader performs better at pump radii below about 200 μm; the thin device soldered to diamond and copper is better at larger pump radii. However, neither of these two scenarios is indefinitely scalable.

For the 2.35 μm device (Figure 2.17), the scenario is quite different: the intracavity heat spreader is predicted to provide the lowest temperature rise up to pump radii of about 400 μm. For larger radii, an idealized thin device would perform better, but the heat spreader outperforms the more realistic thin device case for all modeled pump spot radii. The reason for the distinct differences between the 0.98 μm and 2.35 μm devices is the thermal resistance of the mirror structure: in the 2.35 μm device, the low thermal conductivity of the mirror structure inhibits heat flow directly down to the heat sink. This limits the performance of even an idealized thin device approach; hence, the heat spreader approach, which effectively bypasses this thermal resistance, is favored at all pump spot radii that give a reasonable temperature rise.

In a conventional thin disk laser, power scaling is predicated on a "top-hat" pump profile since this minimizes heat flow in the plane of the disk [32]. However, the pump profile in a VECSEL is often peaked on axis. The average temperature rise in the gain region does not vary significantly between a top-hat pump and a Gaussian pump (see Figure 2.18). On the other hand, the maximum temperature rise is typically 1.5–2 times larger for a Gaussian pump shape. Hence, while differences in the scaling potential of systems pumped with a Gaussian and a top-hat pump can be expected, these differences may not be large. Indeed, it is likely to be easier to obtain good beam quality using a Gaussian pump profile, at least from a mode-overlap perspective, and hence such a profile may be preferable for brightness scaling.

As the modeling presented in this section illustrates, there is no universal approach to thermal management in a VECSEL: where the mirror structure has good thermal conductivity – for example, around 1 μm – a thin device approach and

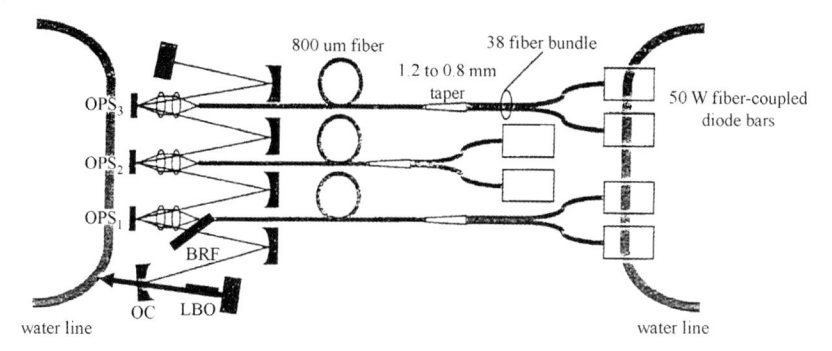

Figure 2.21 Schematic of a high-power frequency-doubled three-chip VECSEL demonstrated by Hunziker et al. [52].

At wavelengths away from 1 µm and away from InGaAs QWs, the current record output power in nearly all cases has been achieved with the use of intracavity diamond heat spreaders – the thermal management technique of choice for low thermal conductivity semiconductor alloys, as described in Section 2.3. The notable exception is the result at 855 nm reported by Beyertt et al. [54] who demonstrated 1.3 W output power with record slope efficiency via in-well pumping, which is discussed in detail in the next section.

At the short-wavelength end of fundamental VECSEL emission, Hastie et al. [55] used a diamond heat spreader bonded to an InGaP/AlInGaP QW gain structure grown on an AlGaAs DBR to achieve >1 W output power at 674 nm with 20% slope efficiency (gain structure details reported in Ref. [56]). This laser was pumped at 532 nm with a frequency-doubled Nd:YVO$_4$ laser; diode pumping of the red VECSEL was not demonstrated until the advent of high-power GaN diode lasers at 445 nm, with VECSEL output power currently limited to a few tens of milliwatts due to the shortened pump absorption length [57].

Beyond 1 µm, dilute nitride materials were developed to target the 1.3 µm communications window (for zero dispersion in optical fiber), using GaInNAs quantum wells with a small fraction of nitrogen to reduce the bandgap of InGaAs. Hopkins et al. [35] demonstrated 0.6 W output power at 1.32 µm from the first VECSEL to use this material system. This was also the first reported use of a diamond heat spreader with a VECSEL to improve on the thermal conductivity of previously used sapphire [33] and SiC [36]. Higher output power using GaInNAs quantum wells has since then been achieved at the shorter wavelength of 1.22 µm by Korpijärvi et al. [58] where the application target is frequency doubling to the red [59].

Moving deeper into the infrared to the next communications window at 1.55 µm (for low attenuation in optical fiber), GaAs-based materials, their reach already extended with dilute fractions of nitrogen, are no longer suitable and materials lattice matched to InP are required. Lindberg et al. [60] used InGaAsP QWs for emission at 1555 nm, employing an intracavity diamond heat spreader to compensate for the low thermal conductivity of this material system. While "standard" diamond heat spreaders used elsewhere have thickness of 0.25–0.5 mm, these authors reported

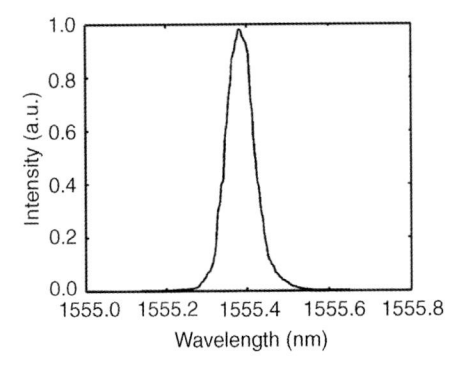

Figure 2.22 Emission spectrum of the InGaAsP-based VECSEL reported by Lindberg et al. [60] with 50 μm thick diamond heat spreader for the dual purposes of thermal management and spectral filtering.

the use of a 50 μm thick heat spreader to improve the spectral purity and wavelength stabilization of the output. Pump power-limited output power of 680 mW (slope efficiency: ~16%) was demonstrated at heat sink temperature of −30 °C, reducing to 140 mW at room temperature. The spectral width of the output was 0.08 nm with a wavelength drift of only 0.03 nm °C^{-1} due to the low thermo-optic coefficient of diamond (see Figure 2.22). The maximum output power demonstrated at this wavelength, however, is the one more recently reported by Rautiainen et al. [61] who achieved 2.6 W at 1560 nm by fusing an InAlGaAs-based gain region (grown on InP) with an AlGaAs DBR, showing significant improvement in performance over monolithically grown structures on InP that suffer higher losses due to low refractive index contrast in DBRs with high thermal resistance.

Toward the shortwave mid-infrared region, antimonide materials have been very successful for the spectral range around 2 μm. Chapter 4 will deal with long-wavelength III-Sb VECSELs in detail, but in terms of power and efficiency, the record is currently held by Burns et al. [62] who demonstrated >5 W at 2 μm with over 33% slope efficiency. Once again, thermal management was achieved with the use of a diamond heat spreader bonded to the intracavity surface of the gain structure.

Also plotted in Figure 2.20 are the best results of the recent work using InAs QDs in the VECSEL gain region. While InGaAs QWs provide better performance at these wavelengths, QDs offer the prospects of lower temperature sensitivity due to their three-dimensional confinement. In addition, the size distribution of the QDs leads to broad gain bandwidth and potentially broad tuning. Chapter 5 deals with QD VECSELs in detail, but results of note are those achieved by Germann et al. [63] who have demonstrated VECSEL emission at 950 nm with the use of submonolayer QDs at 1040 nm using Stranksi–Krastanow (S–K) QDs operating on the excited state and at 1210 nm using S–K QDs operating on the ground state. Improving on the maximum power and slope efficiency, Butkus et al. [64] have demonstrated 4.3 W at 1032 nm using S–K QDs.

Not plotted in Figure 2.20, but certainly worth including in a summary of state-of-the-art VECSEL performance is the first demonstration of a lead–chalcogenide

VECSEL for emission at 5.3 µm [65] (also see Chapter 4). Rahim and coauthors used a PbTe gain region and PbEuTe/BaF$_2$ Bragg mirror grown monolithically by MBE to achieve CW output power of 2 mW at 95 K, and >50 mW under quasi-CW operation with a 3% duty cycle. The authors predict that room-temperature operation should be feasible with improved thermal management, using, for example, crystalline heat spreaders, and lower thresholds might be achieved with the use of QWs.

At the other end of the VECSEL spectrum, microchip structures based on GaN have achieved laser action at 412 nm with optical excitation by a Q-switched frequency-tripled Nd:YAG laser [66]. This material system currently suffers from a lack of lattice-matched alloys with high refractive index contrast for the growth of monolithic Bragg reflectors and therefore dielectric mirrors must be deposited directly onto the sapphire substrate. In addition, short-wavelength pump lasers are not yet available for efficient optical pumping.

As we have seen, watt-level CW output power has been achieved from the visible to the mid-infrared with various groups striving to develop high-brightness sources in application-rich spectral regions after initial demonstrations of VECSEL capability around 1 µm. High output power must of course be achieved with the injection of high pump power; therefore, optical–optical efficiency is as important a target as effective thermal management and power scaling. We will now go on to discuss the CW performance of VECSELs to date with respect to the optical efficiencies achieved using the full variety of QW materials and pump sources available.

2.4.2
Efficiency

The slope (or differential) efficiency is defined as

$$\eta = \eta_{quant} \cdot \eta_{abs} \cdot \eta_{int} \cdot \eta_{out}, \tag{2.6}$$

where η_{quant} is the quantum efficiency $\lambda_{pump}/\lambda_{laser}$, η_{abs} is the pump absorption efficiency, η_{int} is the internal efficiency, and η_{out} is the output efficiency. The absolute limit to the efficiency of an optically pumped device is the quantum efficiency. As previously discussed, semiconductor lasers have the advantage of relatively low sensitivity to pump wavelength and brightness (within certain constraints); therefore, the choice of pump laser is a flexible parameter at the design stage. In reality, most infrared VECSELs are pumped by ubiquitous high-power AlGaAs laser diodes emitting around 808 nm; however, for deeper infrared emission, longer wavelength pump lasers must be used for efficient excitation and to maximize quantum efficiency. In addition, the pump-induced heat deposited in the gain region is inversely proportional to the quantum efficiency – an important consideration for a laser with such thermally sensitive gain. The highest quantum efficiency is achieved by pumping the quantum wells directly [15]; however, in this case the dramatically reduced absorption length necessitates careful structure design to resonate the pump wavelength and perhaps multiple passes of the pump light for high absorption efficiency.

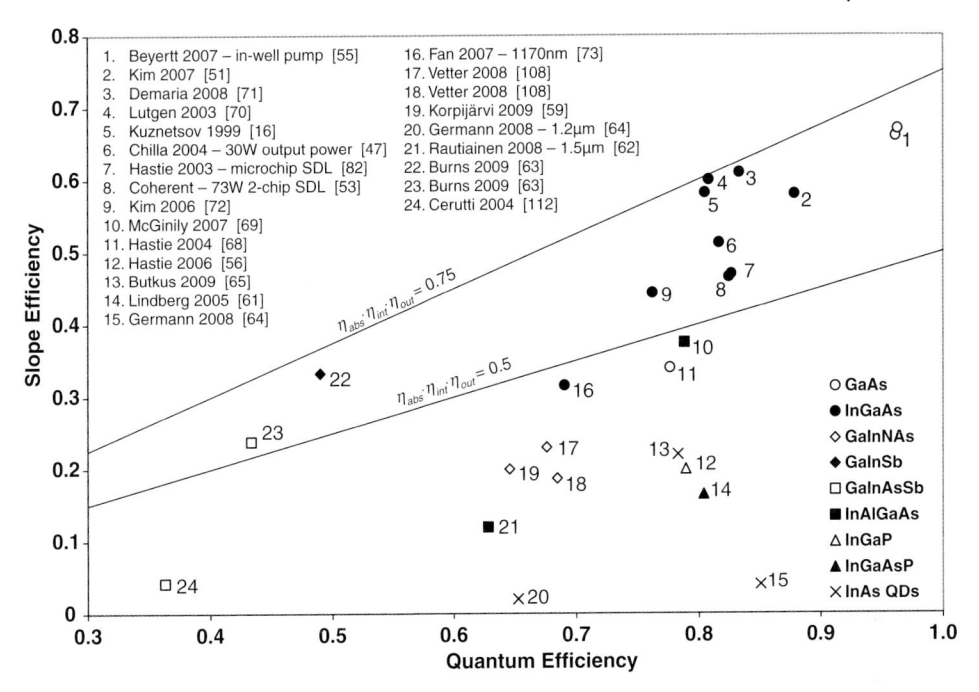

Figure 2.23 VECSEL performance analysis: the slope efficiency of selected VECSELs reported in the literature plotted with quantum efficiency ($\lambda_{pump}/\lambda_{laser}$). For each result, the quantum well composition is indicated [111].

Figure 2.23 shows the slope efficiency achieved for a number of VECSELs reported in the literature plotted with quantum efficiency. Fixed products of the absorption and internal and output efficiencies of 50 and 75% are also plotted herein. The majority of the results reported calculate the slope efficiency based on absorbed pump power, that is, the power incident at the gain structure after surface reflection losses; therefore, the absorption efficiency in most cases may be close to 1 for the purposes of this figure. Pump coupling efficiency is however an important parameter that should not be overlooked and is of paramount importance for a commercially viable laser.

The record slope efficiency demonstrated by a VECSEL was reported by Beyertt et al. [54] who dramatically increased the quantum efficiency to 96% by pumping into the wells of a GaAs QW VECSEL emitting at 865 nm and resonating the 833 nm pump light. Slope efficiency as high as 67% was achieved (data point 1, Figure 2.23) and attributed to the careful design of the active region so that the QWs were positioned at the field antinodes of both the pump and the laser wavelengths and the semiconductor surface was positioned at a field node to minimize losses. While this result demonstrates the efficiency that may be achieved by an optically pumped VECSEL, it must be pointed out that the pump laser was a Ti:sapphire laser, itself a relatively inefficient pump source. Diode pumping (at 660 nm) into the QW barriers of the same material system had previously been demonstrated with less impressive slope efficiency of up to 34% (data point 11, Figure 2.23) [36, 67]; however, it could be argued that the overall system is more practical and efficient. In this case, output power of up to 0.9 W was achieved with the first use of a SiC heat spreader. Later, McGinily et al. [68] demonstrated the use of InAlGaAs QWs with InAlGaAsP strain compensation layers as a novel material system for emission around 850 nm. Here, the use of indium and strain compensation layers prevents the propagation of dislocations into the active region for improved lifetime performance. The output power was pump power limited at 730 mW; however, the slope efficiency was slightly improved over the GaAs QW device at 37% (data point 10).

The best slope efficiencies achieved for barrier-pumped VECSELs have been with the use of InGaAs QWs (see Figure 2.23). High power pump diodes are available around 800 nm allowing for relatively high quantum efficiency >80%; however, this superior material system also offers high internal efficiencies and minimum losses. The bar was set high early on with the report by Kuznetsov et al. who introduced the recognized VECSEL format and demonstrated 58% slope efficiency from their 1 μm device that gave up to just under 700 mW before thermal rollover set in Ref. [16] (data point 5). It was Lutgen et al. who later demonstrated multiwatt output power with high efficiency (data point 4) achieving more than 8 W at 1 μm from their thinned device with a maximum slope efficiency of 60% for a heat sink temperature of 0 °C [69]. The current record for a barrier-pumped VECSEL is held by Demaria et al. [70] who have used careful design of the gain region to demonstrate slope efficiency of 61% for 13.2 W output power at a heat sink temperature of −5 °C, although this reduced slightly to the previous record of 60% when the temperature was increased to 0 °C. More impressive is the pump absorption efficiency of 95% achieved by means of an antireflection coating and a subcavity designed to resonate the pump wavelength as well as the signal wavelength (see Figure 2.24), a technique similar to that used for the group's previous work on quantum well pumping [54]. This VECSEL is, therefore, believed to be the most overall efficient device demonstrated to date.

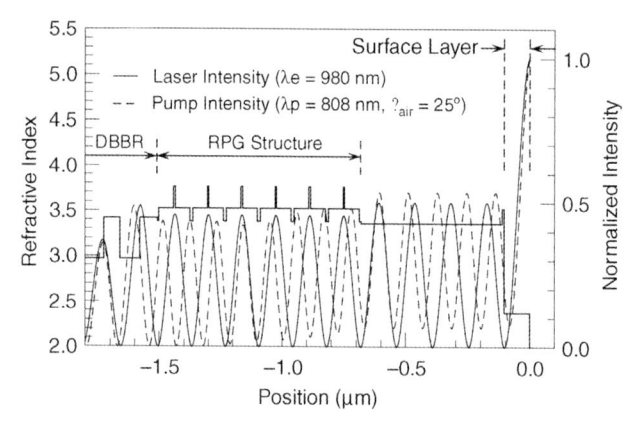

Figure 2.24 Design of the gain region used by Demaria et al. [70] to demonstrate record slope efficiency of 61% and pump absorption efficiency of 95% for a barrier-pumped VECSEL.

Other results of note are those at the low- and high-energy limits of the InGaAs QW spectral emission range. Data points 2 and 9 in Figure 2.23 represent emission wavelengths of 920 and 1060 nm, respectively, both pumped at 808 nm [51, 71]. Further toward the infrared, Fan et al. [72] demonstrated 32% slope efficiency for 7 W output power at 1170 nm (data point 16), an important target for the development of efficient yellow laser sources via frequency doubling.

VECSELs based on other material systems have not yet demonstrated high slope efficiencies, although in the case of antimonides, this is largely due to the unavailability of suitable pump lasers for high quantum efficiency. As indicated in Figure 2.23, the 33% slope efficiency reported by Burns et al. [62] for pump and signal wavelengths of 980 nm and 2 µm is testament to high internal efficiency and near-optimum output coupling.

2.4.3
Tuning

In previous reviews of VECSEL performance, the spectral tuning possibilities have not received as much attention as the output power and efficiency at application-rich wavelengths, although broader tuning is of course generally achieved from a more efficient device. Tunability is an important advantage over many common solid-state lasers at wavelengths where power and efficiency may be otherwise superior. The QW gain is broad. However, in a vertically emitting device, the gain structure design has a significant influence over the laser spectral characteristics; it can limit or increase the gain bandwidth, and hence the tunability, and will in most cases set the temperature dependence of the emission spectrum.

Garnache et al. [73] detailed a method for increasing the overall gain bandwidth and hence the tunability of a VECSEL for intracavity absorption spectroscopy by means of a short antiresonant subcavity. Figure 2.25a shows the design of their VECSEL gain structure and the spatial distribution of the electric field at the signal wavelength. The effective gain is proportional to the product of the material gain and the electric field amplitude squared ($|E|^2$) at the QW positions. The wavelength dependence of $|E|^2$ in the wells plotted in Figure 2.5a shows how the effective gain can therefore be broader than the material gain (dashed curve) for an antiresonant design. The converse is true of a resonant subcavity (Figure 2.25b), although the higher peak effective gain may be desirable for low-threshold operation. Broadening of the effective gain is useful in the development of ultrashort pulse VECSELs and therefore antiresonant structures are generally used for mode locking [74].

By far the most commonly used technique for tuning the emission of a VECSEL is rotation of an intracavity birefringent filter, inserted at Brewster's angle [75]. Within a laser cavity, this is the equivalent of an off-axis birefringent plate between parallel polarizers – rotation of the plate varies the wavelength at which the introduced phase delay results in zero net change of polarization angle and hence varies the wavelength that experiences zero transmission losses at the second polarizer.

Holm et al. [76] were the first to use an antiresonant design to broaden the available tuning range of their GaAs-based VECSEL and to reduce the temperature sensitivity of

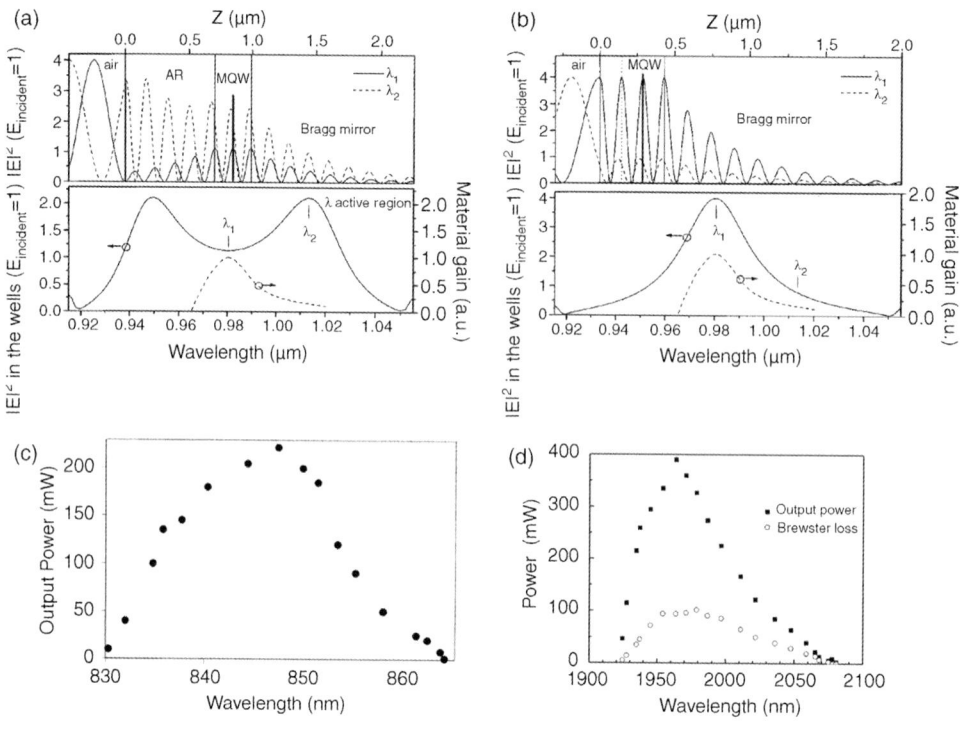

Figure 2.25 Broad spectral tuning of a VECSEL. (a) Design of an antiresonant VECSEL gain structure and electric field calculated by Garnache et al. [73]. Plotted below is the wavelength dependence of the electric field in the wells (solid line) and the material gain (dashed line). (b) Resonant VECSEL gain structure. (c) Tuning curve of a GaAs QW VECSEL with an antiresonant subcavity, tuned by rotation of an intracavity Lyot filter (three-plate quartz birefringent filter) [36]. (d) Tuning curve of an InGaSb QW VECSEL containing three different thicknesses of QW for broad gain; tuned by rotation of an intracavity birefringent filter [77].

the laser. They were able to demonstrate 8.5 nm single-frequency tuning with active stabilization by locking the laser to a reference cavity. Later, Hastie et al. [36] used a SiC heat spreader with the same gain structure to improve the power and efficiency of the laser and subsequently achieved a tuning range of more than 30 nm, corresponding to 460 cm^{-1} (see Figure 2.25c). Figure 2.25d shows the broad tuning of a GaInSb QW-based VECSEL where laser emission over a range of >150 nm was achieved. This represents the current record in terms of wavelength coverage by a single VECSEL. In this case, three different thicknesses of QW were used in the gain region to broaden the gain and to double the tuning range over a similar device with one type of QW [77].

Table 2.4 lists the tuning ranges of selected VECSELs, representative of the main QW alloys. As expected, in general, the tuning range increases with emission wavelength as a small difference in bandgap energy is more significant for lower photon energies. It is, therefore, informative to list the tuning ranges in terms of wave numbers for comparison across the spectral range.

Table 2.4 Broad VECSEL tuning ranges achieved with the use of different QW material systems. In all cases, the laser emission is tuned by rotation of an intracavity birefringent filter [75].

QW material	Range (nm)	Tuning (nm)	Tuning (cm^{-1})	References
InGaP	668–678	10	220	[56]
GaAs	830–863	33	460	[36]
InGaAs	955–995	40	425	[106]
InGaAs	1147–1197	50	364	[72]
GaInNAs	1163–1193	30	216	[107]
GaInSb[a)]	1924–2080	156	390	[77]
GaInAsSb	2189–2318	129	254	[62]

a) Three different thicknesses of QW were used in this device to broaden the tuning range.

2.5 Integration

In this section, ultracompact versions of VECSELs are described. First, microchip format VECSELs and their power scaling capabilities are presented. Possible routes toward direct and "indirect" electrical injection are briefly discussed. Finally, we focus on the fiber-tunable microcavity VECSEL platform. It is shown to be a versatile solution to obtain wavelength-tunable emission with high spectral purity and potentially high power (for a micron-size cavity laser).

2.5.1 Microchip

VECSELs are generally made from discrete, bulk optical elements. While engineering solutions to obtain compact modules exist (see Ref. [78] for a review on such approaches), the optical cavity is still typically several centimeters long and requires active adjustment. A quasi-monolithic and self-aligned VECSEL format, the microchip VECSEL, addresses these issues. It takes advantage of the heat spreader thermal management approach, as discussed in Section 2.3, which consists in liquid-capillary bonding of a thin, transparent high-conductivity crystal to the active side of the semiconductor chip. The outer surface of the heat spreader can be dielectric mirror coated resulting in a highly compact (submillimeter) plane–plane cavity (see Figure 2.26a). This type of VECSEL is inherently alignment free and stabilized by the thermal lens and gain aperture created by optical pumping [79, 80]. This approach can lead to high-brightness microsources [79, 81–83], which emit at the fundamental wavelength and are suitable for array operation [82]. The fundamental microchip VECSEL format can be in principle extended to any wavelength enabled by the semiconductor technology. It has already been demonstrated at 850, 980, 1060, 1300, and 670 nm. The plane–plane cavity geometry, however, has a downside. As Kemp et al. [79] showed, proper choice of the heat spreader thickness, depending on the VECSEL chip structure, is essential to obtain a high-quality beam for a given output

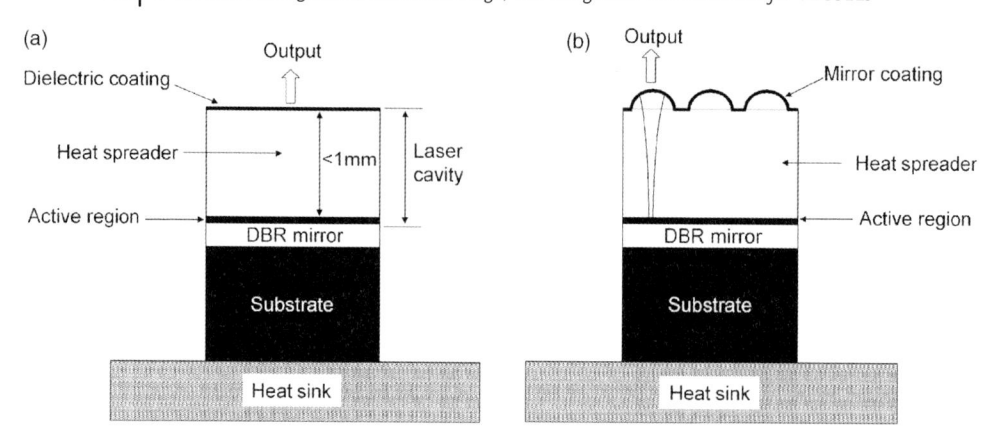

Figure 2.26 (a) Microchip VECSEL schematic having a plane–plane cavity and (b) microlensed microchip VECSEL concept with a plano-concave cavity.

power. Because the cavity stability factor, and consequently the mode size, is mainly determined by the pump-induced thermal lens, keeping the same beam quality as the power is varied is simply not possible [79, 80, 84].

The plano-concave microcavity VECSEL (or microlensed μ-VECSEL) circumvents this problem by rendering the mode matching condition independent of thermal effects [80, 84]. This design makes use of a mirror-coated microlens-patterned heat spreader (see Figure 2.26b), which acts as a stabilizing concave output coupler. For such devices, the microlenses are fabricated in a planar approach [85]. Pillars of resist are first made by photolithography on top of the heat spreader surface. This is followed by a "resist-reflow" step to form resist microlenses. The microlens shape is then transferred into the heat spreader by dry etching. The heat spreader thickness and the microlens radius of curvature (ROC) define the cavity mode size and are therefore important design parameters. Figure 2.27 demonstrates this by showing the evolution of the fundamental mode diameter at a wavelength of 1.06 μm as a function of ROC for two typical diamond heat spreader thicknesses (250 and 500 μm). As a rule of thumb, a shorter heat spreader requires a bigger ROC to obtain a similar mode size.

The first demonstration of such a μ-VECSEL used a 250 μm diamond as a heat spreader and a 1060 nm InGaAs-based chip [80, 84]. Readers are referred to Ref. [86] for specific details about microlenses fabrication in diamond. In this first demonstration, arrays of devices with microlenses ROC ranging from 100 to 1700 μm made a systematic study of the mode size effect on laser performances possible. It was found that when the optical pump overlaps preferentially with the cavity fundamental mode (i.e., when the mode matching condition is fulfilled), microlensed emitters show stable high beam quality ($M^2 \sim 1.1$), even as the power is varied, with efficiency similar to their plane–plane counterparts. It was also noted that if the microlens aperture is small enough compared to the cavity mode size, it acts as a spatial filter. This effect enables single-transverse mode emission even under mode mismatch by preventing the oscillation of higher order modes. Single-transverse mode emission

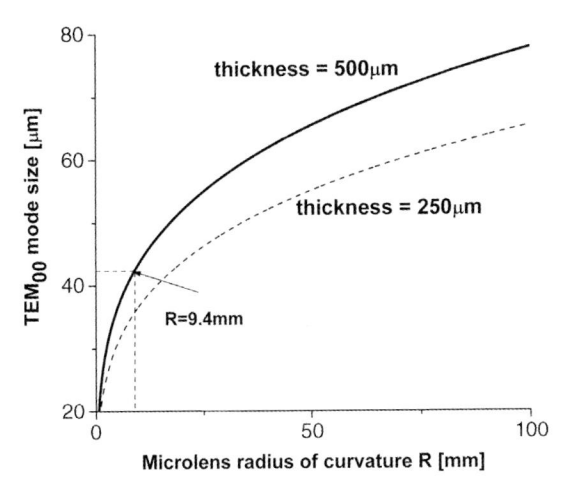

Figure 2.27 Microlensed μ-VECSEL fundamental mode diameter evolution with the microlens radius of curvature for two different heat spreader thicknesses.

from an array of devices optically pumped with a single beam is possible, and has been shown, using this effect, but comes at the price of reduced efficiency [80].

The power of the first generation of microlensed μ-VECSELs (~70 mW) was limited by their relatively small cavity mode size (~10–22 μm diameter in the active region) restricting the choice of optical pump to high-brightness and relatively low-power lasers. A second generation of devices having larger cavity mode size (42 μm diameter) while retaining compactness was devised [87]. For that purpose, the ROC of the microlenses was pushed up to 9.4 mm and the thickness of the diamond heat spreader was doubled to 500 μm. High-power, lower brightness pump laser could then be utilized. The power scaling capabilities of μ-VECSELs were demonstrated [87]. TEM_{00} emission with output powers reaching 230 mW before rollover was obtained (see Figure 2.28a). By defocusing the pump, emission up to 1 W (see Figure 2.28b) could be achieved but at the expense of the beam quality. Technological challenges lie ahead for further power scaling by pushing the ROC of microlenses to even higher values. However, single-transverse mode output powers up to the watt regime are readily anticipated with the present configuration through optimization of the output coupler.

In summary, microchip VECSELs and their microlensed version offer a very compact approach to obtain stable, high-brightness, watt-level, and wavelength versatile laser sources. They are particularly promising for both fixed and addressable array operation.

2.5.2
Pump Integration

While microchip VECSELs are compact and do not require any cavity alignment, the issue of carrier injection still exists. Standard optical pumping offers simplicity and

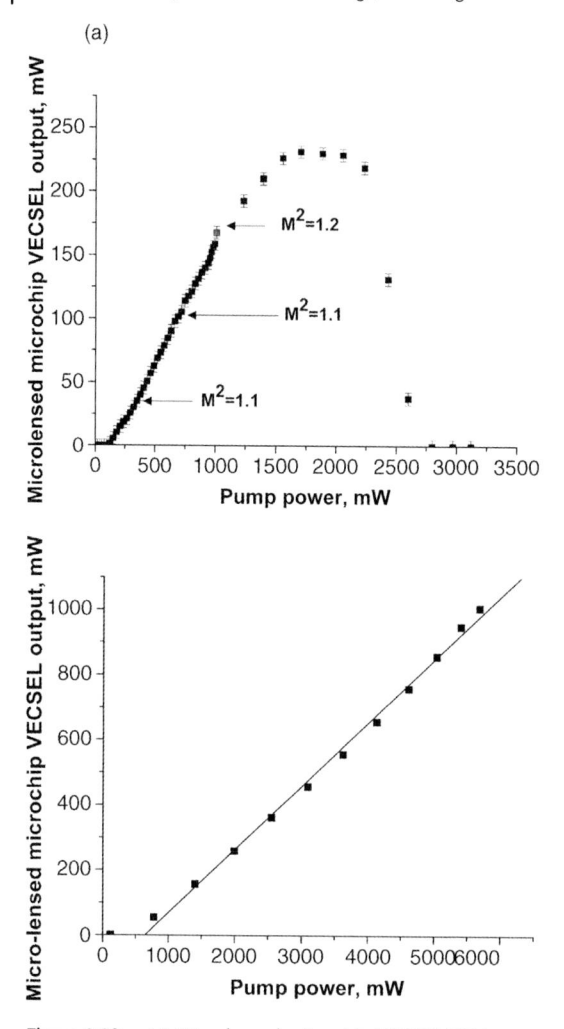

Figure 2.28 (a) Microlensed microchip VECSEL TEM$_{00}$ power transfer and (b) same device operating in multitransverse mode operation, that is, with a bigger pump spot size. Taken from Ref. [87].

great flexibility and can be made relatively compact, but a truly monolithic miniaturized VECSEL would necessitate electrical injection of carriers in the active region. Two methods have been proposed to date: (i) direct electrical pumping and (ii) "indirect electrical injection" through integration of a pump laser within the VECSEL chip. For more information about electrical pumping approaches, the reader is referred to Chapter 7. In the second scheme, the VECSEL structure is coupled to an edge-emitter that acts as the pump. The edge-emitter laser structure is located within the VECSEL external cavity. The pump laser cavity is perpendicular to the VECSEL and the carriers are created through the coupling of the pump mode evanescent field to the VECSEL active region. Such devices pumped by three in-plane lasers over a 400 µm active region have been demonstrated to emit up to 600 mW at 1060 nm [88].

2.5.3
Fiber-Tunable VECSELs

As highlighted in Chapter 1, VECSEL technology is an extension of the VCSEL concept, particularly to provide additional functionality and/or power scaling. As such, tunable VCSELs for which an extended cavity is exploited to control the emission wavelength are considered as a subset of VECSELs. Recent progress achieved with these devices is the subject of this section.

VECSELs are indeed an ideal platform for the realization of extremely small continuously tunable sources with potential applications in spectroscopy and (bio)-sensing and optical communications. In such microlasers, the cavity consists of two mirrors, one being the VECSEL chip itself. The two mirrors are separated by a small air gap (from one to a few tens of microns) and mechanical translation of one with respect to the other changes the oscillating wavelength. The typical cavity length is characterized by a large free spectral range (FSR), which in turn ensures that only one longitudinal mode oscillates.

Extensive work has been reported on electrically and optically driven plane–plane cavity MEMS-mounted mirror-tunable VECSELs [89–91] and, to a lesser degree, on fiber-VECSELs [92, 93]. The latter approach relies on the use of a mirror-coated fiber tip actuated by piezoelectric translation. This configuration simplifies optical pumping of the small cavity: the pump can simply be injected through the fiber tip, which can further act as the output coupler. The inherent fiber coupling also makes fiber-VECSEL suitable for remote interrogation.

If both plane–plane MEMS and fiber-VECSEL typically suffer from limited single-mode output power (maximum of a few milliwatts), fiber-tunable VECSELs bring the flexibility needed to address the problem. Thanks to their geometry, separate optimization of the gain chip and of the output coupling is readily possible. The main power limitation seen in these devices is imposed by thermal rollover or by the appearance of higher order modes. The problem is essentially the same, but on a smaller scale than for the plane–plane microchip VECSEL, as seen in Section 2.5.1. In order to power scale the single-mode output, it is essential to stabilize the cavity mode sizes and to preferentially excite the resulting fundamental mode. To do so, plano-concave microcavities applied to both MEMS and fiber-tunable VECSELs have been used [94–96]. Similarly, microconcave fiber mirrors have been explored for the realization of high-finesse, open-access passive cavities [97, 98]. The process to obtain a microconcave mirror is not, however, straightforward. Furthermore, a microconcave output coupler is far from ideal for optical pumping and for output coupling into single-mode fibers because the efficiency is inherently low.

As an alternative, Laurand et al. [99] have proposed and implemented a fiber-VECSEL that incorporates an intracavity microlens by taking advantage of its open-cavity geometry. The microlens then acts as a guiding element for the cavity modes and allows the use of a flat-facet fiber mirror. This configuration (see Figure 2.29)

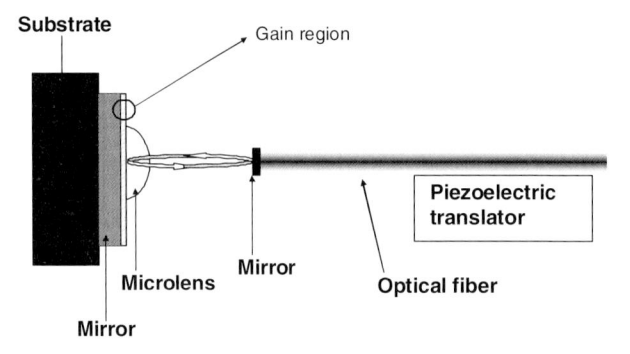

Figure 2.29 Schematic of the fiber-tunable VECSEL platform.

offers simpler fabrication and better fiber-coupling efficiency [100]. Thanks to the focusing microlens, alignment tolerances of the laser cavity are relaxed compared to the plane–plane geometry. The fiber position for which mode matching is optimized depends on the focal of the microlens. Therefore, by controlling this parameter, it is possible to control the air gap length, a promising feature for intracavity spectroscopy experiments. It is also interesting to note that the cavity mode fiber coupling efficiency is maximized at the mode matching position.

As a demonstrator, arrays of transparent polymer microlenses fabricated directly on top of the VECSEL chip have been used. The chip is similar to the one in Section 2.6.1. A typical laser result is shown in Figure 2.30. In this particular case, the laser had a 95% mirror-coated single-mode fiber that was also antireflection coated at the pump wavelength (810 nm). Its microlens had a diameter of 44 μm for a 5.5 μm

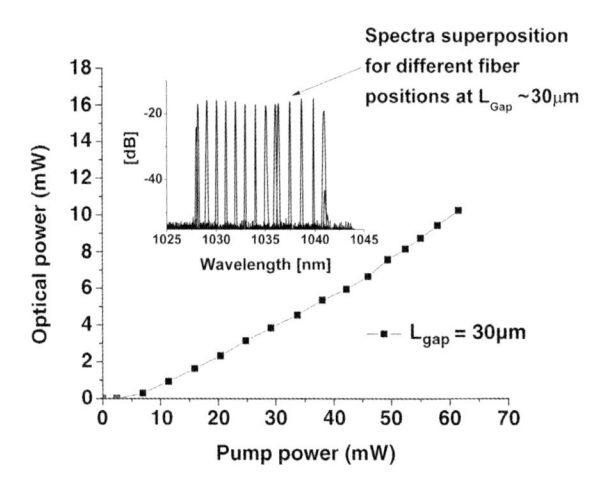

Figure 2.30 Fiber-coupled power transfer of a fiber-tunable VECSEL with a 44 μm diameter intracavity microlens. *Inset*: Shows the superposition of spectrum for different positions of the fiber mirror. L_{gap} represent the optimum air gap length for the given microlens configuration. Results from Ref. [90].

height. The fiber-VECSEL operated around 1030 nm. A single-mode continuous-wavelength tuning over ~13 nm limited by the FSR was demonstrated for an air gap around 30 μm. The fiber-coupled output power was up to 12 mW over that range. Lasing was recorded for air gap lengths of 1 μm and up to more than 100 μm, demonstrating the stabilizing effect of the microlens. This hybrid approach lends itself to further mode control and power scaling, thanks to the flexibility of the fiber technology and microlens design. For example, a very simple way to further increase the output power has been experimentally shown on devices emitting at wavelengths for which rare-earth-doped fibers exist. The principle was demonstrated using a plane–plane 1.55 μm fiber-VECSEL postamplified in an erbium-doped fiber placed in series with the fiber mirror [100]. In such a scheme, there is no compromise on the system simplicity and a single optical pump is used to excite both the VECSEL chip and the doped fiber.

Using combinations of these approaches, fiber-VECSELs hold the promise of yielding single-mode continuously tunable microsources with output powers of several tens of megawatt or more.

2.6
Conclusions

We have seen that the VECSEL format is an unusually versatile laser platform technology. Following a set of "semiempirical" design and optimization procedures outlined in this chapter, together with the use of advanced epitaxial growth techniques and the versatility of III–V alloy semiconductors, fundamental wavelength emission from visible to mid-infrared has been demonstrated at power levels in the multiwatt CW range. Thermal management has been critical to these developments and the specific thermal issues pertinent to these devices have been summarized. In particular, we have highlighted the differences and respective niches of application of the thin device and heat spreader approaches to thermal management. State-of-the-art performance achieved with these approaches has been overviewed. Finally, novel formats of device involving integration and miniaturization have been considered.

References

1 Roxlo, C.B., Putnam, R.S., and Salour, M.M. (1982) Optically pumped semiconductor platelet lasers. *Proc. Soc. Photo Opt. Instrum. Eng.*, **322**, 31–36.

2 People, R. and Bean, J.C. (1985) Calculation of critical layer thickness versus lattice mismatch for Ge_xSi_{1-x}/Si strained-layer heterostructures. *Appl. Phys. Lett.*, **47**, 322–324.

3 Ekins-Daukes, N.J., Kawaguchi, K., and Zhang, J. (2002) Strain-balanced criteria for multiple quantum well structures and its signature in X-ray rocking curves. *Cryst. Growth Des.*, **2**, 287–292.

4 Vurgaftman, I., Meyer, J.R., and Ram-Mohan, L.R. (2001) Band parameters for III–V compound semiconductors and their alloys. *J. Appl. Phys.*, **89**, 5815–5875.

5 Chuang, S.L. (2009) *Physics of Photonic Devices*, 2nd edn, John Wiley & Sons, Inc., New Jersey.
6 Zhang, P., Song, Y., Tian, J., Zhang, X., and Zhang, Z. (2009) Gain characteristics of the InGaAs strained quantum wells with GaAs, AlGaAs, and GaAsP barriers in vertical-external-cavity surface-emitting lasers. *J. Appl. Phys.*, **105**, 053103.
7 Fan, L., Hader, J., Schillgalies, M., Fallahi, M., Zakharian, A.R., Moloney, J.V., Bedford, R., Murray, J.T., Koch, S.W., and Stolz, W. (2005) High-power optically pumped VECSEL using a double-well resonant periodic gain structure. *IEEE Photon. Technol. Lett.*, **17**, 1764–1766.
8 Moloney, J.V., Hader, J., and Koch, S.W. (2007) Quantum design of semiconductor active materials: laser and amplifier applications. *Laser Photon. Rev.*, **1**, 24–43.
9 Sze, S.M. (1985) *Semiconductor Devices: Physics and Technology*, John Wiley & Sons, Inc.
10 Raja, M.Y.A., Brueck, S.R.J., Osinski, M., Schaus, C.F., McInerney, J.G., Brennan, T.M., and Hammons, B.E. (1989) Resonant periodic gain surface-emitting semiconductor-lasers. *IEEE J. Quantum Electron.*, **25**, 1500–1512.
11 Corzine, S.W., Geels, R.S., Scott, J.W., Yan, R.H., and Coldren, L.A. (1989) Design of Fabry–Perot surface-emitting lasers with a periodic gain structure. *IEEE J. Quantum Electron.*, **25**, 1513–1524.
12 Tourrenc, J.P., Bouchoule, S., Khadour, A., Decobert, J., Miard, A., Haimand, J.C., and Oudar, X.L. (2007) High power single-longitudinal-mode OP-VECSEL at 1.55 μm with hybrid metal-metamorphic Bragg mirror. *Electron. Lett.*, **43**, 754.
13 Giet, S., Sun, H.D., Calvez, S., Dawson, M.D., Suomalainen, S., Härkönen, A., Guina, M., Okhotnikov, O., and Pessa, M. (2006) Spectral narrowing and locking of a vertical-external-cavity surface-emitting laser using an intracavity volume Bragg grating. *IEEE Photon. Technol. Lett.*, **18**, 1786–1788.
14 Schmid, M., Benchabane, S., Torabi-Goudarzi, F., Abram, R., Ferguson, A.I., and Riis, E. (2004) Optical in-well pumping of a vertical-external-cavity surface-emitting laser. *Appl. Phys. Lett.*, **84**, 4860–4862.
15 Beyertt, S.S., Zorn, M., Kubler, T., Wenzel, H., Weyers, M., Giesen, A., Trankle, G., and Brauch, U. (2005) Optical in-well pumping of a semiconductor disk laser with high optical efficiency. *IEEE J. Quantum Electron.*, **41**, 1439–1449.
16 Kuznetsov, M., Hakimi, F., Sprague, R., and Mooradian, A. (1999) Design and characteristics of high-power (>0.5-W CW) diode-pumped vertical-external-cavity surface-emitting semiconductor lasers with circular TEM_{00} beams. *IEEE J. Sel. Top. Quantum Electron.*, **5**, 561–573.
17 Zakharian, A.R., Hader, J., Moloney, J.V., Koch, S.W., Brick, P., and Lutgen, S. (2003) Experimental and theoretical analysis of optically pumped semiconductor disk lasers. *Appl. Phys. Lett.*, **83**, 1313–1315.
18 Zakharian, A.R., Hader, J., Moloney, J.V., and Koch, S.W. (2005) VECSEL threshold and output power-shutoff dependence on the carrier recombination rates. *IEEE Photon. Technol. Lett.*, **17**, 2511–2513.
19 Bengtsson, J., Gustavsson, J., Haglund, Å., and Larsson, A. (2008) Trends in cavity designs for vertical cavity lasers – and simulation of their consequences. Numerical Simulation of Optoelectronic Devices Conference, UK, Paper ThPD2.
20 Kim, J.-Y., Cho, S., Lee, J., Kim, G.B., Lim, S.-J., Yoo, J., Kim, K.-S., Lee, S.-M., Shim, J., Kim, T., and Park, Y. (2006) A measurement of modal gain profile and its effect on the lasing performance in vertical-external-cavity surface-emitting lasers. *IEEE Photon. Technol. Lett.*, **18**, 2496–2498.
21 Schulz, N., Rattunde, M., Ritzenthaler, C., Rösener, B., Manz, C., Köhler, K., and Wagner, J. (2007) Effect of the cavity resonance-gain offset on the output power characteristics of GaSb-based VECSELs. *IEEE Photon. Technol. Lett.*, **19**, 1741–1743.
22 Born, M. and Wolf, E. (2002) Basic properties of the electromagnetic field, in

Principles of Optics, 7th edn, Cambridge University Press, Cambridge, pp. 1–74.

23 Macleod, H.A. (2001) *Thin-Film Optical Filters*, Institute of Physics, Bristol.

24 Babic, D.I. and Corzine, S.W. (1992) Analytic expressions for the reflection delay, penetration depth, and absorptance of quarter-wave dielectric mirrors. *IEEE J. Quantum Electron.*, **28**, 514–524.

25 Kemp, A.J., Valentine, G.J., Hopkins, J.M., Hastie, J.E., Smith, S.A., Calvez, S., Dawson, M.D., and Burns, D. (2005) Thermal management in vertical-external-cavity surface-emitting lasers: finite-element analysis of a heatspreader approach. *IEEE J. Quantum Electron.*, **41**, 148–155.

26 Kemp, A.J., Valentine, G.J., Hopkins, J.-M., Hastie, J.E., Smith, S.A., Calvez, S., Dawson, M.D., and Burns, D. (2006) Corrections to "Thermal management in vertical-external-cavity surface-emitting lasers: finite-element analysis of a heatspreader approach". *IEEE J. Quantum Electron.*, **46**, 85.

27 Calvez, S., Clark, A.H., Hopkins, J.M., Macaluso, R., Merlin, P., Sun, H.D., Dawson, M.D., Jouhti, T., and Pessa, M. (2003) 1.3 μm GaInNAs optically-pumped vertical cavity semiconductor optical amplifier. *Electron. Lett.*, **39**, 100–102.

28 Ayers, J.E. (2007) *Heteroepitaxy of Semiconductors: Theory, Growth and Characterization*, Illustrated edn, CRC Press.

29 Hanna, D.C., Sawyers, C.G., and Yuratich, M.A. (1981) Telescopic resonators for large-volume TEM$_{00}$-mode operation. *Opt. Quantum Electron.*, **13**, 493–507.

30 Holm, M.A., Burns, D., Cusumano, P., Ferguson, A.I., and Dawson, M.D. (1999) High-power diode-pumped AlGaAs surface-emitting laser. *Appl. Opt.*, **38**, 5781–5784.

31 Ouvrard, A., Garnache, A., Cerutti, L., Genty, F., and Romanini, D. (2005) Single-frequency tunable Sb-based VCSELs emitting at 2.3 μm. *IEEE Photon. Technol. Lett.*, **17**, 2020–2022.

32 Giesen, A., Hugel, H., Voss, A., Wittig, K., Brauch, U., and Opower, H. (1994) Scalable concept for diode-pumped high-power solid-state lasers. *Appl. Phys. B*, **58**, 365–372.

33 Alford, W.J., Raymond, T.D., and Allerman, A.A. (2002) High power and good beam quality at 980 nm from a vertical external-cavity surface-emitting laser. *J. Opt. Soc. Am. B*, **19**, 663–666.

34 Liau, Z.L. (2000) Semiconductor wafer bonding via liquid capillarity. *Appl. Phys. Lett.*, **77**, 651–653.

35 Hopkins, J.M., Smith, S.A., Jeon, C.W., Burns, D., Calvez, S., Dawson, M.D., Jouhti, T., and Pessa, M. (2004) A 0.6 W CW GaInNAs vertical external-cavity surface-emitting laser operating at 1.32 μm. *Electron. Lett.*, **40**, 30–31.

36 Hastie, J.E., Hopkins, J.M., Calvez, S., Jeon, C.W., Burns, D., Abram, R., Riis, E., Ferguson, A.I., and Dawson, M.D. (2003) 0.5-W single transverse-mode operation of an 850-nm diode-pumped surface-emitting semiconductor laser. *IEEE Photon. Technol. Lett.*, **15**, 894–896.

37 van Loon, F., Kemp, A.J., Maclean, A.J., Calvez, S., Hopkins, J.M., Hastie, J.E., Dawson, M.D., and Burns, D. (2006) Intracavity diamond heatspreaders in lasers: the effects of birefringence. *Opt. Express*, **14**, 9250–9260.

38 Millar, P., Birch, R., Kemp, A.J., and Burns, D. (2008) Synthetic diamond for intracavity thermal management in compact solid-state lasers. *IEEE J. Quantum Electron.*, **44**, 709.

39 Häring, R., Paschotta, R., Aschwanden, A., Gini, E., Morier-Genoud, F., and Keller, U. (2002) High-power passively mode-locked semiconductor lasers. *IEEE J. Quantum Electron.*, **38**, 1268–1275.

40 Lindberg, H., Strassner, M., Gerster, E., Bengtsson, J., and Larsson, A. (2005) Thermal management of optically pumped long-wavelength InP-based semiconductor disk lasers. *IEEE J. Sel. Top. Quantum Electron.*, **11**, 1126–1131.

41 Maclean, A.J., Kemp, A.J., Calvez, S., Kim, J.Y., Kim, T., Dawson, M.D., and Burns, D. (2008) Continuous tuning and efficient intracavity second-harmonic

generation in a semiconductor disk laser with an intracavity diamond heatspreader. *IEEE J. Quantum Electron.*, **44**, 216–225.

42 Kemp, A.J., Hopkins, J.M., Maclean, A.J., Schulz, N., Rattunde, M., Wagner, J., and Burns, D. (2008) Thermal management in 2.3 μm semiconductor disk lasers: a finite element analysis. *IEEE J. Quantum Electron.*, **44**, 125–135.

43 Giet, S., Kemp, A.J., Burns, D., Calvez, S., Dawson, M.D., Suomalainen, S., Härkönen, A., Guina, M., Okhotnikov, O., and Pessa, M. (2008) Comparison of thermal management techniques for semiconductor disk lasers. *Proc. SPIE*, **6871**, 687134.

44 Jacquemet, M., Domenech, M., Lucas-Leclin, G., Georges, P., Dion, J., Strassner, M., Sagnes, I., and Garnache, A. (2007) Single-frequency cw vertical external cavity surface emitting semiconductor laser at 1003 nm and 501 nm by intracavity frequency doubling. *Appl. Phys. B*, **86**, 510.

45 Abram, R.H., Gardner, K.S., Riis, E., and Ferguson, A.I. (2004) Narrow linewidth operation of a tunable optically pumped semiconductor laser. *Opt. Express*, **12**, 5434–5439.

46 Hopkins, J.M., Maclean, A.J., Burns, D., Riis, E., Schulz, N., Rattunde, M., Manz, C., Köhler, K., and Wagner, J. (2007) Tunable, single-frequency, diode-pumped 2.3 μm VECSEL. *Opt. Express*, **15**, 8212–8217.

47 Chilla, J.L., Butterworth, S.D., Zeitschel, A., Charles, J.P., Caprara, A.L., Reed, M.K., and Spinelli, L. (2004) High power optically pumped semiconductor lasers. *Proc. SPIE*, **5332**, 143–150.

48 Rudin, B., Rutz, A., Hoffmann, M., Maas, D.J.H.C., Bellancourt, A.-R., Gini, E., Südmeyer, T., and Keller, U. (2008) Highly efficient optically pumped vertical-emitting semiconductor laser with more than 20 W average output power in a fundamental transverse mode. *Opt. Lett.*, **33**, 2719–2721.

49 Bedford, R.G., Kolesik, M., Chilla, J.L.A., Reed, M.K., Nelson, T.R., and Moloney, J.V. (2005) Power-limiting mechanisms in VECSELs, in *Enabling Photonics Technologies for Defense, Security and Aerospace Applications*, SPIE 5814 edn (eds A.R. Pirich, M.J. Hayduk, E.J. Donkor, and P.J. Delfyett Jr.), pp. 199–208.

50 Härkönen, A., Guina, M., Okhotnikov, O., Rößner, K., Hümmer, M., Lehnhardt, T., Müller, M., Forchel, A., and Fischer, M. (2006) 1-W antimonide-based vertical external cavity surface emitting laser operating at 2-μm. *Opt. Express*, **14**, 6479–6484.

51 Kim, K.S., Yoo, J., Kim, G., Lee, S., Cho, S., Kim, J., Kim, T., and Park, Y. (2007) 920-nm vertical-external-cavity surface-emitting lasers with a slope efficiency of 58% at room temperature. *IEEE Photon. Technol. Lett.*, **19**, 1655–1657.

52 Hunziker, L.E., Shu, Q.-Z., Bauer, D., Ilhi, C., Mahnke, G.J., Rebut, M., Chilla, J., Caprara, A.L., Zhou, H., Weiss, E.S., and Reed, M.K. (2007) Power-scaling of optically-pumped semiconductor lasers. *Proc. SPIE*, **6451**, 64510.

53 Fan, L., Fallahi, M., Hader, J., Zakharian, A.R., Moloney, J.V., Murray, J.T., Bedford, R., Stolz, W., and Koch, S.W. (2006) Multichip vertical-external-cavity surface-emitting lasers: a coherent power scaling scheme. *Opt. Lett.*, **31**, 3612–3614.

54 Beyertt, S.S., Brauch, U., Demaria, F., Dhidah, N., Giesen, A., Kübler, T., Lorch, S., Rinaldi, F., and Unger, P. (2007) Efficient gallium–arsenide disk laser. *IEEE J. Quantum Electron.*, **43**, 869–875.

55 Hastie, J.E., Morton, L.G., Kemp, A.J., Dawson, M.D., Krysa, A.B., and Roberts, J.S. (2006) Tunable ultraviolet output from an intra-cavity frequency-doubled red vertical-external-cavity surface-emitting laser. *Appl. Phys. Lett.*, **89**, 061114.

56 Hastie, J.E., Calvez, S., Dawson, M.D., Leinonen, T., Laakso, A., Lyytikäinen, J., and Pessa, M. (2005) High power CW red VECSEL with linearly polarized TEM_{00} output beam. *Opt. Express*, **13**, 77–81.

57 Smith, A., Hastie, J.E., Foreman, H.D., Leinonen, T., Guina, M., and Dawson, M.D. (2008) GaN diode-pumping of a red semiconductor disk laser. *Electron. Lett.*, **44**, 1195–1196.

58 Korpijärvi, V.-M., Guina, M., Puustinen, J., Tuomisto, P., Rautiainen, J.,

Härkönen, A., Tukiainen, A., Okhotnikov, O., and Pessa, M. (2009) MBE grown GaInNAs-based multi-watt disk lasers. *J. Cryst. Growth*, **311**, 1868–1871.

59 Rautiainen, J., Härkönen, A., Korpijärvi, V.-M., Tuomisto, F., Guina, M., and Okhotnikov, O. (2007) 2.7 W tunable orange–red GaInNAs semiconductor disk laser. *Opt. Express*, **15**, 18345–18350.

60 Lindberg, H., Strassner, M., and Larsson, A. (2005) Improved spectral properties of an optically pumped semiconductor disk laser using a thin diamond heat spreader as an intracavity filter. *IEEE Photon. Technol. Lett.*, **17**, 1363–1365.

61 Rautiainen, J., Lyytikäinen, J., Sirbu, A., Mereuta, A., Caliman, A., Kapon, E., and Okhotnikov, O. (2008) 2.6 W optically-pumped semiconductor disk laser operating at 1.57-μm using wafer fusion. *Opt. Express*, **16**, 21881–21886.

62 Burns, D., Hopkins, J.-M., Kemp, A.J., Rösener, B., Schulz, N., Manz, C., Köhler, K., Rattunde, M., and Wagner, J. (2009) Recent developments in high-power, short-wave mid-infrared semiconductor disk lasers. *Proc. SPIE*, **7193**, 719311.

63 Germann, T.D., Strittmatter, A., Pohl, J., Pohl, U.W., Bimberg, D., Rautiainen, J., Guina, M., and Okhotnikov, O. (2008) Quantum-dot semiconductor disk lasers. *J. Cryst. Growth*, **310**, 5182–5186.

64 Butkus, M., Wilcox, K.G., Rautiainen, J., Okhotnikov, O., Mikhrin, S.S., Krestnikov, I.L., Kovsh, A.R., Hoffmann, M., Südmeyer, T., Keller, U., and Rafailov, E.U. (2009) High-power quantum-dot-based semiconductor disk laser. *Opt. Lett.*, **34**, 1672–1674.

65 Rahim, M., Arnold, M., Felder, F., Behfar, K., and Zogg, H. (2007) Midinfrared lead–chalcogenide vertical external cavity surface emitting laser with 5 μm wavelength. *Appl. Phys. Lett.*, **91**, 151102.

66 Park, S.-H. and Jeon, H. (2006) Microchip-type InGaN vertical external-cavity surface-emitting laser. *Opt. Rev.*, **13**, 20–23.

67 Hastie, J.E. (2004) High power surface emitting semiconductor lasers. PhD Thesis, University of Strathclyde.

68 McGinily, S., Abram, R., Gardner, K.S., Riis, E., Ferguson, A.I., and Roberts, J.S. (2007) Novel gain medium design for short-wavelength vertical-external-cavity surface-emitting laser. *IEEE J. Quantum Electron.*, **43**, 445–450.

69 Lutgen, S., Albrecht, T., Brick, P., Reill, W., Luft, J., and Späth, W. (2003) 8-W high-efficiency continuous-wave semiconductor disk laser at 1000 nm. *Appl. Phys. Lett.*, **82**, 3620–3622.

70 Demaria, F., Lorch, S., Menzel, S., Riedl, M., Rinaldi, F., Rösch, R., and Unger, P. (2008) Design of highly-efficient high-power optically-pumped semiconductor disk lasers. International Semiconductor Laser Conference 2008, Paper WA5.

71 Kim, K.S., Yoo, J.R., Cho, S.H., Lee, S.M., Lim, S.J., Kim, J.Y., and Lee, J.H. (2006) 1060 nm vertical-external-cavity surface-emitting lasers with an optical-to-optical efficiency of 44% at room temperature. *Appl. Phys. Lett.*, **88**, 091107-1–091107-3.

72 Fan, L., Hessenius, C., Fallahi, M., Hader, J., Li, H., Moloney, J.V., Stolz, W., Koch, S.W., Murray, J.T., and Bedford, R. (2007) Highly strained InGaAs/GaAs multiwatt vertical-external-cavity surface-emitting laser emitting around 1170 nm. *Appl. Phys. Lett.*, **91**, 131114.

73 Garnache, A., Kachanov, A.A., Stoeckel, F., and Houdré, R. (2000) Diode-pumped broadband vertical-external-cavity surface-emitting semiconductor laser applied to high-sensitivity intracavity absorption spectroscopy. *J. Opt. Soc. Am. B*, **17**, 1589–1598.

74 Keller, U. and Tropper, A.C. (2006) Passively modelocked surface-emitting semiconductor lasers. *Phys. Rep.*, **429**, 67–120.

75 Bloom, A.L. (1974) Modes of a laser resonator containing tilted birefringent plates. *J. Opt. Soc. Am.*, **64**, 447–452.

76 Holm, M.A., Ferguson, A.I., Burns, D., and Dawson, M.D. (1999) Actively stabilized single-frequency vertical-external-cavity AlGaAs laser. *IEEE Photon. Technol. Lett.*, **11**, 1551–1553.

77 Paajaste, J., Suomalainen, S., Koskinen, R., Härkönen, A., Guina, M., and Pessa, M. (2009) High-power and broadly

tunable GaSb-based optically pumped VECSELs emitting near 2 μm. *J. Cryst. Growth*, **311**, 1917–1919.

78 Calvez, S., Hastie, J.E., Guina, M., Okhotnikov, O., and Dawson, M.D. (2009) Semiconductor disk lasers for the generation of visible and ultraviolet radiation. *Laser Photon. Rev.*, **3**, 407–434.

79 Kemp, A.J., Maclean, A.J., Hastie, J.E., Hopkins, J.M., Calvez, S., Valentine, G.J., Dawson, M.D., and Burns, D. (2006) Thermal lensing, thermal management and transverse mode control in microchip VECSELs. *Appl. Phys. B*, **83**, 189–194.

80 Laurand, N., Lee, C.L., Gu, E., Hastie, J.E., Calvez, S., and Dawson, M.D. (2008) Array-format microchip semiconductor disk lasers. *IEEE J. Quantum Electron.*, **44**, 1096–1103.

81 Hastie, J.E., Hopkins, J.M., Jeon, C.W., Calvez, S., Burns, D., Dawson, M.D., Abram, R., Riis, E., Ferguson, A.I., Alford, W.J., Raymond, T.D., and Allerman, A.A. (2003) Microchip vertical external cavity surface emitting lasers. *Electron. Lett.*, **39**, 1324–1326.

82 Hastie, J.E., Morton, L.G., Calvez, S., Dawson, M.D., Leinonen, T., Pessa, M., Gibson, G., and Padgett, M.J. (2005) Red microchip VECSEL array. *Opt. Express*, **13**, 7209–7214.

83 Smith, S.A., Hopkins, J.M., Hastie, J.E., Burns, D., Calvez, S., Dawson, M.D., Jouhti, T., Kontinnen, J., and Pessa, M. (2004) Diamond-microchip GaInNAs vertical external-cavity surface-emitting laser operating CW at 1315 nm. *Electron. Lett.*, **40**, 935–937.

84 Laurand, N., Lee, C.L., Gu, E., Hastie, J.E., Calvez, S., and Dawson, M.D. (2007) Microlensed microchip VECSEL. *Opt. Express*, **15**, 9341–9346.

85 Daly, D. (2001) *Microlens Arrays*, Taylor and Francis, London.

86 Choi, H.W., Gu, E., Liu, C., Griffin, C., Girkin, J.M., Watson, I.M., and Dawson, M.D. (2005) Fabrication of natural diamond microlenses by plasma etching. *J. Vac. Sci. Technol. B*, **23**, 130–132.

87 Laurand, N., Lee, C.-L., Gu, E., Kemp, A.J., Calvez, S., and Dawson, M.D. (2009) Power-scaling of diamond microlensed microchip semiconductor disk lasers. *IEEE Photon. Technol. Lett.*, **21**, 152–154.

88 Illek, S., Albrecht, T., Brick, P., Lutgen, S., Pietzonka, I., Furitsch, M., Diehl, W., Luft, J., and Streubel, K. (2007) Vertical-external-cavity surface-emitting laser with monolithically integrated pump lasers. *IEEE Photon. Technol. Lett.*, **19**, 1952–1954.

89 Chang-Hasnain, C.J. (2000) Tunable VCSEL. *IEEE J. Sel. Top. Quantum Electron.*, **6**, 978–987.

90 Larson, M.C. and Harris, J.S. (1996) Wide and continuous wavelength tuning in a vertical-cavity surface-emitting laser using a micromachined deformable-membrane mirror. *Appl. Phys. Lett.*, **68**, 891–893.

91 Riemenschneider, F., Maute, M., Halbritter, H., Boehm, G., Amann, M.C., and Meissner, P. (2004) Continuously tunable long-wavelength MEMS-VCSEL with over 40-nm tuning range. *IEEE Photon. Technol. Lett.*, **16**, 2212–2214.

92 Hsu, K., Miller, C.M., Babic, D., Houng, D., and Taylor, A. (1998) Continuously tunable photopumped 1.3-μm fiber Fabry–Perot surface-emitting lasers. *IEEE Photon. Technol. Lett.*, **10**, 1199–1201.

93 Laurand, N., Calvez, S., Dawson, M.D., Jouhti, T., Konttinen, J., and Pessa, M. (2005) Fiber-tunable dilute-nitride VCSEL. *Phys. Status Solidi C*, **2**, 3895–3898.

94 Tarraf, A., Riemenschneider, F., Strassner, M., Daleiden, J., Irmer, S., Hallbritter, H., Hillmer, H., and Meissner, P. (2004) Continuously tunable 1.55-μm VCSEL implemented by precisely curved dielectric top DBR involving tailored stress. *IEEE Photon. Technol. Lett.*, **16**, 720–722.

95 Matsui, Y., Vakhshoori, D., Wang, P., Chen, P., Lu, C.C., Jiang, M., Knopp, K., Burroughs, S., and Tayebati, P. (2003) Complete polarization mode control of long-wavelength tunable vertical-cavity surface-emitting lasers over 65-nm

tuning, up to 14-mW output power. *IEEE J. Quantum Electron.*, **39**, 1037–1048.

96 Bousseksou, A., Kurdi, M.E., Salik, M.D., Sagnes, I., and Bouchoule, S. (2004) Wavelength tunable InP-based EP-VECSEL operating at room temperature and in CW at 1.55 μm. *Electron. Lett.*, **40**, 1490–1491.

97 Steinmetz, T., Colombe, Y., Hunger, D., Hänsch, T.W., Balocchi, A., Warburton, R.J., and Reichel, J. (2006) Stable fiber-based Fabry–Pérot cavity. *Appl. Phys. Lett.*, **89**, 111110.

98 Trupke, M., Goldwin, J., Darquié, B., Dutier, G., Eriksson, S., Ashmore, J., and Hinds, E.A. (2007) Atom detection and photon production in a scalable, open, optical microcavity. *Phys. Rev. Lett.*, **99**, 063601.

99 Laurand, N., Guilhabert, B., Gu, E., Calvez, S., and Dawson, M.D. (2007) Tunable single-mode fiber-VCSEL using an intracavity polymer microlens. *Opt. Lett.*, **32**, 2831–2833.

100 Laurand, N., Calvez, S., Dawson, M.D., Bouchoule, S., Harmand, J.-C., and Decobert, J. (2009) 1.55-μm tunable doped-fiber vertical-cavity surface emitting laser. Conference on Laser and Electro-Optics Europe, Paper CB13.

101 Park, S.-H., Kim, J., Jeon, H., Sakong, T., Lee, S.-N., Chae, S., Park, Y., Jeong, C.-H., Yeom, G.-Y., and Cho, Y.-H. (2003) Room-temperature GaN vertical-cavity surface-emitting laser operation in an extended cavity scheme. *Appl. Phys. Lett.*, **83**, 2121–2123.

102 Schlosser, P.J., Hastie, J.E., Calvez, S., Krysa, A.B., and Dawson, M.D. (2009) InP/AiGaInP quantum dot semiconductor disk lasers for CW TEM$_{00}$ emission at 716–755 nm. *optics express*, **17**, 21782–21787.

103 Lindberg, H., Larsson, A., and Strassner, M. (2005) Single-frequency operation of a high-power, long-wavelength semiconductor disk laser. *Opt. Lett.*, **30**, 2260–2262.

104 Hopkins, J.-M., Hempler, N., Rösener, B., Schultz, N., Rattunde, M., Manz, C., Köhler, K., Wagner, J., and Burns, D. (2008) High-power, (AlGaIn)(AsSb) semiconductor disk laser at 2.0 μm. *Opt. Lett.*, **33**, 201–203.

105 Schulz, N., Hopkins, J.-M., Rattunde, M., Burns, D., and Wagner, J. (2008) High-brightness long-wavelength semiconductor disk lasers. *Laser Photon. Rev.*, **2**, 160–181.

106 Abram, R., Schmid, M., Riis, E., and Ferguson, A.I. (2004) Laser Spectroscopy: Proceedings of the XVI International conference, world of science publishing Co. Pte Ltd, edited by Hamaford, P., Sidirov, A., Bachor, H. and Baldwin, K., 369–372.

107 Vetter, S.L., Hastie, J.E., Korpijarvi, V.-M., Puustinen, J., Guina, M., Okhotnikov, O., Calvez, S., and Dawson, M.D. (2008) Short-wavelength GaInNAs/GaAs semiconductor disk lasers. *Electron. Lett.*, **44**, 1069–1070.

108 Konttinen, J., Härkönen, A., Tuomisto, F., Guina, M., Rautiainen, J., Pessa, M., and Okhotnikov, O. (2007) High-power (>1 W) dilute nitride semiconductor disk laser emitting at 1240 nm. *New J. Phys.*, **9**, 140.

109 Hopkins, J.M., Preston, R.D., Maclean, A.J., Calvez, S., Sun, H., Ng, J., Steer, M., Hopkinson, M., and Burns, D. (2007) High performance 2.2 μm optically-pumped vertical external-cavity surface emitting laser. *J. Mod. Opt.*, **54**, 1677–1683.

110 Lindberg, H., Strassner, M., Bengtsson, J., and Larsson, A. (2004) High-power optically pumped 1550-nm VECSEL with a bonded silicon heat spreader. *IEEE Photon. Technol. Lett.*, **16**, 1233–1235.

111 Cerutti, L., Garnache, A., Ouvrard, A., and Alibert, C. (2003) Low threshold, room temperature laser diode pumped Sb-based VECSEL emitting around 2.1 μm. *Electron. Lett.*, **39**, 290.

3
Red Semiconductor Disk Lasers by Intracavity Frequency Conversion
Oleg Okhotnikov and Mircea Guina

3.1
Introduction

The development of visible light sources has been growing at an unprecedented rate over the past decades and is expected to have a huge impact on television and laser projection technology [1–4]. To achieve the best image quality, red–green–blue (RGB) wavelengths of 620, 532, and 460 nm, respectively, should be used [5]. The primary advantage of semiconductor disk lasers (SDLs) is that they can produce radiation at virtually any wavelength, depending on the composition of the active medium, and can cover the whole visible spectral range after frequency doubling. Their high output power and compact structure, combined with diffraction-limited beam characteristics, make diode-pumped SDLs attractive sources as they allow the projection of sharp points in laser scanning systems [3] and easy homogenization in the case of microdisplays [2]. Finally, direct modulation capability at $\sim$100 MHz frequencies, which enables high-definition, television-standard image projection, is another essential feature of SDLs for laser scanning projection/display applications [3]. Since single-frequency operation of SDLs could be achieved in a simple cavity configuration, spectroscopy is another emerging application of wavelength-tunable SDLs [6, 7]. In the field of biophotonics, the development of small footprint, efficient lasers with emission wavelengths in the UV or blue spectral regions could be conveniently used as replacements for the bulky excitation sources currently used in fluorescence-based bioimaging (microscopy) [8]. The fluorescence-based search for fingerprints in ambient/daylight illumination would benefit from multiwatt, portable, green (510 nm) sources [9]. In medicine, the ability to produce reliable, watt-level, red (630–690 nm) lasers is of great significance, especially for photodynamic therapy, as red lasers are the only viable option for providing sufficient illumination for the treatment of malignant tumors accessible only with optical fibers [10, 11]. Finally, in ophthalmology, yellow emitting (577 nm) sources are in great demand for the treatment of retinal vascular diseases [12, 13].

Unlike SDLs operating at around 1 µm, the development of 1.2–1.3 µm SDLs, required to reach the red wavelength range via frequency doubling, is hindered by the

Semiconductor Disk Lasers. Physics and Technology. Edited by Oleg G. Okhotnikov
Copyright © 2010 WILEY-VCH Verlag GmbH & Co. KGaA, Weinheim
ISBN: 978-3-527-40933-4

lack of practical materials for the fabrication of monolithic gain mirrors. One approach is to use a lattice-matched monolithic structure based on an InP distributed Bragg reflector (DBR). The critical issue associated with InP monolithic structures is the low quality of the DBR, which limits their capability for power scaling. Furthermore, the InP-based quantum wells (QWs) exhibit a low potential depth and reduced thermal conductivity, which again results in poor thermal behavior.

One possible solution to these problems lies in the so-called wafer fusion technology, which is used for combining disparate materials in various optoelectronic devices. This technique allows the integration of semiconductor materials with different lattice constants, which can not be used for highly efficient light emitters if grown monolithically. During the past few years, the wafer fusion technique has been used extensively in the fabrication of vertical-cavity surface-emitting lasers (VCSELs) operating at the telecom wavelengths of 1.3–1.55 µm [14, 15]. Recently, this technique has been successfully applied for the first time to high-power InP-based disk lasers operating at 1.3 and 1.57 µm [16]. Output powers up to 2.7 W at 1.3 µm were observed, which represent the highest powers achieved from SDLs at this wavelength to date. The use of wafer fusion technology in high-power disk lasers is, however, at a very early stage of its development and its potential, for example, in terms of reliability, needs to be further explored.

InGaAs-based quantum dot (QD) ensembles can be used as an alternative to QWs since they exhibit a broad gain spectrum and high temperature stability and can be grown on GaAs/AlAs DBRs [17]. SDLs based on InAs/GaAs submonolayers and InGaAs Stranski–Krastanow (S–K)-grown QDs have both been successfully demonstrated [18]. An output power of 500 mW has been achieved from an optically pumped InGaAs/GaAs QD semiconductor disk laser emitting at 1185 nm. However, the maximum gain of a QD ensemble is typically limited to 4–8 cm^{-1} per QD layer in the vertical configuration, which makes implementation of laser cavities with nonlinear frequency conversion difficult. A substantial increase in the number of QD layers could be a possible solution to this problem; however, a rigorous investigation is necessary to determine whether it is possible to grow large numbers of QD layers without degradation of photoluminescence efficiency.

The approach we have used for developing 1.2–1.3 µm SDLs rests upon the use of high-quality GaAs/AlAs DBRs and the so-called dilute nitride (InGaAsN) QWs, which can be monolithically combined. The operation wavelength of a InGaAsN-based structure is tuned by altering the amount of nitrogen in the composition. The limitation of this approach is that a higher concentration of nitrogen in the QW layers leads to an increase in the amount of point defects, resulting in a higher rate of nonradiative recombination. A recent systematic study of 1.2 µm dilute nitride compounds showed that high-quality InGaAsN can be fabricated, which resulted in the demonstration of a record high power at red wavelengths using intracavity frequency doubling. This section focuses on 1.2 µm SDLs based on dilute nitride semiconductors by reviewing recent achievements at ORC, Tampere University of Technology. The purpose of this chapter is primarily to highlight the progress that is being made in red light generation using dilute nitride semiconductor compounds. Optically pumped 1.2 µm SDLs based on this material system represent a promising solution for bridging the gap in the visible spectrum, which is difficult, if not impossible, to achieve with monolithically grown SDLs using N-free materials.

3.2 SDL with Frequency Doubling

SHG is a technique used to generate new wavelengths, particularly in the visible spectral range, which are difficult to achieve using direct emission from semiconductor diodes. The key aspects of frequency conversion are thoroughly presented in many books and publications. In this section, we provide only a general overview, aimed at familiarizing the reader with the main concepts used in the development of frequency-doubled SDLs.

The lasers used for frequency doubling can operate either in the pulsed or in the continuous-wave (cw) regimes. For lasers operating in pulsed mode, for example, Q-switched or mode-locked regimes, the nonlinear crystals are typically placed outside the laser cavity; the nonlinear conversion process is efficient owing to the high peak powers of the laser output. Conversely, to achieve efficient nonlinear conversion for lasers operating in the cw regime, the nonlinear crystal is usually placed inside the laser cavity, where the intensities are much higher [20]. SDLs typically incorporate an output coupler with only a small percentage of transmission and have excellent capacities for power storage in the cavity; thus, intracavity frequency doubling techniques are a suitable method for frequency conversion with this type of laser. On the other hand, SDL cavities are typically short ($\sim$10 cm), and thus operating the laser in mode-locked regime results in high repetition rate pulses ($\sim$1–100 GHz) with rather modest increase of the peak powers compared to the continuous-wave regime. This makes the intracavity frequency approach the preferred nonlinear conversion technique for SDLs. Detailed descriptions of nonlinear frequency conversion principles and intracavity frequency conversion can be found in Refs [19, 20].

3.2.1 General Principle of Frequency Doubling

Harmonic generation is a nonlinear optical process that occurs in dielectric media characterized by a nonlinear relation between the polarization density and the electric field, $P = \varepsilon_0 \chi E$, where ε_0 is the permittivity of free space and χ is the electric susceptibility of the medium. The relation between P and E is approximately linear for small values of E; however, it starts to deviate from linearity as E increases. This relationship can be expanded in a Taylor series at about $E = 0$ and is usually given as

$$P = \varepsilon_0 \chi E + 2\chi^{(2)} E^2 + 4\chi^{(3)} E^3 + \cdots,$$

where $\chi^{(2)} = d = 1/4a_2$ and $\chi^{(3)} = 1/24a_3$ are coefficients describing the strength of second- and third-order nonlinear effects, respectively. Obviously, a high nonlinear response is crucial for achieving efficient frequency doubling. It is understood that the anisotropy and material dispersion should be taken into account for an accurate description of the nonlinear optical processes. The susceptibility is a third-rank tensor with only two or three components of practical interest. Particularly, in centrosymmetric media, which have inversion symmetry, the second-order nonlinear coefficient d vanishes and, consequently, the lowest order nonlinearity is third order. The second harmonic generation as a second-order nonlinear process thus requires noncentrosymmetric materials that offer nonzero values of d. If the second dielectric response tensor component is large enough, the polarization will have a significant component oscillating at twice the frequency of the applied field, resulting in second harmonic generation or frequency doubling. This can be clearly seen from the second-order nonlinear polarization written as

$$P_2 = 2\varepsilon_0 \chi^{(2)} E^2 = 2\varepsilon_0 \chi^{(2)} E^* E + 2\varepsilon_0 \chi^{(2)} E^2 e^{-i(2\omega)t} + 2\varepsilon_0 \chi^{(2)} (E^*)^2 e^{-i(2\omega)t}.$$

The first time-independent term describes the process known as optical rectification, while the second and third terms comprise a frequency component that is twice the fundamental frequency.

Certain conditions should be fulfilled to realize the efficient energy flow from the original frequency to the second harmonic frequency. Among these, the length of the nonlinear crystal over which fundamental and frequency converted beams interact is a very important parameter in achieving doubling with high efficiency. Due to wavelength-dependent refraction, the first and second harmonic beams travel at different speeds. This causes the second harmonic light generated at different points in the crystal to be out of phase. The resulting destructive interference greatly reduces the conversion efficiency. This problem can be avoided when phase velocities of fundamental and second harmonic waves are equalized. Techniques to maintain the relative phase between the interacting waves must be employed to obtain efficient frequency conversion to second harmonic. Phase mismatching is, therefore, an important aspect of second harmonic generation and originates from the dispersive nature of nonlinear crystals. Typically, only one process can comply with this condition; in particular, the frequency doubling mechanism dominates when the phase matching of the second harmonic generated and fundamental beams is achieved.

Second harmonic beams generated at every local position of the polarization wave interfere constructively inside the nonlinear crystal for lengths less than the so-called coherence length, defined as $l_c = \lambda_\omega/4(n_{2\omega}-n_\omega)$. This length corresponds to the propagation distance over which the fundamental and converted frequency beams acquire a phase difference of π. Propagation over longer distances results in a linear increase in the relative phase difference and, consequently, in a periodic power flow from the fundamental beam to the second harmonic beam and back. Obviously, if the crystal length is an odd multiple of the coherence length ($L_{\text{crystal}} = ml_c, m = 2n+1$), the second harmonic radiation is suppressed.

The first phase matching technique, known as *birefringent phase matching*, relies on the anisotropy of the refractive index and uses the intrinsic birefringence of crystals. Generally, the equalization of phase velocities for fundamental and second harmonic waves in the nonlinear crystal employs the dependence of refractive index on the direction of propagation, polarization, and frequency. The phase matching condition

$\Delta k = 0$ for SHG, expressed using wave numbers for fundamental and frequency converted waves, can be achieved by a proper setting of these parameters, where $\Delta k = 2k_\omega - k_{2\omega} = 4\pi/\lambda_\omega (n_\omega - n_{2\omega})$. The matching condition can be displayed using a graphic representation of the index surfaces of the nonlinear crystal (see, for example, Ref. [21]). The basic phenomenon behind phase matching is a large crystal birefringence, which should be sufficient to compensate the chromatic dispersion. The phase-matched condition then corresponds to the intersection of the index surfaces at the fundamental and harmonic frequencies. Depending on the given nonlinear crystal, this can be achieved when polarization vectors of interacting fundamental beams are parallel (*type I phase matching*) or when these vectors are different (*type II phase matching*).

An effect that results from crystal birefringence and should be accounted for in SHG is known as "Poynting vector walk-off" or spatial walk-off. The large birefringence of an anisotropic material is instrumental in phase matching; however, as a consequence of this birefringence, the wave vector and the energy flow given by the Poynting vector cannot be perfectly collinear. Consequently, walk-off will limit the interaction length of the fundamental beam and reduce the conversion efficiency. To alleviate this effect, the crystal is rotated, a technique known as *critical phase matching*. However, even small deviations from the phase matching angle cause large changes in Δk. This effect can be largely suppressed when the angle between the beam propagation direction and the optical axis of the crystal is set to 90°. This arrangement, called *noncritical phase matching* because of its low sensitivity to angular alignment, is practically performed by tuning the crystal temperature to find the state when all polarization directions are aligned along with crystal axis. Unfortunately, only a limited number of nonlinear materials allow noncritical phase matching configurations. Among them, LBO, BBO, and $LiNbO_3$ are most frequently used for the efficient generation of second harmonics.

Appropriate phase matching based on crystal birefringence provides a strong prerequisite to the optical symmetry of the crystals and can limit the range of usable media with high nonlinear coefficients. The technique called *quasi-phase matching* (QPM) allows the periodic power flow to be prevented by a periodic reversal of the sign of the nonlinear coefficient. QPM based on the spatial modulation of the second-order nonlinearity is an artificial means by which a versatile phase matching can be provided. By introducing a periodic spatial modulation on the sign or strength of nonlinearity, with the period matched to a multiple of the coherence length for the process, one can ensure that the generated fields do not exactly cancel. The QPM principle then allows a cumulative growth of the generated field, as in the case of exact phase matching. Periodic poling is the most practical method for creating spatial modulation in a variety of materials.

When compared with birefringent phase matching, QPM enables nonlinear materials with enhanced nonlinearity and an extended phase matching range to be used and allows noncritical phase matching. Compatibility with waveguide geometry gives an opportunity to maintain small beam cross sections over longer lengths as compared with bulk crystals, where diffraction losses limit the achievable interaction length. Advanced spatial modulation geometry of the nonlinearity, such as aperiodic

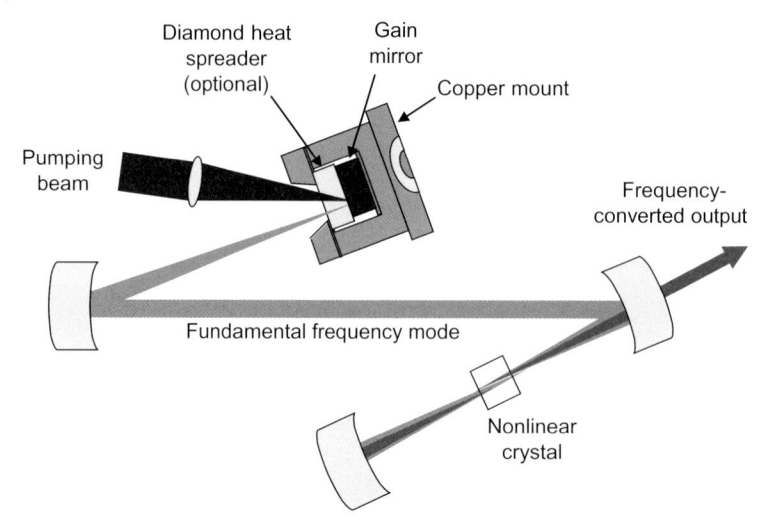

Figure 3.1 Z-type cavity used for frequency conversion with an optically pumped semiconductor disk laser.

patterns and two-dimensional patterns, would further extend the applicability of the technique.

The nonlinear crystal within the laser resonator acts as an output coupler, and the generated beam exits the resonator via a dichroic mirror with high reflectivity at the fundamental wavelength and high transmission at the harmonic wavelength, as illustrated in Figure 3.1. The outcoupling depends on the conversion efficiency and therefore also on the intracavity power. A variety of nonlinear crystals have successfully been used for intracavity frequency converted SDLs. The parameters of the most used crystals are presented Chapter 7.

3.2.2
Power Scaling of SDLs

To achieve efficient generation of high-power frequency-doubled radiation from SDLs, one needs to address operation at the fundamental wavelength and power scaling concepts. Thermal rollover frequently limits the power scaling, making heat dissipation an important aspect in high-power disk lasers. The thin-disk geometry of the gain element allows power scaling because of the virtually one-dimensional heat flow along the beam axis; therefore, the power can be scaled up with beam size [22]. Using intracavity transparent heat spreader capillary bonded to the semiconductor gain element just next to the active layer is an approach that has been successfully demonstrated and is especially useful in long-wavelength SDLs with thick DBRs (see Chapter 2 for details). To boost the SDL power, a number of approaches have been considered, including the concept of multiple gain elements with individual heat spreaders placed in the same cavity for power scaling [23]; details of this technique are described in Chapter 1. The multiple gain cavity can tolerate higher pump power by

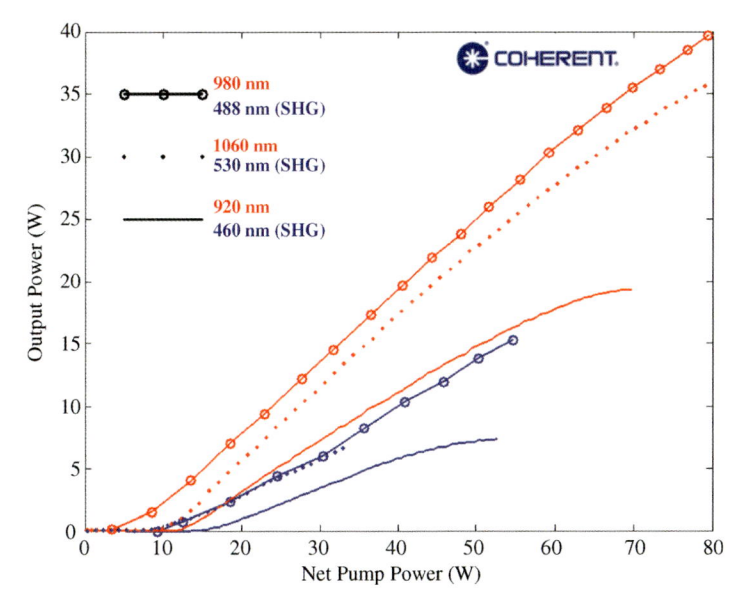

Figure 3.2 Green output from two and three disk SDLs after intracavity frequency doubling. Figure courtesy of Juan Chilla of Coherent.

sharing the thermal load between few gain elements, thus avoiding excessive heating and rollover. Figure 3.2 demonstrates the potential of SDLs with multiple gain elements, with intracavity frequency conversion for power scaling; 55 W of green emission has been obtained with a three-chip design.

Among different techniques used to reduce the thermal load of the gain medium, optimization of the threshold characteristic and slope efficiency is extremely important for the improvement of the output power. It is understood, therefore, that for any given technique used for thermal management, the overall quantum efficiency of the gain plays a key role. The gain element design has a significant impact on various aspects of laser operation, for example, on the efficiency and thermal sensitivity of the output characteristics, especially for the widely used resonant periodic gain (RPG) configuration, where quantum wells are placed at antinodes of the optical standing wave. Spectral detuning between cavity resonance and gain peak with temperature rise may cause rollover in the output characteristics and should, therefore, be properly set to achieve the matching condition at the operation temperature.

The main challenge in the design of short-wavelength SDLs originates from the choice of barrier materials to prevent the degradation of optical efficiency due to poor carrier confinement in the relatively shallow GaInAs/GaAs quantum wells at $\lambda \leq 950$ nm. Alternatively, the performance of short-wavelength SDLs could be improved by using in-well pumping, which could result in better efficiency since the band offset between QW and barrier bandgap energy is not limited by the pump wavelength. The low pump absorption typical for this scheme requires implementation of pump beam recycling [24].

3.3
SDL Frequency Doubled to Red

The 1064 and 940 nm InGaAs/GaAs SDLs with intracavity second harmonic generation to green and blue have been developed very intensively during the past 5–10 years. The best results achieved with single-gain element SDLs frequency converted to green and blue wavelengths are shown in Figure 3.3. On the other hand, accessing the orange–red wavelength range has been addressed just recently, owing to the lack of suitable semiconductor materials for the gain mirror. The purpose of the section is to highlight the details of optically pumped SDLs designed to operate at a specific spectral range around 1.2 μm for further conversion to red emission.

SDL operation at red wavelengths can be achieved via direct emission using the GaInP/AlGaInP/AlGaAs/GaAs material system [25, 26]. Despite recent progress that has led to demonstration of red SDLs with watt-level powers [27], there are several constraints that limit the adoption of direct emission as an effective approach for demonstrating practical SDLs. First, $Ga_xIn_{1-x}P$/AlGaInP quantum wells typically emit within the 640–700 nm wavelength range. Moving toward emission at shorter wavelengths is associated with a significant reduction in carrier confinement, leading to small gain and increased temperature sensitivity. This, in turn, restricts the use of red SDL technology to spectroscopy and generation of UV radiation via frequency doubling; so far attaining an emission wavelength of about 620 nm remains out of reach for SDLs with direct emission. Another important issue to be considered is the pumping source. The requirement to pump at wavelengths shorter than 600 nm leaves only a few options for suitable pump systems. GaInP/AlGaInP SDLs have been typically pumped with cryogenically cooled Ar-ion lasers at 514 nm, or with

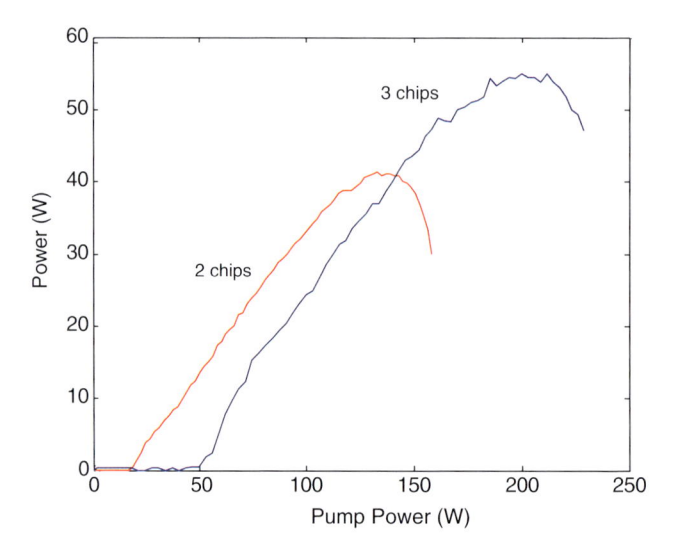

Figure 3.3 Blue and green output from single-disk SDLs after intracavity frequency doubling. Figure courtesy of Juan Chilla of Coherent.

frequency-doubled Nd:YVO$_4$ lasers operating at 532 nm. The size and cost of these pumping systems obliterates the advantages offered by the SDL concept. A much more practical approach for pumping is to use GaN laser diodes with emission at around 475 nm [28]. However, in this case the pump power would be in the range of 1 W or below, which in turn renders the output power from the SDLs to only a few tens of mW. Because of these limitations in obtaining high-power red radiation from SDLs via direct emission, we have to turn our attention towards development of SDLs with emission at around 1200 nm followed by second harmonic generation.

In this section, we will present results concerning the development of SDLs with red–orange emission obtained via intracavity frequency doubling. We will first provide an outlook for the semiconductor technology required to fabricate the gain mirrors with optical emission at 1200–1240 nm. Then, we focus on summarizing the laser results corresponding to operation at fundamental and second harmonic wavelengths.

3.3.1
Dilute Nitride Heterostructures for 1.2 μm Light Emission

The lattice mismatch between InGaAs and GaAs makes it difficult to reach an emission wavelength beyond 1170 nm [29]. The spectral region 1200 nm $< \lambda <$ 1300 nm can be covered by introducing a small amount of nitrogen into the InGaAs/GaAs QWs [30]; a schematic representation of the bandgaps and lattice constants corresponding to InGaAsN is shown in Figure 3.4. We should note that in practice the nitrogen incorporation is usually below 3–4%; however, this is sufficient to reduce the bandgap down to 0.8 eV, corresponding to operation at 1.55 μm [31]. Detailed analysis

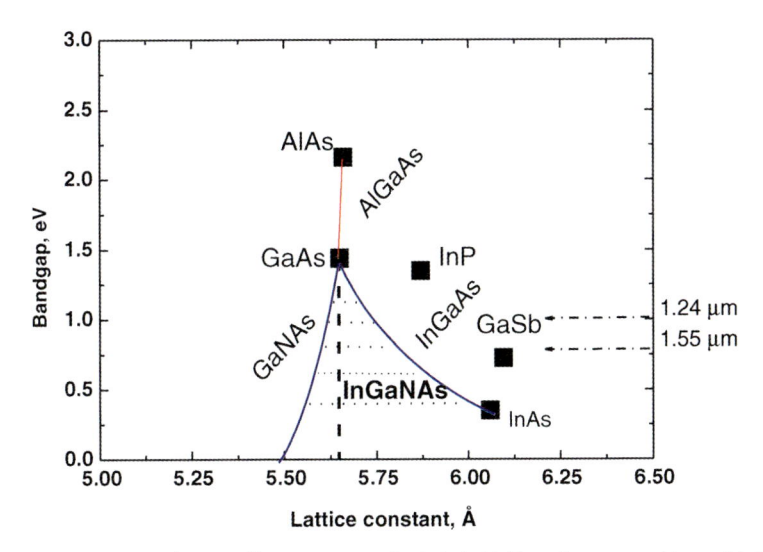

Figure 3.4 Bandgap and lattice constant for InGaAsN alloys. For compositions of In(x) and N(y) satisfying the condition $x \approx 2.8y$, this material system is lattice matched to GaAs.

of the structural, electrical, and optical properties of InGaAsN can be found in Ref. [32]. Similar to InGaAs/GaAs compounds, dilute nitride InGaAsN/GaAs material systems are very useful for developing vertical-cavity devices using high-reflectivity GaAs/AlGaAs distributed Bragg reflectors (DBRs). GaAsSb/GaAs QWs can also be used with GaAs-based DBRs [33]. However, dilute nitride QWs are preferred because of their stronger carrier confinement, which results in higher carrier density and the ability to operate at elevated temperatures.

It should also be noted that the incorporation of nitrogen into the InGaAs lattice is usually accompanied by the generation of nonradiative recombination centers that have detrimental effects on the emission efficiency and long-term device reliability, ultimately limiting the use of dilute nitride lasers in practical applications.

The next section reviews the main technological issues that should be addressed when fabricating dilute nitride heterostructures. It also provides an account of the main results pertinent to high-power operation of dilute nitride-based SDLs.

3.3.2
Plasma-Assisted MBE Growth of Dilute Nitrides

Dilute nitride heterostructures have been synthesized by both metal-organic vapor-phase epitaxy (MOVPE) [34] and molecular beam epitaxy (MBE) [35]. Although the MOCVD technique has advantages in terms of production throughput and maintenance, the highest performance InGaAsN-based heterostructures are routinely fabricated by MBE. MBE is also more favorable for growing devices with vertical geometry, such as gain mirrors for SDLs, because it enables a more accurate thickness control of the deposited layers. The standard technique used to incorporate N is dissociation of atomic nitrogen from molecular nitrogen using a radio-frequency (RF) plasma source attached to the MBE growth chamber [36]. Optimization of the plasma operation is one of the key issues that needs to be addressed to fabricate high-quality dilute nitride heterostructures. The state of the nitrogen plasma depends on the RF power, the flow of N_2, and pressure. The main constituents of the plasma are the molecular nitrogen, atomic nitrogen, and nitrogen ions, each of them having a specific spectral signature that can be used for optimizing the plasma operation [37]. Although the energy of the ions is small, they can cause significant degradation of the optical quality as they impinge on the semiconductor structure during the formation of the quantum wells [38]. The amount of ions can be easily monitored with the ion gauge used for pressure monitoring as a Langmuir probe; the current indicated by the ion gauge is proportional to the amount of ions. Figure 3.5 reveals the typical dependence between the current used to monitor the ions and the plasma settings. As it can be seen, the maximum amount of ions corresponds to small values of flow and large values of RF power. Therefore, plasma is usually operated away from its maximum RF power limits, while the flow is used as a parameter to control the amount of nitrogen incorporated into the structure.

Figure 3.6 shows typical photoluminescence (PL) spectra emitted by InGaAsN QWs grown with different plasma exposures. This figure reveals an unavoidable issue: dilute nitride structures incorporating higher amounts of nitrogen exhibit

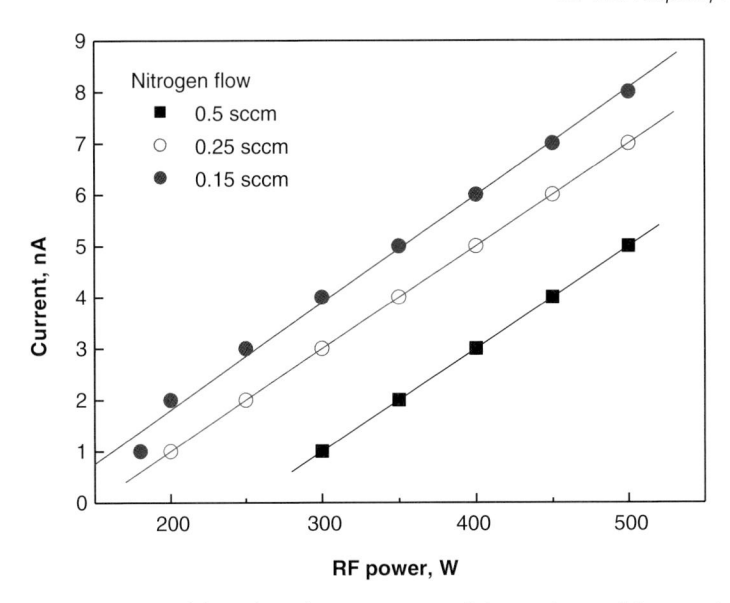

Figure 3.5 Typical dependence between amount of nitrogen ions and the operation condition of the plasma source.

lower photoluminescence efficiencies and larger spectral shifts toward long wavelengths. This degradation of the PL intensity has been the subject of intensive investigations, which revealed that it is associated with the presence of interstitial N [39] and the creation of Ga vacancies [40]. The detrimental effects associated with N

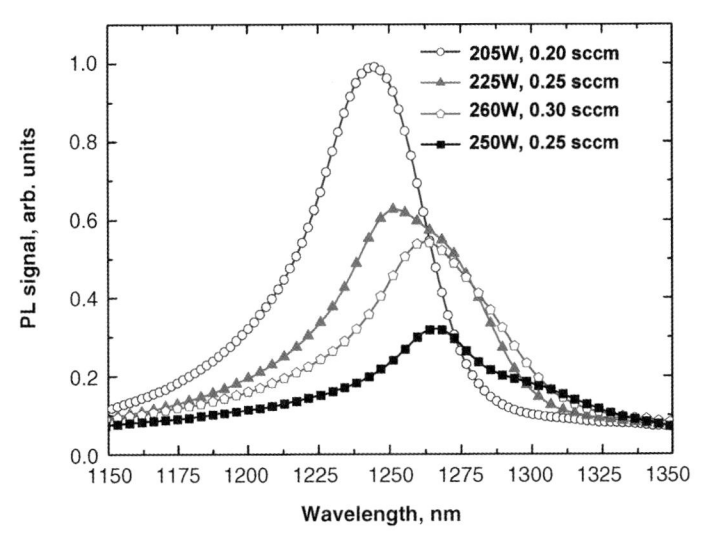

Figure 3.6 Effect of plasma RF-power (W) and nitrogen flow (in percentage of 1 sccm) on the PL spectrum.

incorporation can be partially alleviated by rapid thermal annealing (RTA) that, however, leads to a considerable blue shift of the PL wavelength [41], which should be taken into account to achieve the desired laser performance.

Another growth parameter that calls for careful optimization, and that has a high impact on the optical and structural quality of the dilute nitride QWs, is the N : In ratio. Operation at a given wavelength can be achieved for different N/In mole fractions as the emission wavelength increases with an increase in the concentration of both N and In. However, they affect the optical quality in different ways: increasing the In content is associated with more strain in the structure, while incorporating more nitrogen reduces the strain but potentially decreases the luminescence efficiency, as discussed in the previous paragraph. This optimization becomes more complicated for vertical-cavity devices comprising a large number of QWs, for which the effect of overall strain accumulation becomes significant.

The range of suitable growth temperatures for fabricating high-quality dilute nitrides is narrower than that for growing InGaAs. The typical growth temperature for InGaAsN is $\sim 460\,°C$, whereas for InGaAs QWs is $\sim 520\,°C$. The lower temperature for the growth of dilute nitrides is required to avoid switching from 2D to 3D growth modes [42]. Other important growth parameter affecting the quality of dilute nitrides is the As pressure [43]. Understanding the overall optimization procedure of the various growth conditions makes the fabrication of dilute nitrides challenging. It should also be noted that the growth optimization becomes progressively more difficult when moving to longer wavelengths by adding more N to the InGaAsN compound.

3.3.3
Design and Characteristics of Dilute Nitride Gain Media

To demonstrate some of the aspects discussed above, two structures designed for high-power operation have been grown using the InGaAsN material system. Both structures incorporate a 30-pair GaAs/AlAs DBR grown on an n-GaAs substrate. The structure, shown in Figure 3.7, consists of an active region with 12 InGaAsN quantum wells having a nitrogen content of 1%. The quantum wells, each having a thickness of 7 nm, were distributed over five identical pairs placed at the standing wave antinodes formed within the Fabry–Pérot cavity defined by the DBR and the

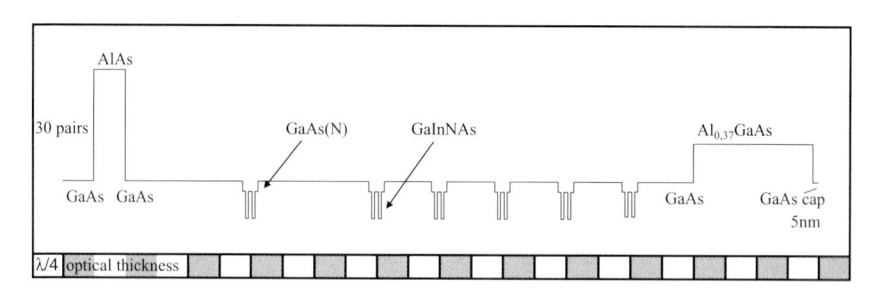

Figure 3.7 Conduction band profile of a gain mirror comprising 12 QWs with 1% N.

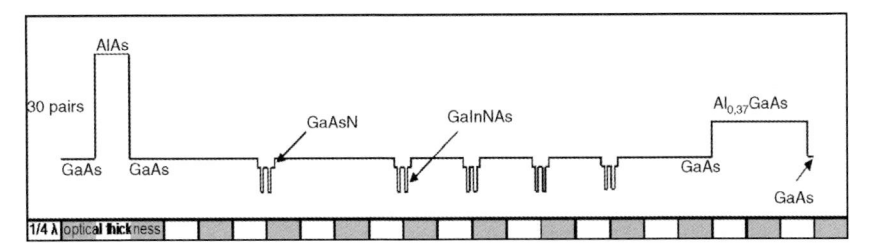

Figure 3.8 Conduction band profile of a gain mirror comprising 10 QWs with 0.6% N.

semiconductor–air interface. The compressive lattice strain caused by the QWs was partially compensated by 4 nm thick tensile-strained GaAsN layers grown on both sides of each QW. These strain compensating layers cause a helpful spectral red shift [44]. A λ-thick $Al_{0.37}Ga_{0.63}As$ window layer was grown on top of the active region to confine the photogenerated carriers within the active region and to avoid non-radiative surface recombination. Finally, a thin GaAs capping layer was grown to prevent oxidation of the window layer.

The alternative structure shown in Figure 3.8 comprises 10 QWs with a nitrogen content of 0.6–0.7%. The actual settings of the plasma source used during the growth of the gain mirror were ~200 W power and 0.2 sccm nitrogen flow. The structure with less N content acquires larger strain because it incorporates a higher amount of In to keep the wavelength constant. However, the reduced N content was confirmed to result in lower pumping threshold and higher output power from the SDLs. The strain for the two structures can be compared from Figure 3.9 showing the XRD (004)-rocking curves corresponding to the two structures.

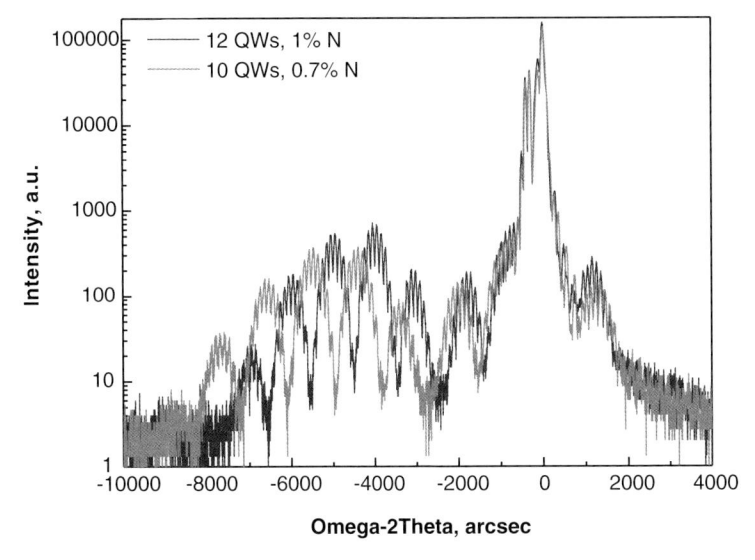

Figure 3.9 XRD (004)-rocking curve measured from the as-grown semiconductor gain mirrors.

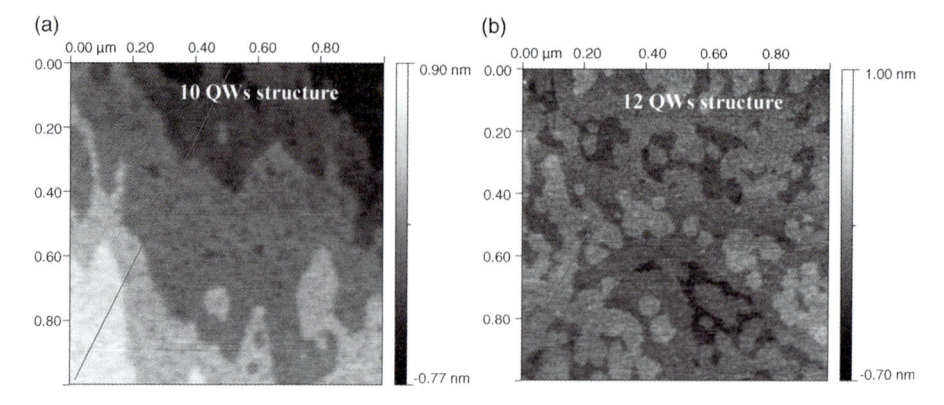

Figure 3.10 AFM pictures demonstrating the high-quality surface. Panel (a) corresponds to the 10 QWs structure and exhibits the presence of larger and more uniform surface domains, which are associated with a lower N content.

For both structures, the growth temperature, measured by an optical pyrometer, was 450 °C for the QWs, 580 °C for the GaAs barriers, and 590 °C for the DBR and the AlGaAs window layer. The structures were *in situ* post-growth annealed at 680 °C for 5 min. Atomic force microscope (AFM) images shown in Figure 3.10 reveal that the surfaces were smooth on the atomic level; it also shows that the structure with 10 QWs exhibits larger surface domains belonging to the same atomic layer. The PL and reflectance spectra measured from an as-grown 10 QWs wafer are shown in Figure 3.11. We should note that the PL signal from the gain mirror peaks at the resonance wavelength of the Fabry–Pérot vertical cavity confining the active region. To avoid the cavity influence on the PL spectrum, a separate PL sample was prepared by removing the substrate and the DBR. The PL spectrum obtained from this sample

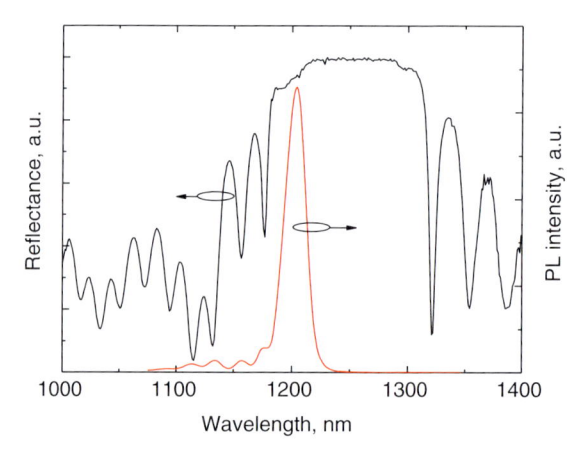

Figure 3.11 Room-temperature reflectivity and PL spectra of the as-grown sample.

was blue shifted by 20 nm compared to the PL signal measured from the gain mirror, which determines the spectral detuning between the QWs peak emission and the resonance of the subcavity.

A distinct difference between the two structures was observed when they were used as SDL gain mirrors. Though each structure supports more than 1 W of output power at 1220 nm, the SDL with the gain chip comprising 12 QWs with more nitrogen content exhibits a significantly higher threshold and smaller power compared to the structure with 10 QWs and less nitrogen [45]. The performance of the SDL with a 10 QW gain chip is described in Sections 3.3.4 and 3.3.5.

3.3.4
Performance of 1220 nm SDL

Chips of 2.5×2.5 mm^2 were cut from the as-grown wafers and capillary bonded with water to a $\sim 3 \times 3 \times 0.3$ mm^3 type IIa natural diamond heat spreader to ensure efficient heat removal from the gain region. The bonded chip was fixed between two copper plates with indium foil to ensure good thermal and mechanical contact. The pump light and the 1220 nm signal were coupled through a circular aperture in the top copper plate. The mounted samples were attached to a water-cooled copper heat sink; note that at this point the gain chip was not antireflection coated. Two different cavity configurations were used: a linear cavity and a V-shaped cavity, which are schematically shown in Figures 3.12 and 3.13, respectively.

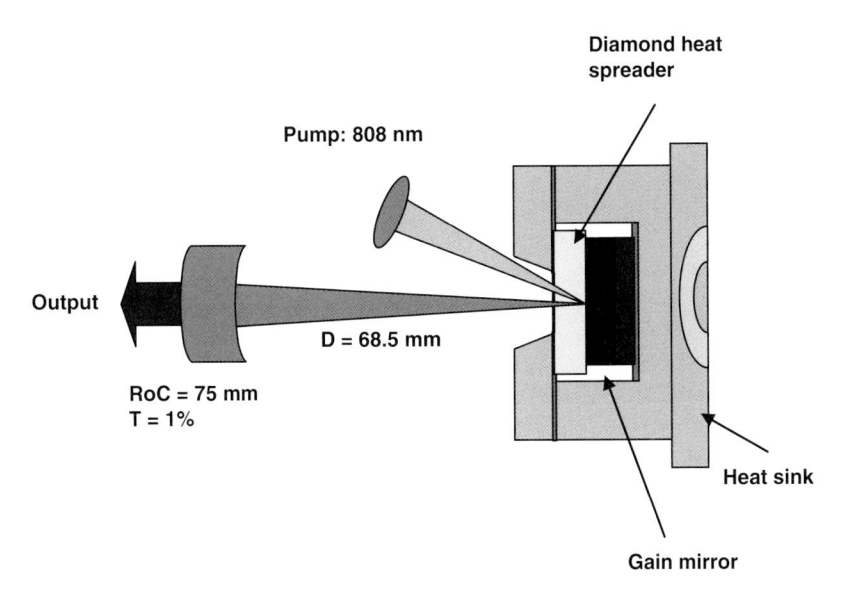

Figure 3.12 Schematic representation of the linear SDL. RoC, radius of curvature; T, transmission of a mirror at the operating wavelength; D, separation between elements.

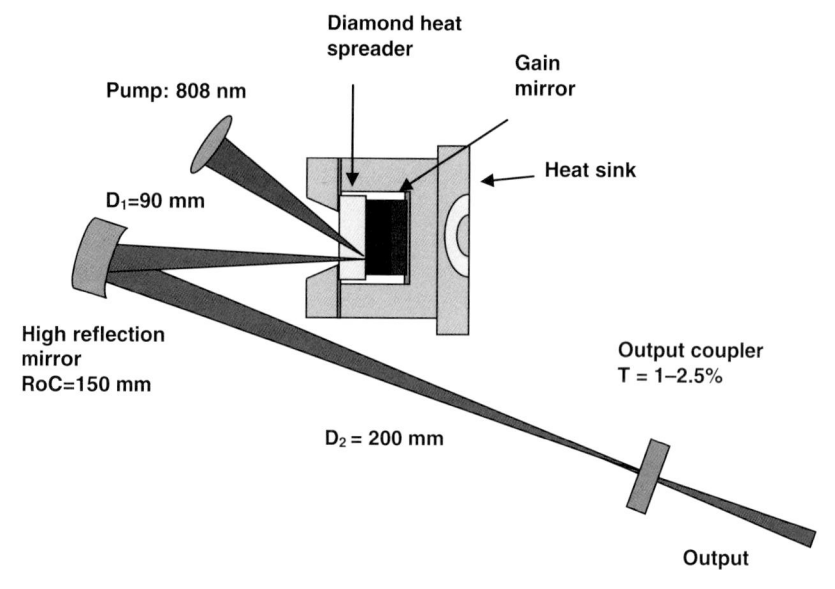

Figure 3.13 Schematic representation of the V-shaped cavity.

The linear cavity consisted of the reflective semiconductor gain chip and a curved output mirror with a radius of curvature (RoC) of 75 mm and a transmission of 1% at the signal wavelength. The output coupler was placed at a distance of 68.5 mm from the gain mirror. The mode diameter on the gain mirror was set to ~180 μm.

The V-shaped cavity consisted of the gain mirror, the high-reflectivity curved mirror with a radius of curvature of 150 mm placed at a distance of 90 mm from the gain mirror, and a planar output coupler placed at 200 mm from the curved mirror. The 808 nm pump beam was focused on a spot with a diameter of approximately 180 μm, comparable to the beam size used in the linear cavity. Both cavities were designed to match the pump and the fundamental cavity mode sizes on the gain mirror.

The output power characteristics for the linear cavity are shown in Figure 3.14. The output power at 8 °C with 20 W of pump power reaches a value of 3.14 W, corresponding to a differential efficiency of 20%. The dependence of the threshold pump power on the temperature is shown in the inset of Figure 3.14. The minimum threshold pump power was observed at a mount temperature of 10 °C. The slight decrease in threshold when the temperature was raised from 8 to 10 °C could be explained by a closer match between the peak gain wavelength and the Fabry–Pérot cavity resonance of the gain chip at 10 °C. The emission spectra at the maximum pump power are shown in Figure 3.15 for two mount temperatures. The multiple spikes in the spectra originate from the resonant modes of the intracavity etalon formed by the diamond heat spreader.

The output characteristics for the SDL with the V-shaped cavity are shown in Figure 3.16. Using a 2.5% output coupler, 3.5 W has been achieved at a pump power

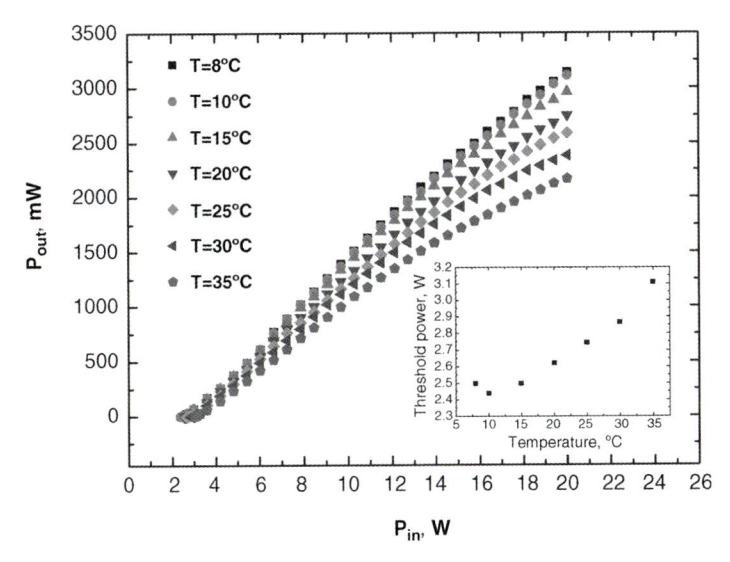

Figure 3.14 Output characteristics of the SDL with linear cavity. The inset shows the temperature dependence of the thresholds pump power.

of 20 W. The threshold pump power was about 2.7 W and the differential efficiency was estimated to be ~20% for the mount temperature of 15 °C.

To scale up the output power, the gain chip was antireflection coated and the diameter of the pump spot on the gain mirror was increased to 290 μm. The V-shaped cavity parameters were changed to $D_1 = 170$, $D_2 = 265$, and RoC = 250 mm to ensure overlap between the pump beam and the cavity mode. The resulting output characteristics are shown in Figure 3.17.

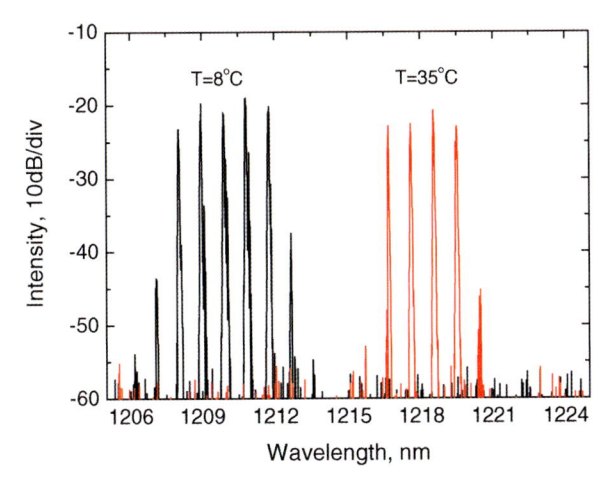

Figure 3.15 Output spectra for the SDL with linear cavity for two temperatures and 5 W of pump power.

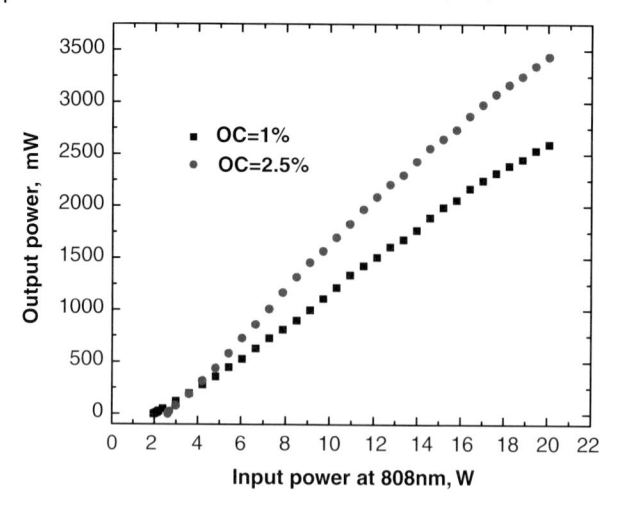

Figure 3.16 Output characteristics of the SDL with V-shaped cavity for two values of the output coupler.

3.3.5
SDL Intracavity Light Conversion to Red–Orange

The 1220 nm InGaAsN gain chips described above have been used in an SDL for orange–red laser radiation by frequency conversion using an intracavity scheme [46]. The frequency doubling was studied with a Z-type cavity formed by the gain mirror

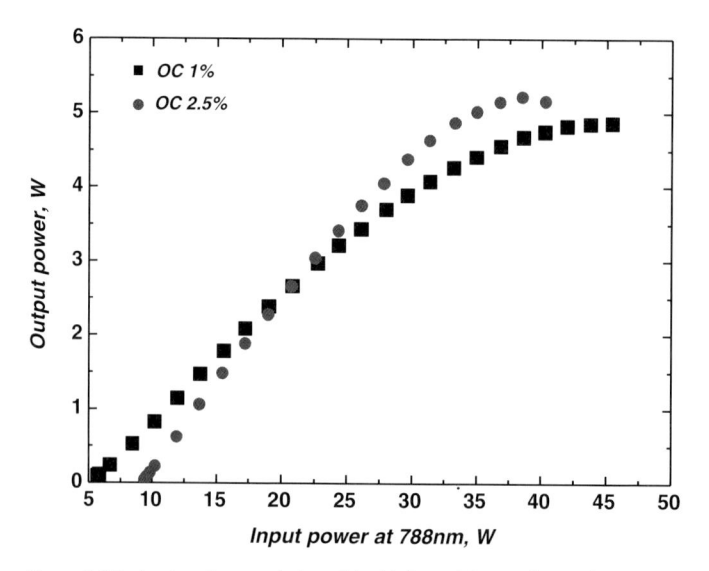

Figure 3.17 Lasing characteristics of the V-shaped SDL with 1 and 2.5% transmission of output couplers. The mount temperature was set to 15 °C.

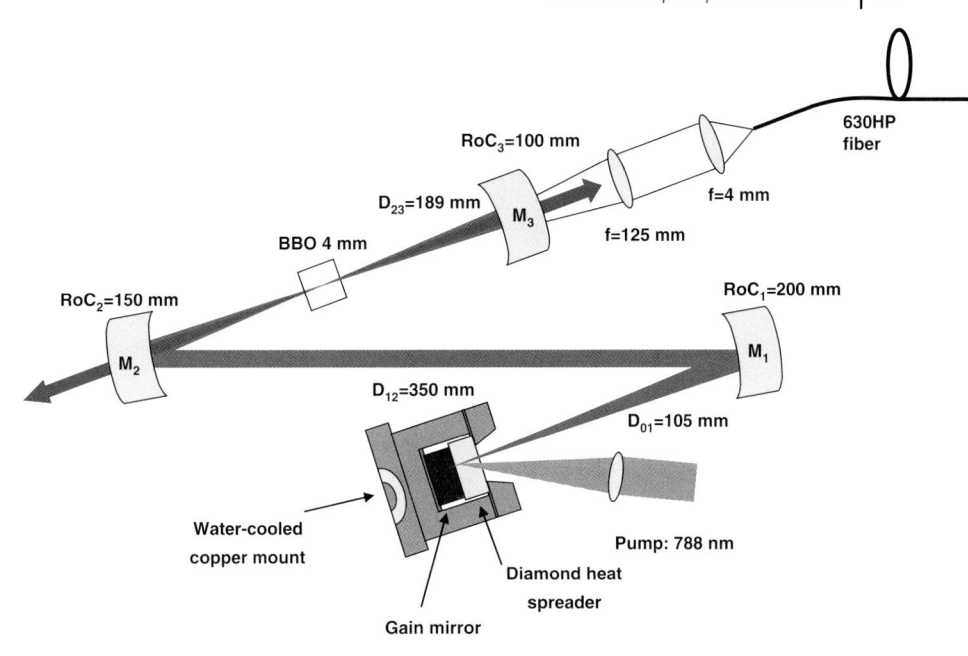

Figure 3.18 Z-type laser cavity consisting of three curved mirrors and the gain chip. The output light was coupled to an optical fiber located behind mirror M_3. RoC, radius of curvature; D, distance between mirrors.

and three curved mirrors, as shown in Figure 3.18. All curved mirrors are highly reflective for the fundamental wavelength of 1.2 μm and have transmission over 90% for the second harmonic radiation. The gain medium was pumped with a fiber-coupled 788 nm pump diode laser at an angle of 35° to the surface normal of the gain mirror. The cavity was designed such that the laser mode matched the 290 μm diameter pump spot at the gain mirror. A 4 mm long nonlinear BBO crystal was placed between mirrors M_2 and M_3 in the waist of the cavity transverse mode. The waist diameter was calculated to be approximately 160 μm with a Rayleigh length of 18 mm. The BBO crystal is critically type I phase matched and antireflection coated for 1220 and 610 nm on both facets. Red radiation was coupled out through mirrors M_2 and M_3. Using the gain chip capable of producing 5 W at the fundamental wavelength (Figure 3.17), 4.5 W of orange–red radiation has been achieved at visible wavelengths [47]. The spectrum of red emission is shown in Figure 3.19a separated from residual infrared power with a dichroic beam splitter.

A narrow-line spectrum was achieved by inserting a 25 μm solid glass etalon in the cavity. The etalon was placed between mirrors M_1 and M_2. Tilting the etalon enabled a discrete tuning of the operation wavelength around 1.2 μm with a step of 0.9 nm corresponding to the free spectral range of the intracavity diamond. The tuning range at the fundamental wavelength was 16 nm and it was limited by the free spectral range of the etalon. The corresponding tuning range in the visible region was 8 nm, as shown in Figure 3.19b.

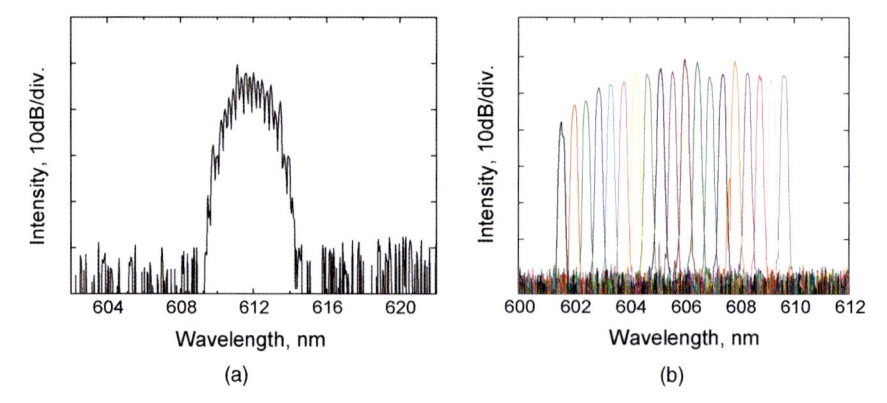

Figure 3.19 The spectra of 5 W laser output frequency converted to red (a) without and (b) with etalon that allows 8 nm tuning band.

The beam profile of the red emission was studied with a CCD camera. The transverse beam profiles are shown in Figure 3.20a. The beam was found to be elliptical with a Gaussian profile in both directions. The ellipticity is likely to originate from the spatial walk-off in the BBO crystal. Pumping the gain medium at an angle to the surface normal may also contribute to noncircular beam shape formation. To quantify the quality for the elliptical beam, we have launched the red emission from output to a fiber with single-mode guiding at visible wavelength and determined the coupling efficiency for different powers. The coupling efficiency into the fiber was 78% at low power and gradually decreased to 70% with an increase in the output power. Coupling efficiency as a function of the output power is shown in Figure 3.20b. Efficient coupling to a single-mode fiber indicates that despite the elliptical shape, the output beam corresponds to a nearly single-transverse mode with a very moderate beam quality degradation with increase in output power.

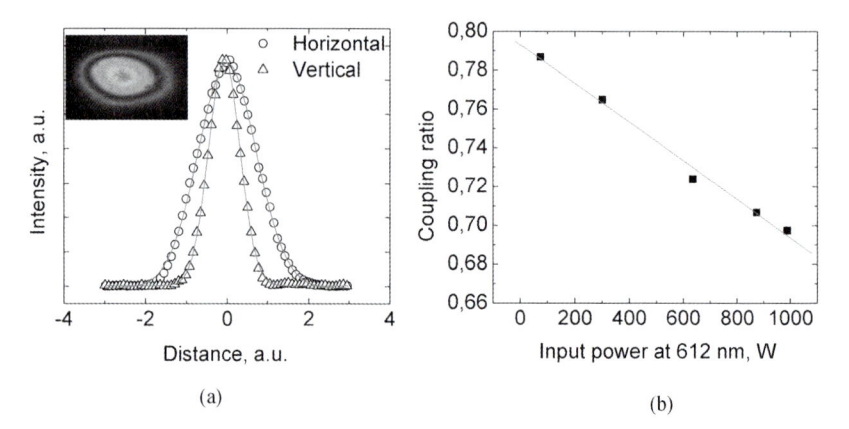

Figure 3.20 (a) Transverse beam profile of the red beam, with a Gaussian fit in vertical and horizontal directions. *Inset*: Corresponding picture of the beam. (b) Coupling efficiency to a single-mode fiber as a function of output power from single laser output port.

3.4
Conclusions

Intracavity frequency doubling is particularly attractive for implementation in semiconductor disk lasers owing to their low loss and small output coupling, which in turn results in high intracavity power and high conversion efficiencies.

Though laser sources generating radiation over the whole visible range are of great interest, the development of orange–red emitters producing light in the 550–620 nm band is particularly challenging as compared with blue–green sources, primarily due to lack of suitable gain materials. We have reviewed here the main results concerning the development of red SDLs based on dilute nitride semiconductor compounds. The study proves that InGaAsN/GaAs quantum well heterostructures are a promising approach for the development of SDLs emitting at 12XX nm for frequency doubling to red.

Acknowledgments

We would like to thank Ville-Markus Korpijärvi, Janne Puustinen, and Pietari Tuomisto for fabricating and characterizing the dilute nitride gain mirrors. We gratefully acknowledge Jussi Rautiainen and Antti Härkönen for building the lasers and performing the experiments on frequency doubling. Lasse Orsila, Jari Nikkinen, Antti Tukiainen, Kimmo Haring, and Charis Reith are acknowledged for their various contributions. The main results presented here have been obtained within the EU FP6 project "NATAL" (IST-NMP 016769).

References

1 Chilla, J.L.A., Zhou, H., Weiss, E., Caprara, A.L., Shou, Q., Gorvokov, S.V., Reed, M.K., and Spinelli, L. (2005) *Proc. SPIE*, **5740**, 41; Chilla, J., Shu, Q., Zhou, H., Weiss, E., Reed, M., and Spinelli, L. (2007) Recent advances in optically pumped semiconductor lasers. *Proc. SPIE*, **6451**, 645109-1.

2 Jansen, M., Carey, G.P., Carico, R., Dato, R., Earman, A.M., Finander, M.J., Giaretta, G., Hallstein, S., Hofler, H., Kocot, C.P., Lim, S., Krueger, J., Mooradian, A., Niven, G., Okuno, Y., Patterson, F.G., Tandon, A., and Umbrasas, A. (2007) Visible laser sources for projection displays. *Proc. SPIE*, **6489**, 648908.1–648908.6.

3 Schmitt, M. and Steegmüller, U. (2008) Green laser meets mobile projection requirements. *Optics and Laser Europe*, p. 17.

4 Van Kessel, P.F., Hornbeck, L.J., Meier, R.E., and Douglass, M.R. (1998) A MEMS-based projection display. *Proc. IEEE*, **86**, 1687.

5 Masters, A. and Seaton, C. (2006) Laser-based displays will deliver superior images. *Laser Focus World*, **42**, S9–S11.

6 Wieman, C.E. and Hollberg, L. (1991) Using diode lasers for atomic physics. *Rev. Sci. Instrum.*, **62**, 1.

7 Kumagai, H. (2004) Development of a continuous-wave, deep-ultraviolet, and single-frequency coherent light source: challenges toward laser cooling of silicon. *IEEE J. Sel. Top. Quantum Electron.*, **10**, 1252.

photoluminescence and structural properties of GaInNAs/GaAs quantum wells grown by molecular beam epitaxy. *J. Cryst. Growth*, **281**, 249.

44 Pavelescu, E.-M., Peng, C.S., Jouhti, T., Li, W., Pessa, M., Dumitrescu, M., and Spanulescu, S. (2002) Effects of insertion of strain-mediating layers on luminescence properties of 1.3 μm GaInNAs/GaNAs/GaAs quantum-well structures. *Appl. Phys. Lett.*, **80**, 3054.

45 Korpijärvi, V.-M., Guina, M., Puustinen, J., Tuomisto, P., Rautiainen, J., Härkönen, A., Tukiainen, A., Okhotnikov, O.G., and Pessa, M. (2009) MBE grown GaInNAs-based multi-watt disk lasers. *J. Cryst. Growth*, **311**, 1868.

46 Rautiainen, J., Härkönen, A., Korpijärvi, V.-M., Tuomisto, P., Guina, M., and Okhotnikov, O.G. (2007) 2.7 W tunable orange–red GaInNAs semiconductor disk laser. *Opt. Express*, **15**, 18345.

47 Rautiainen, J., Härkönen, A., Korpijärvi, V.-M., Puustinen, J., Orsila, L., Guina, M., and Okhotnikov, O.G. (2009) Red and UV generation using frequency-converted GaInNAs-based semiconductor disk laser. Paper presented at CLEO/IQEC, Baltimore, MD, May 31–June 5.

4
Long-Wavelength GaSb Disk Lasers

Benno Rösener, Marcel Rattunde, John-Mark Hopkins, David Burns, and Joachim Wagner

4.1
Introduction

Shortly after the first demonstration of an optically pumped semiconductor disk laser (SDL) based on GaAs, a great deal of effort was made in extending the wavelength range by using different material combinations within the semiconductor structure. These activities eventually led to the development of SDLs based on group-III antimonides. As already shown for diode lasers, using semiconductor alloys such as GaInSb or GaInAsSb in quantum well (QW) layers enables coverage over a broad range of emission wavelengths in the short-wavelength end of the mid-infrared (MIR) spectrum between 1.9 and 3.3 μm. While early Sb-based SDLs exhibited moderate output power at wavelengths around 2.1 μm, progress continuously advanced and finally included milestones like multiwatt operation and maximum emission wavelengths of up to 2.8 μm. Transferring the SDL technology to the III-Sb material system thus proved to be particularly advantageous in realizing high-performance semiconductor-based laser sources in the MIR. In this way, III-Sb SDLs could significantly expand the range of existing laser technologies (e.g., doped-dielectric solid-state lasers) in this wavelength range.

There are numerous applications that can benefit from the development of MIR SDLs. First, this wavelength range is particularly important as many technically relevant gases exhibit specific absorption lines at wavelengths between 2.0 and 3.0 μm. These features corresponding to molecular rotational–vibrational oscillations can be readily investigated by MIR lasers, the latter thus being able to serve as powerful tools in gas sensing [1]. Besides, there are also various applications that can take advantage of the genuine feature of SDLs to provide high output power at excellent beam quality. In particular, MIR SDLs have been discussed as laser sources for applications such as free-space optical communications, standoff detection, and even infrared countermeasures [2]. Furthermore, the spectrum of possible applications includes employing III-Sb-based lasers as a master oscillator or pump source in solid-state lasers to improve their efficiency and functionality.

Semiconductor Disk Lasers. Physics and Technology. Edited by Oleg G. Okhotnikov
Copyright © 2010 WILEY-VCH Verlag GmbH & Co. KGaA, Weinheim
ISBN: 978-3-527-40933-4

This chapter provides a broad overview of the field of Sb-based SDLs, discusses the basic design issues, and introduces recent key developments. The structure of the chapter is as follows: the first two sections cover the basic properties of the group-III antimonide materials, which are commonly used in state-of-the-art MIR SDLs, as well as several aspects of the structural layout and the growth process of the laser chips. A review of past progress and recent developments in the field of high-performance Sb-based SDLs will be presented subsequently. Following this, the spectral properties of Sb-based SDLs will be discussed with special focus on tunability and single-frequency operation. As an outlook, concepts for future SDLs emitting at wavelength beyond 3 μm will finally be presented.

4.2
The III-Sb Material System

The quaternary semiconductor $Al_xGa_{1-x}As_ySb_{1-y}$ is the ideal material for barrier, window, and absorbing layers in III-Sb-based semiconductor disk lasers. Furthermore, efficient distributed Bragg reflectors (DBRs) can be fabricated using $AlAs_{0.08}Sb_{0.92}$/GaSb layer pairs. Because of the large width of the individual DBR layers as well as of the whole layer stack, they are almost exclusively grown lattice matched to the GaSb substrate; this can be achieved by adding a small amount of As to the AlGaSb to form $Al_xGa_{1-x}As_ySb_{1-y}$ with $y = 0.08x$ [3]. The direct bandgap for AlGaAsSb at 300 K, lattice matched to GaSb, is given by $E_g(\Gamma) = 2.297x + 0.727 (1-x) - 0.48x(1-x)$ eV [3].

For the active layers, $Ga_{1-x}In_xAs_ySb_{1-y}$ is used that has a direct bandgap for all alloy compositions and is lattice matched to GaSb if the condition $y = 0.913x$ is satisfied [3, 4]. Using a quaternary material for the active layer adds an additional degree of freedom for the design of QWs compared to a ternary material (such as the well-known GaInAs): by changing the composition, two of the relevant parameters bandgap E_g, strain ε_{zz}, and conduction (valence) band offsets ΔE_C (ΔE_V) can be adjusted individually within certain limits. To illustrate this property of the GaInAsSb material system, Figure 4.1 shows the band edge profile for $Ga_{1-x}In_xAs_ySb_{1-y}$ grown on GaSb as a function of the In content x for three cases with different As contents y and thus different strain ε_{zz}. Starting with GaSb on the left side of each plot, the bandgap is decreased with increasing In content in all cases. Also shown are the band edges of AlGaAsSb lattice matched to GaSb with an Al content of 30% as a reference. The latter represents a typical barrier material, used by many different groups in their GaSb-based lasers [5, 6]. All material data for this calculation are taken from Ref. [3].

Without the addition of As in Figure 4.1a, that is, for $Ga_{1-x}In_xSb$, the layers are compressively strained with $\varepsilon_{zz} = +0.063x$ and due to strain effects, a type I band alignment is formed with regard to the chosen barrier material ($Al_{0.3}Ga_{0.7}As_{0.02}Sb_{0.98}$) in the whole composition range with the heavy hole (hh) band as the topmost valence band. But due to the increasing strain upon increasing In content, the maximum emission wavelength achievable with this ternary compound

4.2 The III-Sb Material System

$Ga_{1-x}In_xAs_ySb_{1-y}$

(a) compressive strained $y = 0$ (b) lattice matched $y = 0.913 x$ (c) tensile strained $y = 1.7 x$

$\varepsilon_{zz} = + 0.063 x$ $\varepsilon_{zz} = 0$ $\varepsilon_{zz} = - 0.063 x$

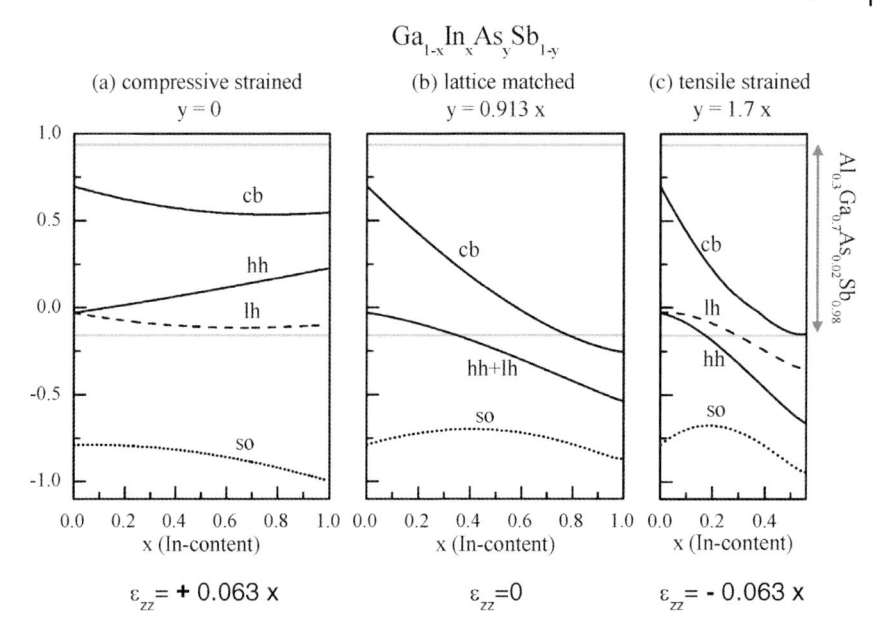

Figure 4.1 Relative positions of the conduction band (cb), heavy hole (hh), light hole (lh) valence band, and split-off (so) valence band versus In content for the active region QW material $Ga_{1-x}In_xAs_ySb_{1-y}$ grown on GaSb; for comparison, conduction and valence band edges of the $Al_{0.3}Ga_{0.7}As_{0.02}Sb_{0.98}$ barriers are also indicated (gray lines). Three representative cases are shown: (a) compressively strained QW material with $y = 0$, (b) lattice-matched material with $y = 0.913xi$, and (c) tensile-strained material with $y = 1.7x$. For the strained GaInAsSb in (a) and (c), the strain ε_{zz} is given by $+0.063x$ and $-0.063x$, respectively.

is limited by the critical layer thickness for pseudomorphic growth of GaInSb to around 2.1 μm. This limit depends on details of the active region such as the number and width of the QWs and on the strain in the surrounding barrier material (i.e., using tensile strained barriers, the accumulated strain in multi-QW systems can be reduced, leading eventually to a higher tolerable compressive strain in the individual QWs [7]).

For material lattice matched to GaSb with $y = 0.913x$ (Figure 4.1b), the band offset in the valence band ΔE_V is reduced with increasing In content x and leads to a type II band alignment with respect to this barrier material for $x > 0.35$. In Figure 4.1c, a tensile-strained material is shown, with higher As than In concentration ($y = 1.7x$) and a strain given by $\varepsilon_{zz} = -0.063x$. In this case of tensile strain, the light hole band (lh) forms the topmost valence band and the band alignment is type I again only for low In concentrations ($x < 0.3$). Therefore, compressively strained GaInAsSb is favorable to allow sufficient confinement of the holes in the active QWs. This is especially important for GaSb-based lasers with longer wavelength toward 3 μm (see Section 4.6).

Furthermore, recent theoretical and experimental studies have shown that the band structure in compressively strained GaInSb is almost ideal to achieve high optical gain [8]. The electron and hole effective masses are rather low and almost

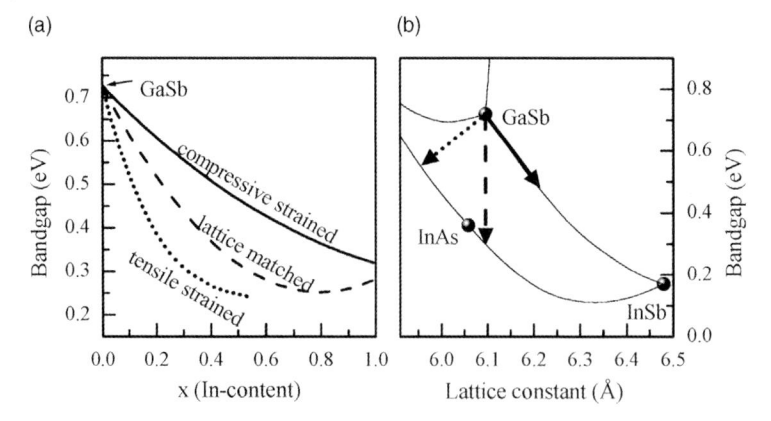

Figure 4.2 Bandgap E_g dependence of the In content x for the three different cases of compressively strained (solid line), lattice-matched (dashed line), and tensile-strained (dotted line) $Ga_{1-x}In_xAs_ySb_{1-y}$ on GaSb as displayed in Figure 4.1. In (b), the bandgap versus lattice constant relation is illustrated for these three different cases.

identical, making GaInSb QWs more favorable in terms of high material gain than, for example, the well-studied GaInAs material system.

For the three different cases of compressively strained, lattice-matched, and tensile-strained material discussed above, the bandgap E_g of GaInAsSb is plotted versus the In concentration in Figure 4.2a and the bandgap versus lattice constant relation is illustrated in Figure 4.2b. In the case of compressively strained GaInSb, the bandgap decreases at a lower rate with increasing In content than that for the other two cases (Figure 4.2a), that is, the addition of As leads to an additional reduction of the bandgap. This is due to bowing and strain effects, as compressive strain increases E_g. However, the addition of As decreases the band offset in the valence band ΔE_V (see Figure 4.1). Note that for all calculations shown in Figures 4.1 and 4.2, a pseudomorphic growth of the GaInAsSb layer is assumed despite the fact that for strained material, the critical layer thickness approaches zero for increasing In content.

4.3
Epitaxial Layer Design and Growth of III-Sb Disk Laser Structures

To a great extent, Sb-based SDLs benefit from the proven design concepts that had already been extensively investigated during the development of GaAs-based VCSELs and SDLs. However, there are also some distinct differences that become evident on comparing SDL concepts and technology in both material systems. Although realizing SDLs based on GaAs has become commonplace, specific challenges have to be faced in developing high-performance Sb-based SDLs. These difficulties mainly center on thermal management issues, however, there are specific challenges regarding the growth and processing technology of Sb-based SDLs that require careful consideration.

In the following sections, the basic layer design developed for the state-of-the-art MIR SDLs will be reviewed. The aforementioned difficulties as well as methods to mitigate them will then be discussed.

4.3.1
Basic Structural Layout

At first glance, there are only few structural differences between the layer designs of GaAs- and Sb-based SDLs. To date, as for GaAs-based SDLs, light amplification in Sb-based SDLs originates in type I QWs embedded in a microcavity terminated by a highly reflective distributed Bragg reflector and the interface between the topmost semiconductor layer and the surrounding air. In most laser structures, the QW gain is strongly enhanced by utilizing the concept of resonant periodic gain (RPG) [9] – this derives from VCSEL design that was subsequently transferred to GaAs-based SDLs.

The QW layers in Sb-based SDLs can either consist of ternary or quaternary alloys. For emission in the wavelength range below 2.1 µm, $Ga_{1-x}In_xSb$ QWs have been favored. In such QWs, the bandgap energy (starting at 0.72 eV for In-free GaSb) is continuously reduced by increasing the In content. Simultaneously, the amount of compressive strain increases, thus continuously reducing the maximum layer thickness that can be achieved without causing a relaxation in the crystalline structure. As mentioned in Section 4.2, the addition of arsenic allows independent control of the bandgap energy and the induced strain. A certain amount of compressive strain is, however, mandatory since in lattice-matched $Ga_{1-x}In_xAs_ySb_{1-y}$, the valence band offset with respect to the barrier layers continuously decreases with increasing In content, which would ultimately result in a type II band alignment in the QWs. In current SDL designs for the wavelength range between 1.9 and 2.8 µm, QWs having thickness between 8 and 10 nm are used. At a given pump power and output coupler reflectivity, the number of QWs in the active region can be optimized for the intended output power. The active region designs of high-power Sb-based SDLs typically contain more than 10 QWs (placed at the antinodes of the internal standing-wave pattern to increase the modal gain). Furthermore, for optimized high-power operation, the maximum of the gain spectrum at room temperature is detuned with respect to the spectral position of the microcavity resonance – this will be discussed in detail in Section 4.4.

All Sb-based SDLs demonstrated so far have been optically pumped. For this purpose, high-power diode lasers emitting at a wavelength around 1 µm are typically used as a pump source. In conventional SDL designs, the pump light is absorbed within both the barrier layers and the QWs. To ensure a high pump absorption, the barriers consist of either GaSb or $Al_xGa_{1-x}As_ySb_{1-y}$ layers grown lattice matched to the GaSb substrate. For $Al_xGa_{1-x}As_ySb_{1-y}$ barriers, the Al content is chosen to enable sufficient pump absorption while simultaneously providing a high valence band offset with respect to the QW layers. Such a high valence band offset is desirable since experimental results indicate that type I Sb-based QW heterostructures can suffer from the thermionic emission of holes [10]. As hole delocalization negatively impacts

Table 4.1 Important thermal, optical, and electronic constants of different barrier materials.

	GaSb	$Al_{0.3}Ga_{0.7}As_{0.02}Sb_{0.98}$	$AlAs_{0.08}Sb_{0.92}$
Thermal conductivity ($W\,m^{-1}\,K^{-1}$)	33	10.5	9.8
Absorption coefficient at $\lambda = 980\,nm$ (cm^{-1})	3.59×10^4	1.55×10^4	Transparent
Absorption cut-off wavelength (nm)	1710	1130	540
Valence band offset in $Ga_{0.74}In_{0.26}Sb$ QWs (meV)	59	187	391
Valence band offset in $Ga_{0.65}In_{0.35}As_{0.1}Sb_{0.9}$ QWs (meV)	10	140	340

the efficiency and threshold carrier density, it should be avoided, if possible. To this end, $Al_xGa_{1-x}As_ySb_{1-y}$ barrier layers with an Al content of $x = 0.3$ are usually employed.

In Table 4.1, some basic properties of GaSb, $Al_{0.3}Ga_{0.7}As_{0.02}Sb_{0.98}$, and $AlAs_{0.08}Sb_{0.92}$ barriers are compared. It can be seen that GaSb barriers offer a comparatively high thermal conductivity and a high absorption coefficient for pump wavelengths around 1 μm. Compared to SDL structures using $Al_{0.3}Ga_{0.7}As_{0.02}Sb_{0.98}$ barriers, the total barrier layer thickness can be significantly reduced when using GaSb barriers since pump radiation can be almost completely absorbed within a shorter optical path length. This, of course, also reduces the path length for the heat flow from the active region to the heat sink. In addition, heat extraction also benefits from the higher thermal conductivity of GaSb. However, all these features are negated due to the significantly lower valence band offset to the QWs – and the situation only gets worse at longer wavelengths where quaternary QWs are used.

In a more advanced approach, active region designs incorporating $AlAs_{0.08}Sb_{0.92}$ barriers have been demonstrated [11]. In the III-Sb material system, type I QWs embedded between such barrier layers exhibit the highest valence band offset possible. However, such barriers cannot be pumped by conventional AlGaAs or GaInAs diode lasers as the bandgap energy far exceeds the photon energy of such pump lasers. However, these novel structures can be utilized in schemes where direct pumping of the QWs is employed, and this concept will be discussed in more detail in Section 4.4.3.

The microcavity containing the QWs is terminated by a distributed Bragg reflector on one side. In Sb-based SDLs, the DBR is typically a sequence of $AlAs_{0.08}Sb_{0.92}$–GaSb layer pairs. However, in comparison to GaAs-based SDLs that usually feature ≈30 AlGaAs–GaAs layer pairs, in Sb-based SDLs, this number is significantly reduced due to the high refractive index contrast ($\Delta n = 0.7$ at $\lambda = 2.3\,\mu m$) afforded by the III-Sb material system. A suitably high reflectivity can therefore be readily achieved with only 20 $AlAs_{0.08}Sb_{0.92}$–GaSb layer pairs.

Furthermore, the high index contrast also enhances the mirror bandwidth to >300 nm for a DBR centered at 2.0 μm. The number of layer pairs of Sb-based SDL structures reported so far varies between 18.5 and 23.5 layer pairs, where the noninteger number of layer pairs accounts for the requirement to terminate on an $AlAs_{0.08}Sb_{0.92}$ layer. Also, as $AlAs_{0.08}Sb_{0.92}$ exhibits the highest bandgap energy in the III-Sb material system, strong confinement of photogenerated electrons in the neighboring barrier layers is provided. Similarly, a $AlAs_{0.08}Sb_{0.92}$ layer (or an $Al_xGa_{1-x}As_ySb_{1-y}$ layer having a high Al content) is usually used as a window layer on top of the active region of Sb-based SDLs.

Although high refractive index contrast can be achieved in $AlAs_{0.08}Sb_{0.92}$–GaSb DBRs, the total thickness is still considerably greater than that of AlGaAs–GaAs DBRs used in SDLs emitting at ≈ 1 μm since the total thickness is directly proportional to the emission wavelength. For a 20 layer pair $AlAs_{0.08}Sb_{0.92}$–GaSb DBR, the thickness is ≈ 6 μm at a wavelength of 2.0 μm. Since Sb-based semiconductor alloys exhibit a relatively poor thermal conductivity, heat extraction through the DBR becomes problematic in MIR SDLs. More important, the growth of MIR SDL structures is not straightforward as the total thickness can be up to 10 μm resulting in very long growth times, and so unavoidable layer inhomogeneities. This becomes even more of an issue at wavelengths approaching 3 μm. To circumvent these difficulties, alternative approaches can be considered. For example, at 1.55 μm, InP-based SDLs incorporating the so-called hybrid mirrors have been successfully realized [12]. These mirrors deploy a reduced layer stack of a conventional DBR in conjunction with a terminating metal layer (typically gold). The layer stack can be formed either by epitaxially grown semiconductor or by dielectric layers deposited by plasma-enhanced chemical vapor deposition, for example. Due to the broadband reflectivity of the underlying metal coating, the number of DBR layers can be significantly reduced to only a few layers in such mirrors. Therefore, hybrid mirrors could play an even more important role as the wavelength of Sb-based SDLs is increased. However, in contrast to the GaAs material system, the availability of a well-developed substrate removal technique has not been established in the III-Sb system. One possible solution to this may be the growth of III-Sb-based active regions on $AlAs_{0.08}Sb_{0.92}$–GaSb DBRs directly on GaAs substrates; this novel approach will be discussed in Section 4.4.4. Sb-based laser structures grown on this platform would then benefit from the well-established GaAs processing technology that includes a proven substrate removal process.

4.3.2
Sample Growth and Post-Growth Analysis

Sb-based semiconductor lasers are usually grown by molecular beam epitaxy (MBE) [13]. In contrast to GaAs-based semiconductor devices, metal-organic vapor-phase epitaxy (MOVPE) [14] has not been widely used for the growth of Sb-based semiconductor lasers, although MOVPE has become established in the production of Sb-based thermophotovoltaic devices, for example. Sb-based SDL structures are grown on (100)-oriented undoped GaSb substrates; however, there have been recent efforts to realize III-Sb active regions on GaAs substrates (see Sections 4.3.4 and 4.4.4).

The epitaxial growth is performed using elemental Ga, In, and Al evaporated from conventional effusion cells as group-III elements. For high layer quality, dimeric antimony and arsenic molecules have to be used as the group-V species. Since tetrameric group-V elements are predominantly evaporated within the effusion cells, cracker cells have to be employed [15]. Within a cracker cell, tetrametric group-V elements are dissociated into As_2 and Sb_2 at elevated temperatures.

After the growth of the laser structures, the epitaxial layers are routinely analyzed by a variety of characterization techniques. These characterizations provide valuable information for accurate readjustment of the MBE growth parameters. This closed-loop design/growth/calibration process is indispensable in the development and realization of high-performance SDLs. A complete post-growth analysis of SDL structures typically includes the following characterizations: high-resolution X-ray diffraction (HRXRD), secondary ion mass spectroscopy (SIMS), photoluminescence (PL) measurements as well as reflectance spectroscopy.

Aided by HRXRD measurements, the crystalline structure of the MBE grown samples can be assessed. For example, Figure 4.3 shows the recorded HRXRD profile of a 2.25 µm emitting SDL structure where the 004 reflection of the GaSb substrate and the almost perfectly lattice-matched barrier and window layers can be clearly seen. Also the diffraction peaks arising from the compressively strained GaInAsSb MQW active region can be identified. From a detailed analysis of the measured HRXRD profiles in conjunction with simulated diffraction profiles (see lower curve

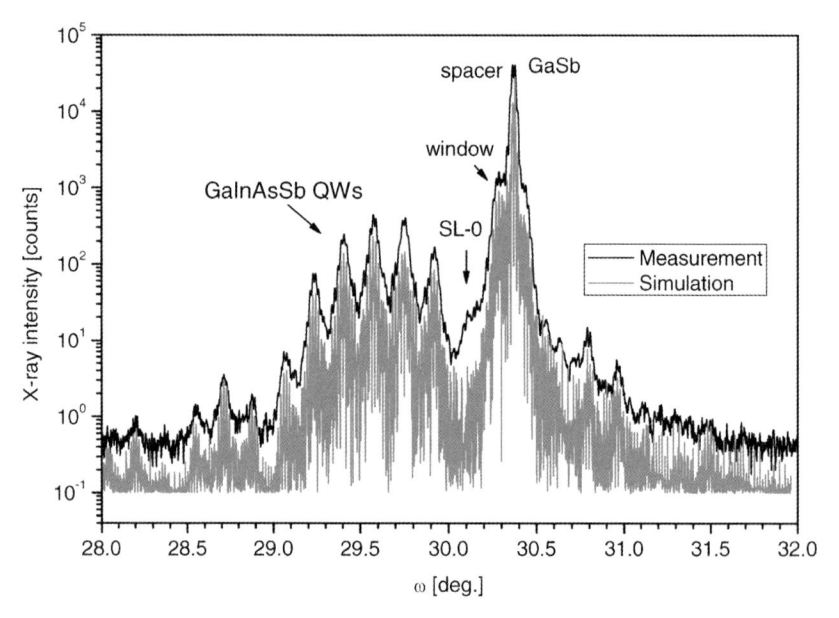

Figure 4.3 HRXRD profile (004 reflection) of a GaSb-based SDL structure designed for an emission wavelength of 2.25 µm.

in Figure 4.3), the individual thickness of the QW and barrier layers can be determined along with any individual lattice mismatch perpendicular to the growth plane $(\Delta a/a_0)_\perp$. From the latter value, the strain ε_{zz} can be determined and thus the strain effects on the band structure can be calculated.

From the HRXRD data, it can be concluded that the residual lattice mismatch of the barrier layers of the structure is below $(\Delta a/a_0)_\perp < 1 \times 10^{-3}$, indicating excellent lattice matching and high crystalline quality. A typical calibration procedure for the growth of AlGaAsSb lattice matched to GaSb is to initially grow AlGaSb with the targeted Al content and then to add As to the alloy by increasing the As flux up to the point where lattice matching is achieved.

From the data shown in Figure 4.3, a lattice mismatch $(\Delta a/a_0)_\perp$ of 2.7% was calculated for the quaternary $Ga_{1-x}In_xAs_ySb_{1-y}$ QWs. This value corresponds to a compressive strain (ε_{zz}) of 1.3%. In contrast to the amount of strain, the layer composition of the QWs cannot be determined from the HRXRD analysis alone. Additional input is needed, such as the Ga-to-In concentration ratio, which can be obtained from group-III flux calibrations. The concentrations of the group-V elements As and Sb can then be calculated from the lattice mismatch $(\Delta a/a_0)_\perp$ as obtained from the HRXRD. An alternative approach is to use the PL emission wavelength as an additional input parameter, along with the lattice mismatch, to characterize the QW active layer. Based on the PL emission wavelength and the lattice mismatch, the QW material composition can be deduced, if quantization effects within the QWs are taken into account. Unsurprisingly, this approach requires a sound knowledge of the material data in the (AlGaIn)(AsSb) material system. An example of such data can be found in Ref. [3].

Another important characterization tool is SIMS, with which the thickness of the individual epitaxial layers within the SDL structure can be determined to an accuracy of 5–10%. This characterization instrument should not be underestimated since deviations from the intended layer thickness can be recognized quickly, especially for comparatively thick layers. In particular, it can be verified that the barrier layers exhibit the correct overall thickness, thus ensuring that the resonance condition is fulfilled at the targeted wavelength. Although the QWs cannot be completely resolved by SIMS measurements, the respective position of the QW groups within the microcavity can be determined. In this way, and importantly, verification of the RPG layout can be confirmed.

In comparison to SIMS measurements, systematic deviations in layer thickness can be analyzed more precisely using reflectance measurements, thus yielding exact recalibration factors for the MBE growth. Reflectance spectra (see Figure 4.4) for Sb-based SDL structures are recorded using a Fourier transform infrared (FTIR) spectrometer system incorporating a thermal-emitting light source and a pyroelectric detector. Similar to HRXRD measurements, quantitative information can only be obtained if the spectra can be accurately simulated. This, of course, requires in-depth knowledge of the optical constants, including the refractive index and its dispersion. Measured reflectance spectra can then be reconstructed through systematically fitting layer thickness with suitable transfer matrix method simulation software. Typical reflectance spectra of a 2.25 μm SDL structure measured at different heat sink

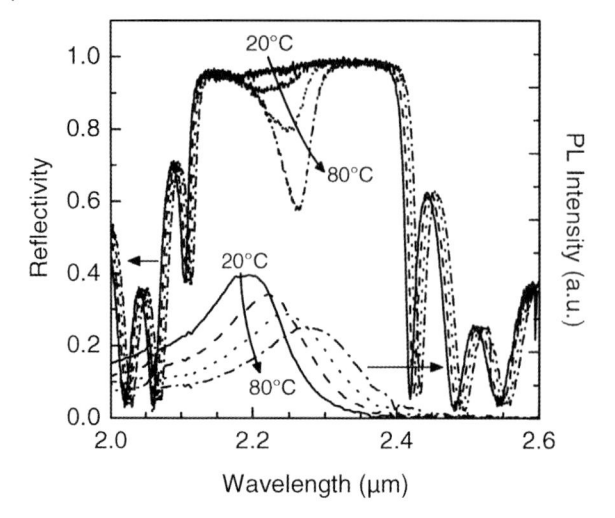

Figure 4.4 Reflectivity and edge PL emission spectra of a 2.25 μm SDL structure for several heat sink temperatures – 20, 40, 60, and 80 °C.

temperatures are shown in Figure 4.4. Within the stop band of the DBR, a clear absorption dip is visible. The spectral position of this reflectance minimum does not necessarily coincide with the resonance wavelength of the microcavity. In fact, the absorption dip results from a superposition of the QW absorption and the microcavity resonance. Since both shift at a different rate with temperature, a change in the shape and spectral position of the dip can be observed within the set of spectra displayed in Figure 4.4. From such temperature-dependant measurements, the spectral position of the microcavity resonance at room temperature can be derived more precisely.

The peak gain wavelength of the quantum wells can be investigated by means of photoluminescence measurements. For this characterization method, the beam from a low-power ($\approx$10 mW) Nd:YVO$_4$ laser emitting at a wavelength of 1064 nm is focused on the surface of an SDL chip. The PL emitted from the edge of the SDL chip is then collected by the FTIR spectrometer. In this way, the shape of the PL spectrum corresponds to the spectral distribution of the QW gain without interference from the DBR or microcavity. Typical PL spectra recorded for different heat sink temperatures are also shown in Figure 4.4. As it can be seen in Figure 4.4, the maximum of the PL emission spectrum is red shifted at increasing temperature. From these spectra, a temperature-tuning coefficient of 1.25 nm K^{-1} was calculated, which corresponds to a tuning rate of 2.5 cm^{-1} K^{-1} in wave numbers. These values are typical for QWs in the (AlGaIn)(AsSb) material system. In contrast, the temperature shift of the microcavity resonance is determined by the temperature-dependent change in the refractive index and as such exhibits a different temperature-dependent tuning behavior. For the Sb-based SDL structures, the tuning coefficient of the resonance is roughly 0.3 nm K^{-1}, which is significantly lower than the tuning rate of the gain peak. In SDL structures optimized for high-power continuous-wave (cw) operation, the QWs are designed such that the gain peak at room temperature is blue shifted with respect to the spectral position of the cavity resonance to account for this differential temperature-dependent tuning [16]. In this way, the gain peak and the cavity resonance coincide at a higher temperature and the SDL is thus optimized for high-power operation with a concomitant higher thermal load. The PL and reflectance spectra shown in Figure 4.4 were recorded for a laser structure optimized for an emission wavelength of 2.25 μm. At room temperature, the gain curve of this structure peaks at a wavelength of 2.19 μm corresponding to an offset of 60 nm from the microcavity resonance.

4.3.3
Epitaxial Design of In-Well-Pumped SDLs

As an alternative approach to the barrier-pumped SDL structures mentioned above, the concept of in-well pumping has attracted significant attention in recent research on Sb-based SDLs [11, 17]. The objective of this approach is to reduce the difference between the pump wavelength and the laser wavelength – also known as the quantum defect. In most SDL concepts, the gain medium is pumped by using standard commercially available GaAs-based diode lasers emitting either at 808 or 976 nm typically used as pump sources in Nd-doped solid-state lasers or Erbium-doped fiber amplifiers. For long-wavelength SDLs, however, the considerable quantum defect becomes a serious issue since the corresponding energy is converted into waste heat that can severely limit the SDL performance. For this reason, the employment of alternative pump lasers capable of reducing the quantum defect has been recently investigated. For in-well pumping, the pump wavelength is chosen such that pump radiation is only absorbed in the QW layers, whereas negligible absorption takes place in the barrier layers (see Figure 4.5). This concept has been investigated in

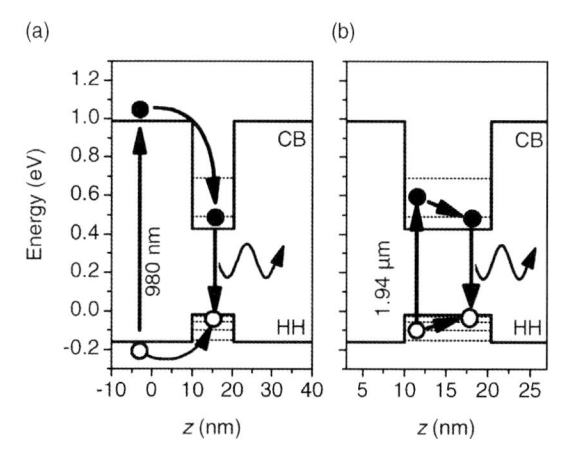

Figure 4.5 Band edge profile (solid lines) and energy levels (dotted lines) of a $Ga_{0.65}In_{0.35}As_{0.10}Sb_{0.90}$ QW (emission wavelength: ≈2.35 μm) embedded between $Al_{0.30}Ga_{0.70}As_{0.02}Sb_{0.98}$ barrier layers. (a) Electronic transitions for barrier pumping at a pump wavelength of 980 nm. (b) Electronic transitions for in-well pumping at 1.94 μm.

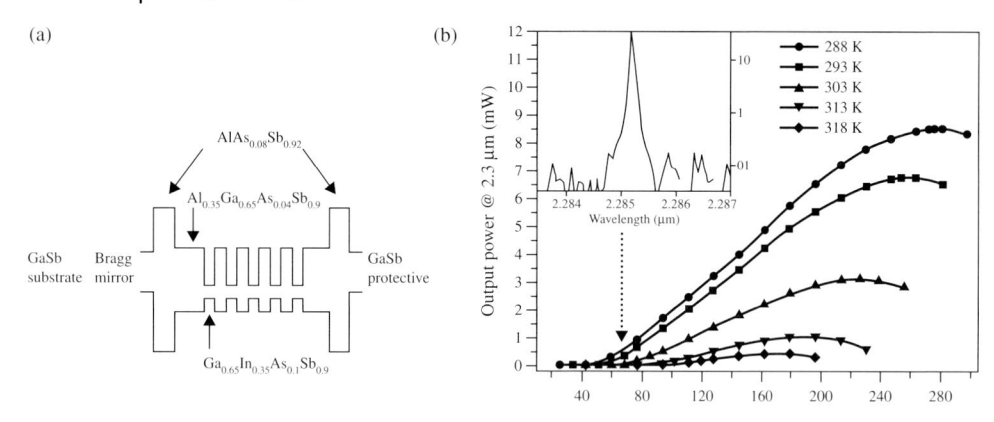

Figure 4.9 (a) Schematic energy bandgap diagram of the SDL structure realized by Cerutti et al. [31]. The microcavity used in this design had a total thickness corresponding to one wavelength and incorporated five QWs. (b) Cw output power versus incident pump power at various heat sink temperatures using an output coupler with $R = 99.2\%$. Inset: Normalized laser spectrum at $T = 288$ K and at low power.

Fresnel losses and the limited pump absorption ($\approx 80\%$) into account, the differential efficiency increased to $\approx 8.1\%$.

Considering the limited performance of these devices, it became clear that an improved heat extraction scheme was required to achieve output powers in the watt range. For InGaAs-based SDLs, one route to high-power operation is to directly mount the mirror side to a heat sink. For this purpose, the active region is grown prior to the DBR layers and the substrate is then removed after mounting. No attempts to transfer this technology to Sb-based SDLs have been reported so far. This can be partially explained by a lack of a proven substrate removal technology for III-Sb materials grown on GaSb. However, in-depth thermal modeling has revealed that mounting the mirror side down is inappropriate for Sb-based SDLs [29], as even with the substrate removed, the considerable thermal resistance of the DBR would impede efficient heat removal from the active region. Using hybrid mirrors with broadband metal coatings may somewhat improve the performance of such an approach; however, the use of intracavity heat spreaders has been proven to be a successful heat extraction technique for Sb-based SDLs.

4.4.2
Sb-Based SDLs Using Intracavity Heat Spreaders

The first results on Sb-based SDLs employing intracavity heat spreaders were reported by Schulz et al. in 2006 [32]. Within the scope of these experiments, an SDL structure emitting at 2.35 μm was investigated. This structure incorporated a DBR composed of 21.5 GaSb–AlAsSb layer pairs followed by a microcavity consisting of nine 10 nm $Ga_{0.64}In_{0.36}As_{0.10}Sb_{0.90}$ QWs embedded between $Al_{0.30}Ga_{0.70}As_{0.02}Sb_{0.98}$ barrier layers. To achieve more than 90% absorption of the pump light, the active region was designed to have a thickness of $6 \times \lambda/2$ (where λ equals

2.35 μm). The microcavity was terminated by an $Al_{0.85}Ga_{0.15}As_{0.06}Sb_{0.94}$ top window to ensure efficient confinement of carriers generated in the active region. The laser chip was pumped by a Nd:YVO$_4$ laser emitting at 1.064 μm. When the laser chip was mounted substrate side down to a heat sink, cw performance at room temperature (18 °C) similar to the SDL in Ref. [31] was achieved. In this case, output powers in the mW range were obtained at a differential efficiency (with respect to the incident pump power) of ≈4% and low threshold pump power densities of 0.73 kW cm^{-2} (for an output coupler reflectivity of 99.4%) were observed. A considerable increase in maximum output power to 0.6 W (at a heat sink temperature of −18 °C) was then achieved when a synthetic polycrystalline diamond heat spreader was liquid capillary bonded [33] to the surface of the SDL chip. As in InGaAs SDLs, these results prove the potential of this mounting technology to upscale the output power.

Following this result, high-power Sb-based SDLs were reported by a number of other groups. In 2006, Härkönen *et al.* published results on a 2.0 μm SDL capable of emitting ≈1 W cw output power at a heat sink temperature of 5 °C [34]. In this case, the SDL chip consisting of 15 $Ga_{0.78}In_{0.22}Sb$ wells of 8 nm thickness between GaSb barrier layers was bonded to a single-crystalline natural diamond heat spreader. The microcavity was terminated by a 18-pair GaSb–AlAsSb DBR and a AlAsSb confinement layer, respectively. This Sb-based SDL was pumped by a fiber-coupled diode laser emitting at 790 nm. During laser operation, a differential efficiency with respect to the incident pump power of ≈6% was measured at a heat sink temperature of 20 °C for an output coupler reflectivity of 98%. Taking into account the pump spot diameter given in Ref. [34], the threshold pump density under these experimental conditions amounted to 4.7 kW cm^{-2} (Figure 4.10).

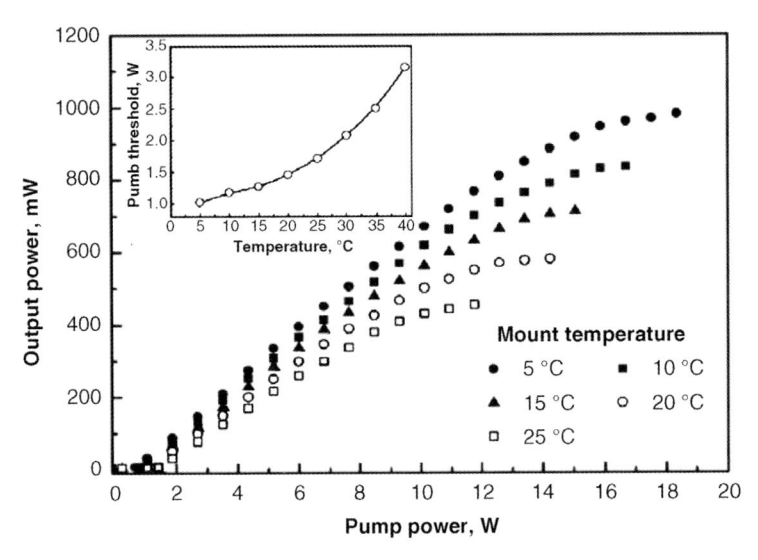

Figure 4.10 Light output characteristics of the 2.0 μm SDL presented in Ref. [34] for different submount temperatures. An output coupler reflectivity 2% was used. *Inset*: Threshold pump power in dependence of the submount temperature.

Using this SDL chip, first mode-locking experiments with Sb-based SDLs were performed [35]. Here, the SDL chip was synchronously pumped at various harmonics of the fundamental cavity frequency ($\approx$460 MHz). In this way, a pulse length of $\approx$160 ps was achieved at a pulse repetition frequency of 2.8 GHz. Slope efficiencies of up to 14% were measured at a heat sink temperature of 15 °C – this increase over Ref. [34] was largely due to the reduced quantum defect afforded by the use of a 1.57 µm pump source.

Recent developments have been characterized by the continued optimization of the underlying Sb-based structures with a view to further increasing both the output power and the efficiency. In 2008, the first results on laser operation of Sb-based SDLs with a multiwatt output power were published [36, 37]. A key feature of these SDL structures was the precise control of the temperature-dependent modal gain spectrum (the product of the material gain and the longitudinal confinement factor Γ). As mentioned in Section 4.3.2, the spectral position of the maximum material gain red shifts with increasing temperature at a significantly higher rate than the microcavity resonance. However, through careful spectral control of the relative room temperature positions of the material gain maxima and the microcavity resonance, the modal gain characteristic can be optimized for any active region temperature. For the Sb-based structures reported in Refs [36, 37], the room temperature gain maxima were intentionally blue shifted with respect to the microcavity resonance by approximately 60 and 40 nm, respectively. From the measured temperature tuning rates of approximately 1.25 and 1.0 nm K^{-1}, respectively, for the gain maxima and 0.3 nm K^{-1} for the microcavity resonance, these offset values correspond to an optimum operating temperature T_{res} of around 70 °C. Such elevated temperatures are readily achieved within the active region during cw operation.

The introduction of a negative gain peak offset at room temperature results in an optimized temperature characteristic of the threshold pump power. This can be seen in Figure 4.11 where the threshold pump power density at different heat sink temperatures is displayed for the SDL structure in Ref. [37]. For this SDL structure,

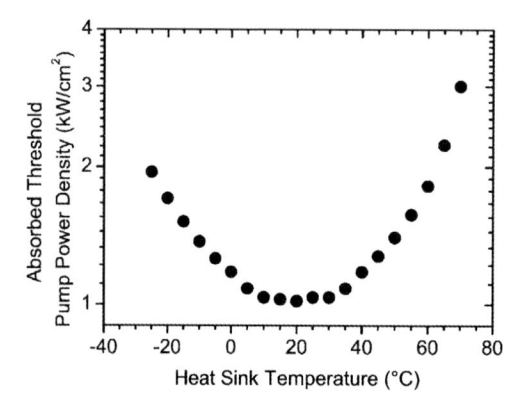

Figure 4.11 Absorbed threshold pump power density as a function of heat sink temperature recorded for a 2.25 µm SDL with an output coupling mirror reflectivity of 95% [37].

minimum threshold pump power and minimum temperature sensitivity of the threshold pump power are observed close to room temperature. Generally, the temperature characteristic of the threshold pump power is determined by the temperature dependence of the modal gain and of various loss mechanisms. On one hand, the temperature tuning of the cavity resonance and the wavelength of maximum QW gain tend to maximize the modal gain at T_{res} (see above). On the other hand, the decrease in the peak value of the QW gain and the increase of the various loss mechanisms with temperature tend to increase the threshold power monotonically with rising temperature. The combination of both effects leads to a minimum in the threshold power at a temperature T_{min} well below T_{res} [38]. For the SDL structure in Ref. [37], the minimum threshold pump power density of 1.02 kW cm^{-2} was recorded at a heat sink temperature $T_{min} = 20\,°C$ for an output coupler reflectivity of 95%.

Power transfer curves for different heat sink temperatures are shown in Figure 4.12b In these experiments, the gain medium was operated in a two-mirror cavity and a 980 nm fiber-coupled diode laser was used as the pump source. When using an optimum output coupler reflectivity of 95%, a maximum output of 3.4 W was measured at $-10\,°C$ for an absorbed pump power of 21 W. From the linear part of the power transfer curve measured at $-10\,°C$, a differential efficiency with respect to the incident pump power of 19.0% was calculated. Since approximately 20% of the incident pump power was reflected by the gain medium and the heat spreader, this corresponds to a differential efficiency with respect to the absorbed pump of 24%. From this value, a differential quantum efficiency of 54.5% can be derived. For higher heat sink temperatures, the differential efficiency with respect to the incident pump power slightly decreased to $\approx 18\%$ at $20\,°C$; nevertheless, a maximum output power of $>1.6\,W$ at $20\,°C$ could still be observed.

Almost simultaneously to the publication of these results on 2.25 μm Sb-based SDLs, the state-of-the-art at 2.0 μm emission wavelength was redefined with a report on new high-performance SDLs also capable of multiwatt output power [36]. Similar to the design presented in Ref. [34], compressively strained ternary QWs were used in this structure (10 $Ga_{0.74}In_{0.26}Sb$ layers of 10 nm thickness). However, the QW layers were embedded between $Al_{0.30}Ga_{0.70}As_{0.02}Sb_{0.98}$ barrier layers replacing the GaSb layers in Ref. [34]. The microcavity exhibited a total optical length of $6\lambda/2$ and the whole structure included a DBR consisting of 22.5 GaSb–AlAsSb layer pairs. As in the above-mentioned 2.25 μm SDLs, the spectral position of the QW gain maximum was intentionally blue shifted with respect to the microcavity resonance at room temperature. Figure 4.12a shows the power transfer characteristic recorded for different heat sink temperatures of this SDL chip bonded to a diamond heat spreader. Here, the gain medium was operated in a three-mirror resonator and pumped by a 980 nm fiber-coupled diode laser. When the submount was water/glycol cooled to $-15\,°C$, a maximum output power up to 6 W was recorded from this 2.0 μm SDL [39]. At this temperature, a slope efficiency with respect to the incident pump power of 26% was calculated. At a heat sink temperature of $21\,°C$, a maximum output power of 3.3 W was measured before the onset of thermal rollover. At this temperature, the slope efficiency was reduced somewhat to 22%.

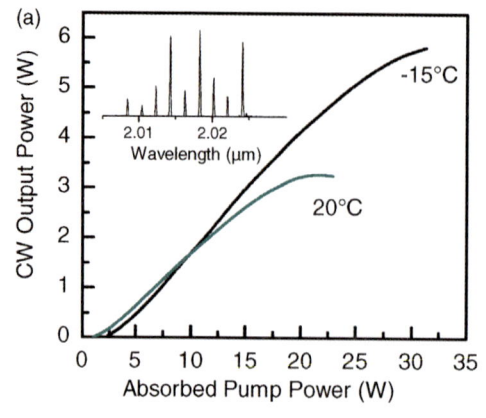

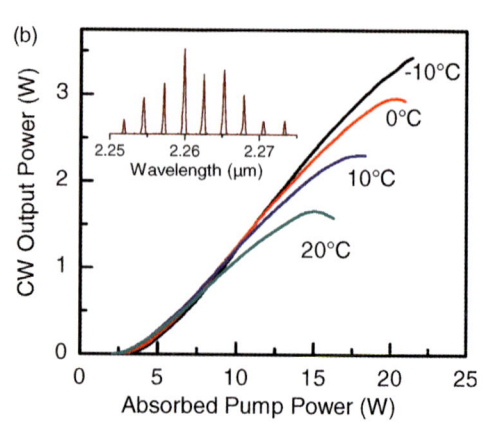

Figure 4.12 (a) Cw output power characteristics of a GaSb-based SDL emitting at 2.0 μm for several heat sink temperatures [36]. (b) Cw output power characteristics of a GaSb-based SDL emitting at 2.25 μm for several heat sink temperatures [37]. The insets in both graphs show typical emission spectra. In both (a) and (b) cases, the pump wavelength was 980 nm.

More recently, comparable output power characteristics at 2.0 μm emission wavelength were reported [40] for an SDL structure similar to that presented in Ref. [34]. The active region of this structure incorporated 15 $Ga_{0.80}In_{0.20}Sb$ QWs embedded between GaSb barrier layers. Here it is noteworthy that the microcavity was not resonant at the design wavelength of this structure. As discussed extensively in Ref. [41], such an "antiresonant" design results in a lower modal gain compared to resonant designs; however, the modal gain bandwidth is increased, which can result in an enhanced tunability (see Section 4.5). For the SDL presented in Ref. [40], diamond intracavity heat spreaders were used for efficient heat extraction. Furthermore, the gain chip was pumped using a 980 nm fiber-coupled diode laser. When operated in a three-mirror cavity, with an output coupler reflectivity of 98%, up to 3.5 W was recorded at a heat sink temperature of 20 °C. Similar to Ref. [36], a slope efficiency of approximately 20% was observed at this temperature. Published simultaneously with these results, a new design approach using QW layers with different thicknesses in a single structure was demonstrated [40]. This structure designed for enhanced tunabilty will be discussed in more detail in Section 4.5.

In the context of high-power Sb-based SDLs, it is worthwhile to mention some recent experiments in which the possibilities of power scaling have been further explored.

While most of the experiments described in this section have been carried out in cw operation, there have also been investigations on pulsed pumping of Sb-based SDLs. In Ref. [42], a low-cost pulsed diode laser emitting at 905 nm was used as the pump source. This diode laser is capable of producing up to 75 W "on-time" output power in a square pulse of up to 200 ns duration at a duty cycle of <0.1%. In these experiments, it was shown that significant and rapid heating of the SDL chip takes place, however, even so up to 16 W peak output power could be obtained for a 2.0 μm

SDL [39]. These sources are now being used as pump sources for broadly tunable Cr:ZnSe solid-state lasers [43].

With regard to further power scaling, promising results of multichip Sb-based SDLs have recently been published [44]. In these experiments, the output power of an SDL was almost doubled when two SDL chips were used in a single resonator. In this way, up to 3.3 W output power at a submount temperature of 20 °C was recorded for a dual-chip SDL emitting at 2.25 µm and over 6 W of output power was possible from a dual-chip 2.0 µm SDL at room temperature [45].

4.4.3
In-Well-Pumped Sb-Based Semiconductor Disk Lasers

The SDLs discussed previously are referred to as "barrier-pumped" devices, that is, the absorption of the short-wavelength (≈ 1 µm) pump radiation predominantly takes place in the barrier layers surrounding the quantum wells. Typically, in the thick barrier layers, the incident pump radiation can be almost completely absorbed on a single pass. However, the quantum defect is generally high for barrier-pumped SDLs (e.g., the quantum defect amounts to $\approx 57\%$ in a 2.3 µm SDL pumped at 980 nm). This large proportion of waste pump power causes strong device heating that severely limits high-efficiency cw SDL operation. The use of in-well optical pumping, where the pump photons are only absorbed, in the QWs instead of in the barrier layers (see Section 4.3.3) can be used to significantly reduce the quantum defect and, as such, the parasitic heating. For this purpose, pump sources emitting at wavelengths closer to the SDL emission wavelength have to be employed. As discussed in Section 4.3.3, in-well-pumped SDLs typically feature an optimized layer design to ensure sufficient pump absorption within the QWs. Such a design typically involves a dual-band DBR as well as a doubly resonant microcavity.

In-well-pumped Sb-based SDLs were first presented in 2006 by Schulz et al. [17]. This SDL structure was optimized for a pump wavelength of 1.92–1.98 µm and an emission wavelength of 2.35 µm, corresponding to a quantum defect of only $\approx 17\%$. The active region of the structure incorporated 15 compressively strained $Ga_{0.64}In_{0.36}As_{0.10}Sb$ QW layers of 10 nm thickness surrounded by $Al_{0.30}Ga_{0.70}As_{0.02}Sb_{0.98}$ barrier layers. The microcavity was designed to exhibit resonances both at the laser and the pump wavelengths for the intended angle of incidence. In addition, a high pump absorption and a high overlap of the laser mode with the individual QWs were simultaneously achieved by placing the QWs in regions of maximum spatial overlap between the pump and the laser standing-wave pattern (see Figure 4.6). The layer structure included a two-band DBR consisting of 24.5 GaSb–AlAsSb pairs. For initial experiments, a thulium-doped fiber laser emitting at 1.96 µm was used as a pump source. The highest pump absorption measured within the structure was 60% for an angle of incidence of 35° at 1.96 µm. However, after the attachment of a SiC heat spreader, the pump absorption was reduced to 25% due to the reduced contrast in refractive index at the interface between the gain chip and the heat spreader. By refocusing the pump light reflected from the surface of the SDL chip through the

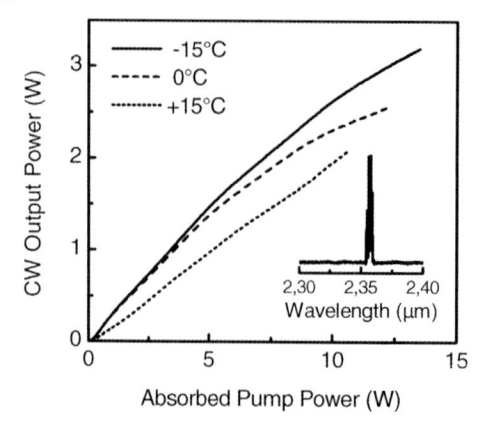

Figure 4.13 Heat sink temperature-dependent cw power transfer characteristics of an in-well-pumped SDL employing a SiC intracavity heat spreader. The inset shows a laser spectrum recorded at a heat sink temperature of −15 °C and an absorbed pump power of 10 W.

active region, a total pump absorption of 40% was achieved. In Figure 4.13, the cw power transfer characteristics of this in-well-pumped SDL is shown. It can be seen that there is no indication of a thermal rollover within the range of available pump power – thus reflecting the beneficial impact of the significantly reduced quantum defect on the laser performance. At a heat sink temperature of 15 °C, a slope efficiency with respect to the absorbed pump of 20% was observed.

For the in-well-pumped active region designs, the barrier material composition is more flexible since it does not have to be optimized to provide high absorption of the pump radiation. Therefore, barrier material compositions exhibiting a larger bandgap energy and capable of significantly enhancing the electron and hole confinement in the QWs can be employed. In the (AlGaIn)(AsSb) material system, the highest lattice-matched alloy bandgap energy is achieved for $AlAs_{0.08}Sb_{0.92}$ barriers. As discussed in Section 4.3.1, typical QWs embedded between such barrier layers exhibit a valence band offset approximately 330 meV higher than QWs between GaSb barriers. This provides much stronger hole confinement, significantly reduces carrier leakage within the QWs, and consequently improves laser performance in terms of reduced threshold pump power, increased efficiency, and reduced temperature sensitivity.

The laser operation of a 2.25 μm Sb-based in-well-pumped SDL structure employing $AlAs_{0.08}Sb_{0.92}$ barrier layers and 13 10 nm $Ga_{0.66}In_{0.34}As_{0.08}Sb_{0.92}$ within the active region has recently been demonstrated [11]. Apart from the barrier layers and a GaSb window layer terminating the active region, the basic structural layout was similar to that of the in-well-pumped SDLs described above. Again, the SDL chip design was optimized for pumping with a Tm:fiber laser emitting at 1.96 μm and therefore featured a dual-band DBR as well as a doubly resonant microcavity. Improved heat extraction was again enabled by using an intracavity SiC heat spreader.

As mentioned previously, the main advantage of using $AlAs_{0.08}Sb_{0.92}$ barrier layers arises from an increased hole confinement due to the increased valence band offset in the QWs. However, there are further benefits of using such barriers in combination with $Ga_{0.66}In_{0.34}As_{0.08}Sb_{0.92}$ QWs and a GaSb window layer as this can also lead to a strong resonant enhancement of the QW gain and absorption. Compared to the in-well-pumped SDL structure of Ref. [17] for which a longitudinal confinement factor of $\Gamma = 6$ was determined for the pump and laser microcavity resonances, for the SDL structure using $AlAs_{0.08}Sb_{0.92}$ barriers the confinement factor increased to ≈ 10. The enhanced optical confinement can be attributed to the larger difference in the refractive index between the various active region layers. As a result of this increased confinement, a pump absorption of more than 50% could be achieved with this active region concept without using any external pump retroreflection. In comparison to the in-well-pumped SDL with $Al_{0.3}Ga_{0.7}As_{0.02}Sb_{0.98}$ barriers, the pump absorption was increased by a factor of 2, even though the number of pump-absorbing QWs had been reduced.

Figure 4.14 shows the cw output power as a function of the absorbed power pump for several values of the output coupling mirror transmission. The maximum slope efficiency obtained was 32.1% at an output coupler transmission of 10.3% and a heat sink temperature of 20 °C, being significantly higher than that of 20% at 15 °C for the previous in-well-pumped SDL reported in Ref. [17]. In particular, the efficiency with respect to the incident pump power is more than doubled as a result of the higher pump absorption combined with the higher efficiency of the SDL structure itself.

In conclusion, the first in-well-pumped Sb-based SDLs have shown very promising performance. Exploitation of microcavity resonances for an enhancement of the QW gain and pump absorption has enabled high-performance laser operation and high single-pass absorption of the pump radiation. The in-well-pumped active region designs, employing $AlAs_{0.08}Sb_{0.92}$ barrier layers, have been shown to be particularly

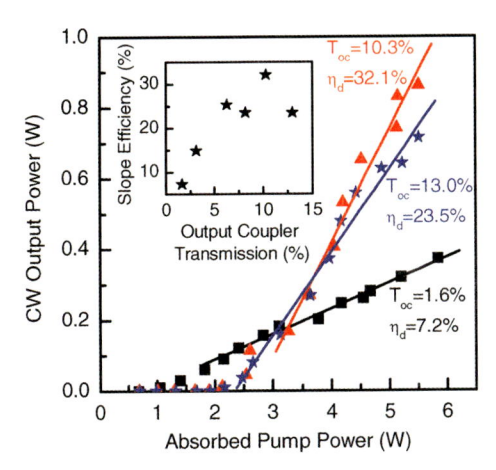

Figure 4.14 Cw output power of the in-well-pumped SDL with $AlAs_{0.08}Sb_{0.92}$ barriers for output coupling mirror transmissions T_{oc} of 1.6% (black squares); 10.3% (red triangles); and 13% (blue asterisks), respectively, at a heat sink temperature of 20 °C. The full lines indicate the linear fit functions used to obtain the respective slope efficiencies, η_d. The values for η_d and T_{oc} are displayed at the side of each power curve. In the inset, the measured slope efficiency is displayed as a function of the output coupling mirror transmission.

advantageous for increased device performance. Although significant improvements have also been recently reported for barrier-pumped Sb-based SDLs, the concept of in-well pumping is still of particular interest, especially when targeting longer emission wavelength of 3 μm and above. For future compact, in-well-pumped SDL modules, Sb-based diode lasers can be used as pump sources. The performance of such diode laser has advanced over the past couple of years and should still continue to improve in the near future.

4.4.4
Sb-Based Semiconductor Disk Lasers on GaAs Substrates

As an initial step toward the realization of Sb-based SDLs on GaAs substrates, active regions for photoluminescence studies have been grown on both GaAs and GaSb substrates by Balakrishnan et al. [28]. These active regions designed for a target wavelength around 2 μm incorporated QWs formed by $Ga_{0.75}In_{0.25}Sb$ layers between $Al_{0.3}Ga_{0.7}Sb$ barrier layers. Photoluminescence spectra recorded for test samples on each substrate are shown in Figure 4.15. Since the peak PL intensity recorded with the active region grown on a GaAs substrate is even higher than the PL intensity of the same structure grown on a GaSb substrate, it has been concluded that excellent structural quality can be achieved with the IMF growth technique. Data from photomodulated reflectance measurements performed with these active regions on GaAs and GaSb furthermore delivered some evidence that a defect state present in the structures on GaSb is strongly suppressed in the structures grown on GaAs [46].

After these initial experiments, a similar active region was grown on top of GaAs–AlGaAs DBRs (center wavelength: 2 μm) on GaAs substrates [47]. Within the laser setup used in the following characterization, a 980 nm fiber-coupled diode laser module served as pump source. For efficient heat extraction, SiC heat spreaders were bonded to the diced chips. Power output characteristics of such an SDL are shown in Figure 4.16. It can be seen that cw operation is readily achieved at submount

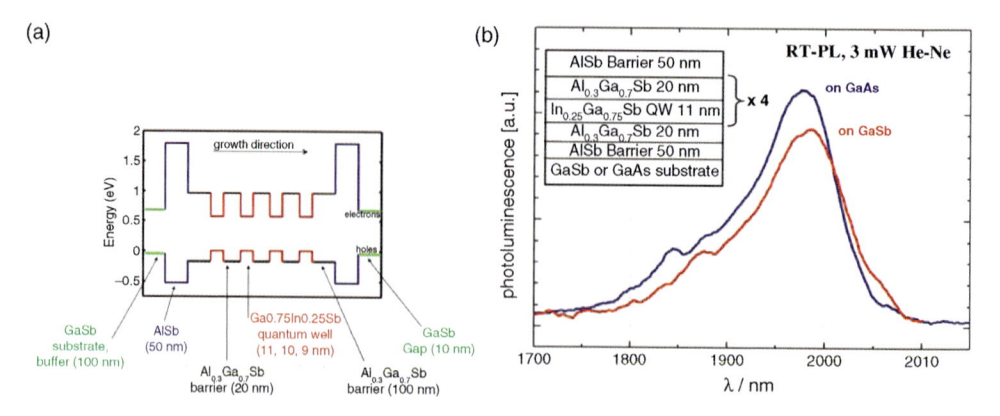

Figure 4.15 (a) Schematic cross section of active material grown either on GaAs or on GaSb substrate using the respective sublayers. (b) PL spectra recorded with both active region structures (graphs taken from Ref. [28]).

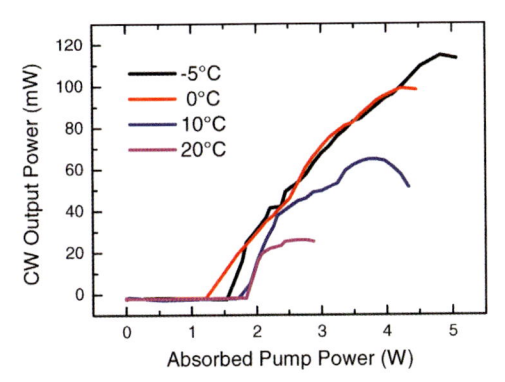

Figure 4.16 Output power curves at different submount temperatures recorded for a Sb-based active region structure grown on top of a AlGaAs–GaAs DBR on a GaAs substrate. An output coupler reflectivity of 99% was used in this laser setup.

temperatures of up to 20 °C. With these first III-Sb SDLs grown on GaAs substrates, a maximum output power of approximately 110 mW was achieved at a submount temperature of −5 °C. Compared to the still more mature III-Sb SDLs grown on GaSb substrates, there is certainly still some need for further optimizations, particularly when considering the comparatively high threshold pump power (approximately 11 kW cm^{-2} at 20 °C for a output coupler reflectivity of 99%) and the relatively low slope efficiency. Nevertheless, these first results are very promising, particularly when considering the prospect of a simplified heat sinking in such structures. As mentioned in Section 4.3.4, Sb-based SDLs on GaAs substrates could benefit from substrate removal processes that have already been established in the GaAs system. This technology as well as the smaller costs of the GaAs substrates would enable to produce these Sb-based SDLs in higher quantities at a reasonable pricing.

4.5
Tunable, Single-Frequency Lasers

Much like a conventional solid-state laser geometry, the SDL lends itself well to the precise optical, spectral, and temporal control via the ability to easily insert additional optical elements into the extended cavity [48]. The tailorability this offers is crucial for a wide variety of applications, including the sensitive detection and spectroscopy of industrially significant atmospheric chemicals. This is especially true for longer wavelength operation in the mid-IR where many molecules exhibit strong overtone and fundamental absorptions [49]. The ability to provide tunable narrow-line and single-frequency operation over a wide spectral region in a compact and relatively low-cost laser is a key advantage of SDL technology for many applications in this field. Indeed, the first >2 µm single-frequency SDLs were used for the demonstration of methane detection [51] and subsequently for water [1], carbon dioxide [1], and carbonyl sulfide [52]. Figure 4.17 shows the absorption spectra of many

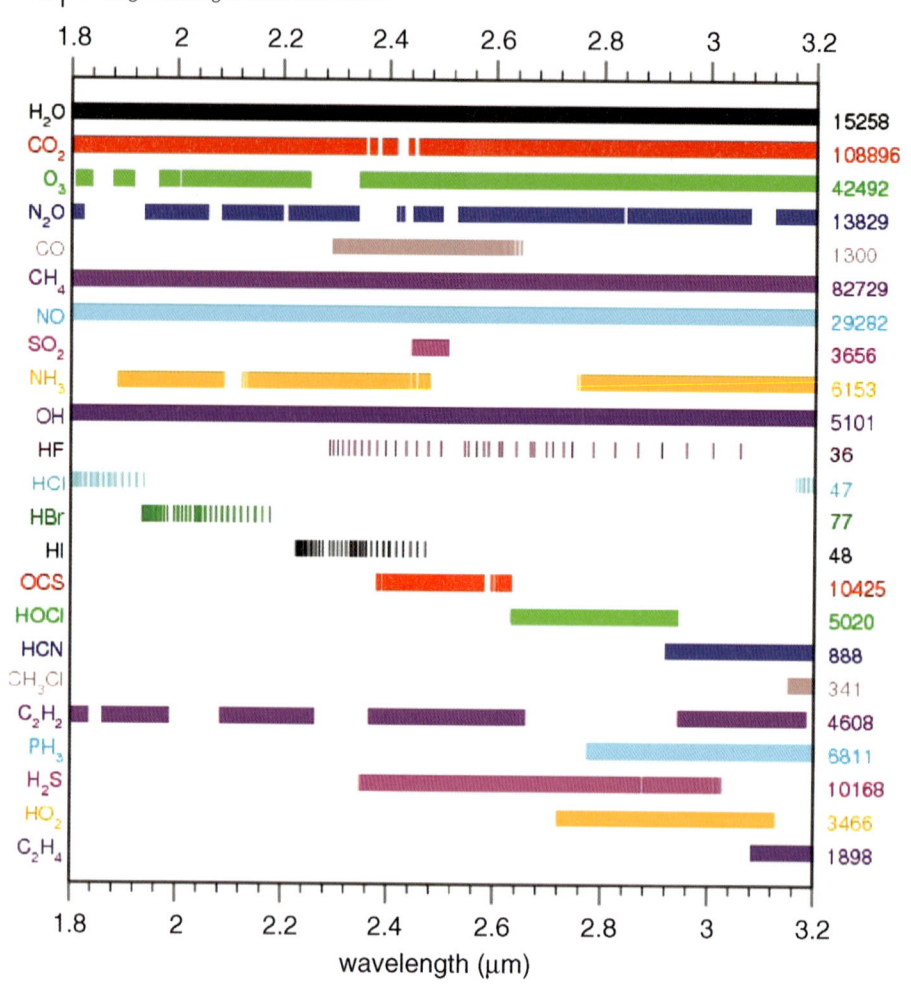

Figure 4.17 Absorption lines for selected chemicals in the 1.8–3.2 μm waveband taken from the Hitran 2004 database [53] – the numbers on the right correspond to the number of discrete absorption lines.

atmospherically significant chemicals in the 2–3 μm spectral region, including CO, CH_4, HF, and HBr.

4.5.1
Tunability

The wide gain bandwidth associated with multiple quantum well (MQW) gain media, coupled with the structural design and extended cavity of the SDL, results in a laser format that offers the potential of wide tunability and broad bandwidth operation. This is a major advantage over many conventional solid-state gain media that exhibit relatively narrow emission bandwidths. The wide bandwidths have been shown to be highly advantageous for compact and potentially low-cost intracavity laser absorption spectroscopy (ICLAS) systems [54]. Traditionally, however, the majority of SDLs have been optimized for output power performance at a specific wavelength by design. This is achieved via the use of a resonant subcavity (formed between the bottom Bragg mirror and the air/semiconductor interface) and the specific distribution of the quantum wells within the structure to exploit this resonance at a particular wavelength by placing the wells at the antinodes of the intracavity standing-wave field (the so-called resonant periodic gain already described). The use of such structural enhancements to the modal gain, however, impact on the spectral width of the modal gain [55]. Despite this, Sb-based SDLs have been shown to exhibit relatively large tunability of over 170 nm (equivalent to 400 cm^{-1} in wave numbers), which compares very well to the tunability of shorter wavelength SDLs. A more detailed summary of SDL tunability can be found in section 2.4.3.

There are a number of standard methods to tune laser resonators containing broadband gain media, including the use of angle-tuned diffraction gratings [56, 57], etalons [58], and birefringent filters [59]. In SDLs, however, the internal losses must be minimized as the thin active region results in a moderate or low single-pass gain, and as such methods such as intracavity diffraction gratings that introduce high losses can be ruled out. Typically, in SDLs, tunability has been implemented via the use of a birefringent tuning element placed at Brewster's angle within the laser cavity, for example, a quartz plate. The plate then acts as a polarization rotator placed between two analyzers (provided by the Brewster surfaces) and provides a wavelength-dependent loss function. Low losses occur at wavelengths where no net polarization rotation is accumulated relative to the tangential plane of the resonator. The rotation of the birefringent filter (BFR) then allows wavelength tuning. The wavelength selectivity can be increased while still maintaining a wide free spectral range by using a multiple plate birefringent filter (Lyot filter), which is particularly useful for broadband gain media.

In this manner, the tuning performance of a selection of GaSb SDL structures (as already described) designed to emit at wavelengths of 1.9, 2.0, 2.25, and 2.3 μm (denoted samples 1–4, respectively) was assessed. The SDL chips were bonded to diamond heat spreaders and mounted in cooled brass compression mountings. These were then placed in a three-mirror cavity consisting of a 100 mm radius of curvature high-reflecting mirror and a plane/wedged output coupler (see Figure 4.18).

The laser performance for a range of output coupler reflectivities, from 91 to 99%, was then assessed. The optical pump was a $\lambda = 980$ nm, 100 μm core fiber-coupled diode laser. In all cases, tuning was promoted by a 2 mm single-plate quartz BRF in the long arm of the resonator. In Figure 4.19, the typical free-running power spectrum of the SDL is reproduced, where the characteristic spectral modulation induced by the heat spreader acting as an etalon can clearly be seen. In this case, the so-called diamond etalon was ≈250 μm thick consistent with the resultant spectral modulation period of ≈3 nm at 2.0 μm. The rotation of the BRF about its normal allowed a tuning range of >100 nm. In addition, varying together the heat sink temperatures and output coupler reflectivity, the maximum wavelength tuning range

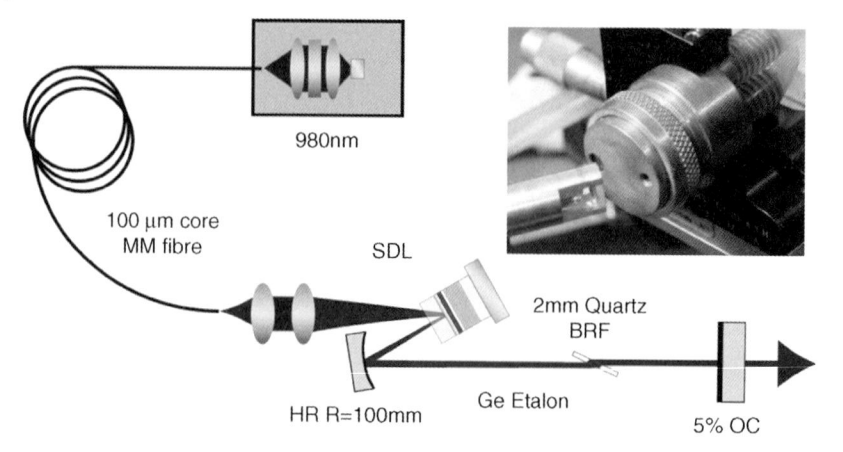

Figure 4.18 Schematic of the three-mirror 2.3 μm SDL resonator arrangement showing the 980 nm fiber-coupled diode laser pump and the 2 mm quartz intracavity birefringent filter (BRF) used for tuning, the inset photograph shows the water-cooled brass mounding ring housing the SDL chip and diamond heat spreader.

for any given SDL was extended to 150–170 nm. Figure 4.20 summarizes the spectral tuning performance obtained for four SDL samples – the individual device characteristics are summarized in Table 4.2.

To asses the full tuning range of any given sample, the SDL is tested in a variety of configurations. Obtaining the longest wavelength for each device requires operating the device at the highest possible temperature, whereas the shortest wavelengths are

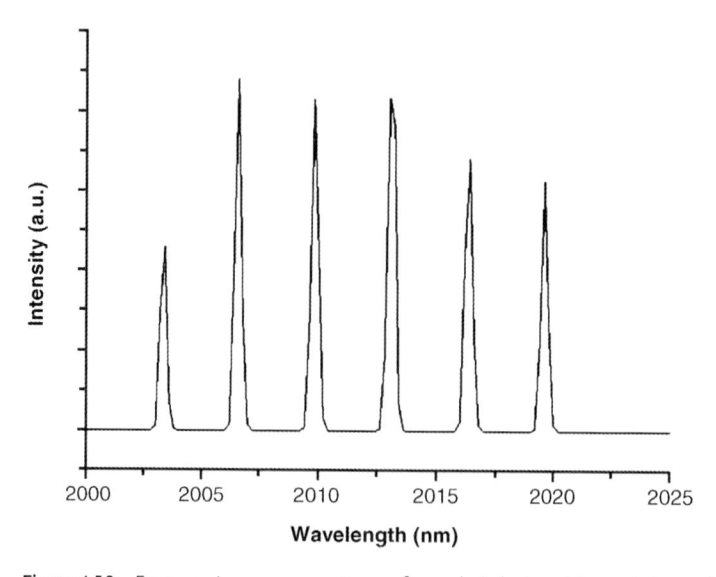

Figure 4.19 Free-running power spectrum of sample 2 designed for emission at 2.0 μm showing the Fabry–Pérot etalon spectral modulation imposed by the plane/plane ≈250 μm intra cavity diamond heat spreader.

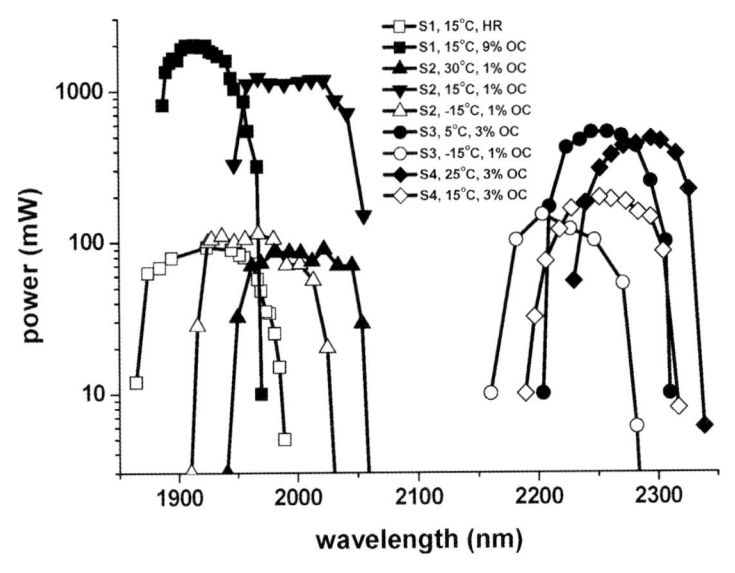

Figure 4.20 Tuning ranges of four 2.X μm SDL samples S1–S4 with central emission wavelengths of 1.9, 2.0, 2.25, and 2.3 μm, respectively. Various pump power, output coupler reflectivities and heat sink temperature combinations are employed to explore the limits of tuning.

obtained for the coolest temperatures. At these extremes of the tunable range, where the system gain is compromised, lower output coupling is used, resulting in lower output powers. Using this technique, tuning ranges of 150–170 nm with output powers >10 mW have been achieved from a single resonant microcavity 2.0 μm SDL – when optimized for higher power output ≈1 W, a tunable range of ≈70 nm was obtained.

First experiments were tried using similar antiresonant microcavity SDLs in an attempt to further increase these tuning ranges; however, performance was limited by the reflectivity of the available cavity optics. In general, such widely tunable resonant SDLs are preferable, as the microcavity enhancement of the gain results in lower threshold pump powers – also it could be argued that microcavity enhancement

Table 4.2 Summary of the tuning performance of the four SDL samples.

Design wavelength (μm)	Cavity length	Quantum wells	Position of resonance at RT (μm)	Tuning range, $\theta_{heatsink} =$ 15–20 °C (nm)	Tuning range, $\theta_{heatsink} =$ −15–30 °C (nm)
1.9	$6 \times \lambda/2$	$10 \times Ga_{0.80}In_{0.20}Sb$	1.89	125	NA
2.0	$6 \times \lambda/2$	$10 \times Ga_{0.74}In_{0.26}Sb$	1.99	154.4	170
2.25	$7 \times \lambda/2$	$10 \times Ga_{0.67}In_{0.33}As_{0.11}Sb_{0.89}$	2.25	135	170
2.3	$6 \times \lambda/2$	$9 \times Ga_{0.64}In_{0.36}As_{0.10}Sb_{0.90}$	2.45	70	NA

of the absorption in the unpumped regions of the chip will lead to a more definite spatial mode aperturing effect, and as a result, improved spatial modal control.

The reasons for the large tunability observed are manifold but can be ascribed to a combination of the increased relative bandwidth of the individual quantum wells and the high gain available in the strained GaSb-based structures, the relatively strong barrier absorption resulting in short (in wavelength terms) active region lengths, and the subsequent smearing out of the RPG effect due to the clustering of the QWs into groups of three or more at the field antinodes. In addition, there is a significant weakening of the subcavity resonance due to the employment of the high thermal conductivity intracavity heat spreader that acts as a coarse index matching layer between the semiconductor structure and the extended cavity. Despite the impressive tuning performance, significant improvements can be expected through further optimization of the SDL structure – for example, exploring a variable thickness [40] and position of the QWs by spoiling the subcavity resonance by way of AR coating on the SDL structure and/or by ensuring the thickness of the active region corresponds to an integer number of half-wavelengths (i.e., the so-called antiresonant structure) [55]. Also, by intentionally clustering a reduced number of quantum wells, the modal gain bandwidth of the SDL may be significantly increased beyond the intrinsic bandwidth of the material, as shown previously by Garnache and coworkers [55].

Such antiresonant design has been most recently implemented in a Sb-based SDL structure. The authors of Ref. [40] used a special design consisting of three separate gain sections containing different width $Ga_{0.8}In_{0.2}Sb$ QWs (3×6.5, 3×9.5, and 2×16 nm, respectively) in GaSb barriers. The gain regions were separated by two thin $AlAs_{0.08}Sb_{0.92}$ blocking layers to equalize pumping of the wells and prevent carrier diffusion between regions. The structure is shown in Figure 4.21. By placing this structure in a three-mirror "V cavity" and using a BRF, the authors were able to obtain a tuning range of 156 nm (a twofold increase over a similar structure without well variation) and a maximum output of approximately 0.4 W (Figure 4.22).

4.5.2
Single-Frequency Operation

In addition to providing a relatively broad gain bandwidth, SDLs represent an ideal source for narrow-line and single-frequency operation. SDLs in fact favor these

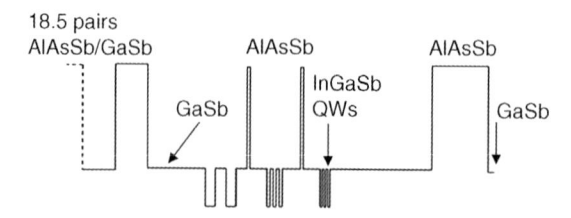

Figure 4.21 Schematic of the structure optimized for broadly tunable operation incorporating 3×6.5, 3×9.5, and 2×16 nm GaInSb QWs.

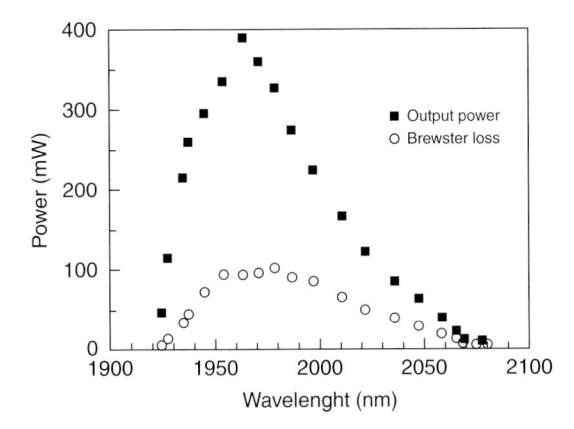

Figure 4.22 Output power of the broadly tunable structure of Figure 4.21 at different wavelengths. Brewster losses refer to reflection losses from the birefringent filter.

modes of operation over their VCSEL and edge-emitting counterparts due to the combination of the thin (few microns), homogeneously broadened semiconductor gain region and the relatively long (hundreds of millimeters) and high-finesse extended resonators that effectively suppress spatial hole burning. In addition, the wavelength-selective, distributed gain (or RPG) structures commonly employed also act to filter the output spectrum. Single-frequency operation naturally requires oscillation on a single-transverse mode and a single-polarization mode; in the SDL, the transverse mode behavior is strictly controlled by the external resonator and polarization switching is effectively suppressed by the use of a polarization-selective cavity element (usually a birefringent filter). Evidence indicating that gain anisotropy along different crystal axes and that by using an elliptical pump spot in the active region (introduced by the off-axis pumping often employed in SDLs) [51] has also been suggested to induce single-polarization oscillation. It is therefore the oscillation of the remaining longitudinal modes that will tend to dominate the spectral behavior of the SDL. The authors of Refs [60, 61] show that there is a characteristic time t_c for a laser based on an ideally homogeneously broadened gain medium (such as an SDL) to collapse to SLM operation in the absence of any disturbances of the order or less than this time. Here t_c is proportional to the curvature of the modal gain and cavity loss rate and inversely proportional to the longitudinal mode spacing.

In practice, however, when single-frequency operation is not the primary function of the SDL setup and chip design, either the long (approximately hundreds of millimeters) cavities with reduced longitudinal mode spacing and/or the clustering of the quantum wells (3–5) at the antinodes of the intracavity field, which acts to reduce the selectivity of the RPG structure by flattening the modal gain, tend to support broader bandwidth operation. Any source of perturbation, and therefore a longitudinal mode coupling mechanism in the laser coupled with the broad gain bandwidth from individual QWs, will tend to increase the bandwidth and force multilongitudinal mode operation with linewidths covering many hundreds of longitudinal modes. Such perturbations can include high-frequency acoustic noise, amplitude noise caused by mode competition in the multimode pump laser, and polarization noise induced by residual birefringence in the heat spreader [62] or other intracavity elements. Therefore, to practically promote and preserve narrow-line and single longitudinal mode operation, intracavity frequency-selective elements and active cavity length stabilization are required [48]. The typical forms of frequency selection utilized are the same types of intracavity filter elements used to implement broad tuning, as described previously; however, to obtain the desired selectivity, a combination of filters is often used.

Typically, coarse linewidth narrowing down to a small number of longitudinal modes is achieved with a single-plate BRF or indeed with a Lyot filter to further increase the selectivity. Additional etalons of varying thickness may be then added to the cavity to further enhance the wavelength filtering such that only one longitudinal mode can oscillate [63, 64]. The use of a birefringent etalon has also been shown to promote high-quality single-frequency oscillation while at the same time reducing the complexity of the cavity locking mechanics and electronics [65]. The first demonstration of a single-frequency SDL, a GaAs-based VECSEL emitting around 850 nm [48], employed a three-plate birefringent tuning element and the higher finesse mode selection of a thin 500 μm etalon to provide SLM operation. The cavity was subsequently side fringe locked to a 300 MHz reference cavity by controlling the cavity length via a piezo-actuated folding mirror and feedback electronics. The resulting linewidth of the laser of $\approx$4 kHz (derived from the locking error signal) was limited by acoustic noise in the optomechanics and the locking circuitry. The use of BRFs and etalons has been repeated often for different material structures and wavelength regimes to provide significant linewidth reduction in these lasers and/or single-frequency operation [60, 66, 74].

More recently, the use of volume Bragg gratings (VBGs) has been shown to significantly narrow the linewidth of SDLs to include only a few longitudinal modes [67]. Here, a narrow band reflector is holographically written in a photorefractive material and used as the wavelength-selective output coupler in an SDL while introducing minimal loss. The locking of a Sb-based SDL to a narrow line (measurement resolution limited to $\approx$75 MHz) at a wavelength of 2021.3 nm has recently been reported using a VBG [68]. Figure 4.23 shows the setup of the 2.0 μm SDL with VBG output coupler and the resulting output spectrum. The VBG locked the SDL output to this mode for the entire power range of the oscillating SDL. Using a heat sink temperature of 18 °C (12 °C), a maximum output power of 0.45 W (0.55 W) was reached.

For the case of a VBG, however, the polarization remains uncontrolled without additional elements that limit its suitability for a number of applications. Also, the use of high reflecting gratings (HRGs) – where a DBR is fabricated on a subwavelength grating – has been shown to exhibit significant linewidth narrowing and single-frequency operation while crucially controlling the polarization in the cavity [69]. Limited tuning in the VBG and HRG cases is also available in these sources by angle tuning of the gratings.

In spite of the above argument, it has been shown that if sufficient care is taken to remove sources of mode competition such as employing a single-transverse mode

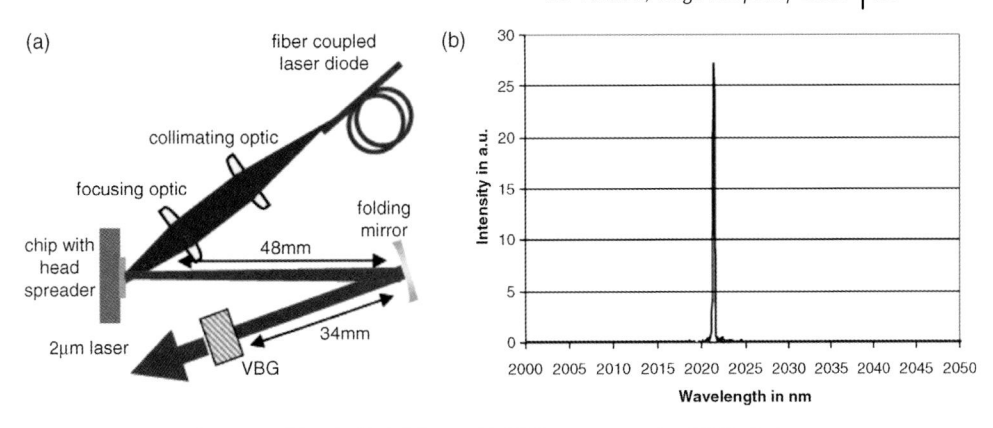

Figure 4.23 (a) Schematic of the GaSb disk laser with VBG output coupler. (b) Typical output spectrum for the VBG stabilization [68].

pump [51], or reducing the number of longitudinal modes by reducing the cavity length [51, 61, 70–73], or by carefully controlling the spatial mode properties of the SDL [74], then the single-frequency operation can be encouraged without the use of significant intracavity elements. It is worthy of note that the SDLs described in Ref. [48] also exhibited self-single-frequency operation; however, here the QWs were distributed individually throughout a relatively long (14 $\lambda/2$) active region, thus enhancing the selectivity of the RPG design.

In general, it has been shown that the Sb-based SDL structures described above can also exhibit robust, spontaneous high-power single-frequency operation (>0.5 W) when the laser cavity and pump are carefully aligned for single-spatial mode operation ($M^2 < 1.1$) [71]. A careful matching of the pump and laser modes encourages single-transverse mode oscillation, since reabsorption in unpumped areas discourages higher order transverse modes. Here, no attempt is made to shorten the cavity, but the longer operating wavelength results in a larger mode spacing that is, of course, advantageous. Mode selection is also aided by the presence of the weak etalon effect provided by the thin intracavity heat spreader, which having a free spectral range of ≈3 nm acts to filter the net gain of the SDL (see Figure 4.19). The use of a thin heat spreader with larger FSR in conjunction with a short ≈4 mm cavity has also been shown to promote single-frequency operation without stabilization control [70].

This mode of operation, however, has limited use in a practical laser as the laser output wavelength is generally fixed by the intracavity diamond etalon. To improve the tunability while retaining the heat spreader function, wedged antireflection-coated heat spreaders have been employed [75]. For the broadest and stable mode-hop-free SLM tunability (>100 GHz) without filtering, it has also been shown that it is beneficial to remove all sources of unintentional frequency selection such as the residual intracavity etalon by AR-coating the semiconductor surface (when no heat spreader is used) and the weak extracavity etalon formed between the back surface of the substrate and the DBR by wedging the substrate (≈3°) [61]. The specific wavelength control mechanisms such as etalons and precise piezoelectric cavity length control, as discussed above, can then be introduced to ensure tunable single-frequency operation [61].

4.5.3
Experimental Results of a 2.3 µm Single-Frequency SDL

An SDL chip designed to operate at 2.3 µm, after the manner described in Section 4.3, was placed in a short two-mirror cavity arrangement. When close to optimum pump/cavity mode overlap was achieved, the laser output exhibited relatively large intensity fluctuations indicated in Figure 4.24.

However, with careful optimization of the pump/cavity mode overlap, an abrupt increase in output power and a reduction in intensity noise (Figure 4.24) are observed – indicating a significant reduction in longitudinal mode competition. This was accompanied by a significant spectral narrowing of the laser – switching from operating on a number of diamond etalon modes (Figure 4.24a) to only one etalon mode (Figure 4.24b). True single-frequency operation in this configuration was confirmed using a scanning Fabry–Pérot etalon with a free spectral range of $\approx$12 GHz (cavity mode spacing $\approx$3 GHz, see Figure 4.25).

Despite the lack of active stabilization, such careful cavity alignment promoted single-frequency operation without requiring any extra cavity elements. This mode of

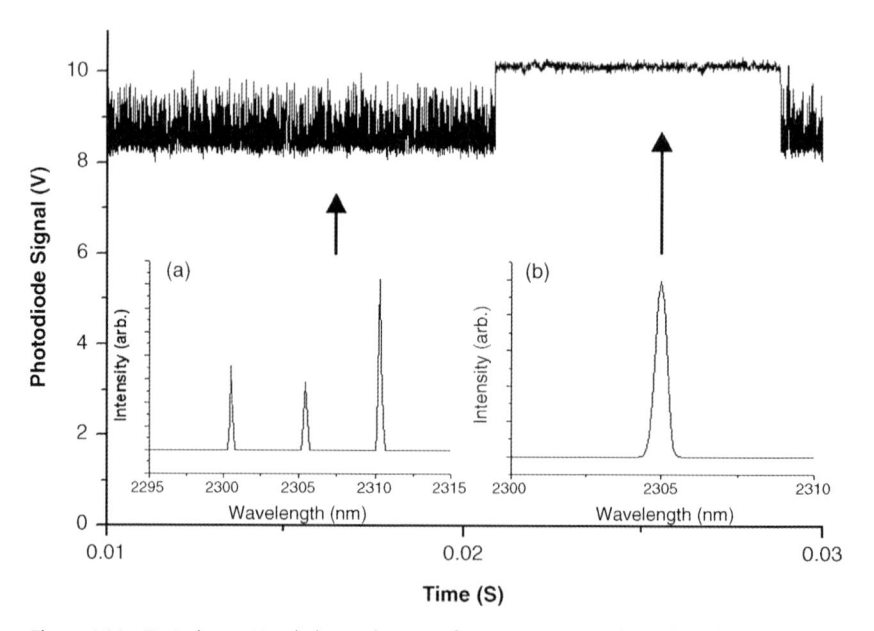

Figure 4.24 Typical transition behavior between free-running (a) multimode and (b) SLM operation (see inset spectra) with careful cavity alignment (resolution limited to 0.44 nm by the grating spectrometer – Jobin Yvon HR460).

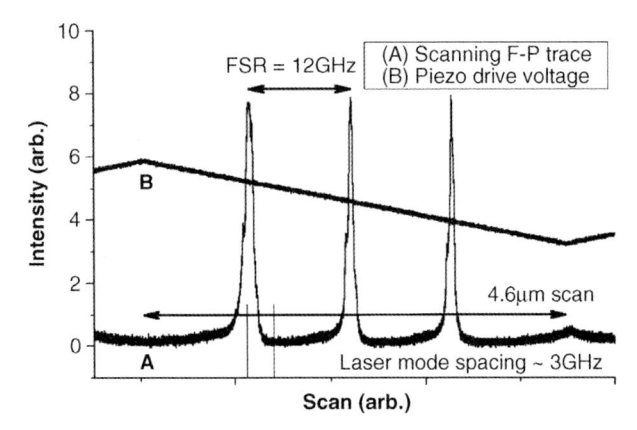

Figure 4.25 Trace from a scanning Fabry–Pérot interferometer (A) with a free spectral range (FSR) of 12 GHz showing the piezoelectric drive voltage (B) and the relative position of the adjacent longitudinal cavity mode (vertical lines at the bottom) of a self-single-frequency Sb-based SDL emitting at 2.3 μm.

operation was found to be stable over periods of many minutes, indicative of the enhanced gain extraction and lower loss of this operating mode. This free-running single-frequency state was observed for substantial output powers of up to 680 mW at −15 °C. As mechanical instabilities would inevitably frustrate this state, intracavity bandwidth controlling elements must be utilized for long-term stable single-frequency operation.

In a three-mirror cavity arrangement featuring a 2 mm quartz BRF (Figure 4.18), the laser could be readily tuned, and SLM operation obtained, over a tuning range of 70 nm from 2260 to 2335 nm (limited only by the reflection bandwidth of the cavity optics – see Figure 4.26).

In a manner similar to Ref. [48], the resonator was then locked to a Fabry–Pérot reference cavity by fixing the output coupler to a piezoelectric transducer coupled to a simple feedback circuit employing the convenient side fringe locking technique [76] (see Figure 4.27). In this way, the time-averaged (over 1 s) SLM linewidth was narrowed to <4 MHz, the measured linewidth being limited by the short ($l \approx 12.5$ mm) reference cavity and noise in the locking electronics. While further linewidth narrowing to kHz levels may be expected by using longer reference cavities having higher finesse (see, for example, Holm et al. [48]), for the majority of mid-IR sensing applications, such linewidths are appropriate and indeed the compact, robust nature of this configuration is desirable.

Sb-based SDLs have been shown to operate over a wide spectral range in the mid-IR spectral region. At present, by way of compositional tuning of the QWs, the range of 1.9–2.8 μm has been explored. Furthermore, any given Sb–SDL can be tuned over a wavelength range typically around 125–175 nm throughout this range – using a suitable intracavity component, typically a single-plate birefringent filter. Thermal management by way of high-conductivity heat spreaders is essential for high-power

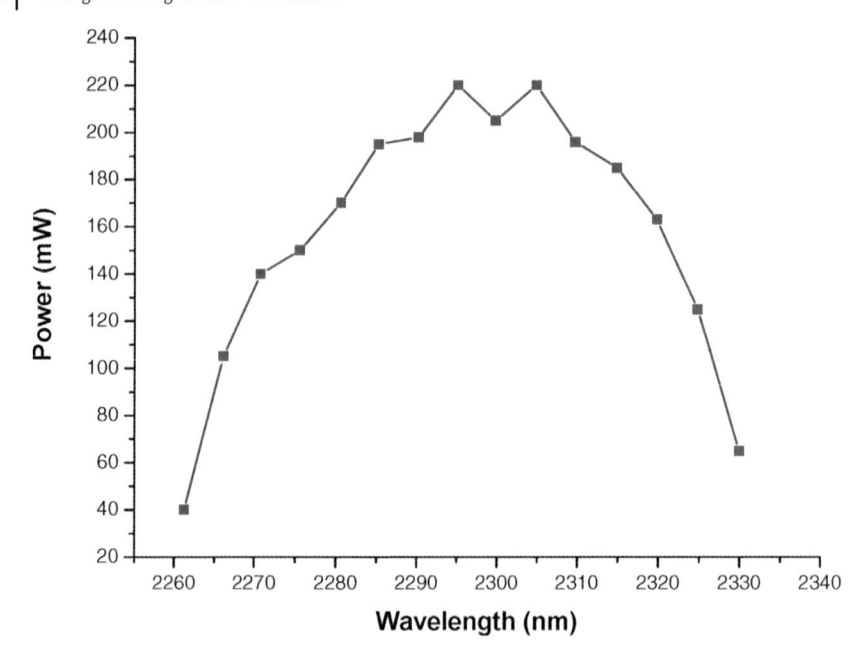

Figure 4.26 Typical tuning characteristic obtained at moderate output powers. Each point was optimized for stable SLM operation.

operation, and output powers approaching 10 W have so far been obtained. Single-frequency operation at these elevated output powers is also readily obtained through suitable wavelength controlling elements and electromechanical stabilization to a reference cavity. The Sb-based SDLs therefore represent a flexible, compact optical source suitable for a wide range of existing and emerging applications in the mid-IR spectral region.

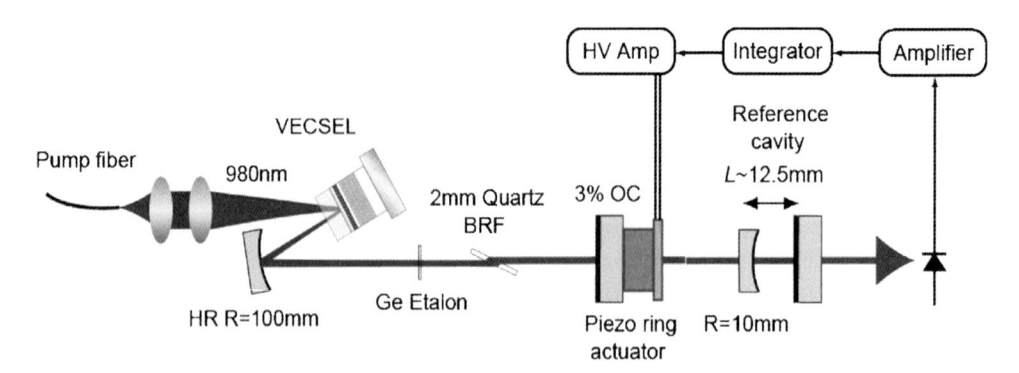

Figure 4.27 Schematic of the cavity arrangement and electronics used for the locking and line narrowing of the single-frequency SDL operation.

4.6
Disk Lasers At and Above 3 μm Wavelength

The (AlGaIn)(AsSb) material system is also well suited to realize lasers emitting well above 3 μm. In fact, the very first III–V laser of this material family, an InAs homojunction diode laser, operated at 3.1 μm at low temperatures back in 1963 [77]. Later on, double-heterostructure lasers employing InAsSb/AlGaAsSb layer sequences grown on GaSb or InAsSb/InAsPSb layers on InAs substrates were fabricated, emitting in the 3–4 μm range with a maximum operating temperature of around 170 K in pulsed operation [78–80]. More recently, GaInAsSb/AlGaAsSb type I diode lasers have been operated at room temperature with emission wavelength up to 3.36 μm [81]. For even longer emission wavelength, a type II band structure in the active region can be used, as in the so-called W-structure, first introduced in 1995 [82]. In this concept, GaSb-based optically pumped edge-emitting lasers were realized, operating over the very broad spectral range of 2.4–9.3 μm [83]. The performance of these W-lasers, although notable at low temperatures (80–150 K), however remains comparatively poor when operated at room temperature [82–84].

Given this work on GaSb-based edge-emitting lasers, it seems only a question of time until the first GaSb-based disk laser operated above 3 μm will be demonstrated (to date, the maximum emission wavelength demonstrated is 2.8 μm, for a cw SDL operating at room temperature [85]). Indeed, the long-wavelength GaSb-based disk lasers have one prominent advantage compared to GaSb-based diode laser– they are undoped and so the optical absorption due to free carriers is greatly reduced. This optical loss mechanism is predominant in GaSb-based diode lasers [86] and is expected to increase further with increasing emission wavelength [87].

Another active region concept for long-wavelength semiconductor laser is of course the quantum cascade laser, but due to the selection rules for the dipole transitions, surface-emitting QCLs are not feasible.

Finally, it is noteworthy that the disk laser format has also been successfully transferred to IV–VI lead–chalcogenide-based devices in 2007 [88] with the demonstration of a PbTe/Pb$_{1-x}$Eu$_x$Te/BaF$_2$ structure, grown by MBE on BaF$_2$ substrate. The emission wavelength was 5 μm and the laser was operated around 95 K; however, more recently, room-temperature operation was demonstrated at 4.5 μm with 6 mW output power in pulsed operation (100 ns pulse width) [89]. These lead–chalcogenide SDLs also utilize a type I active region, thus the emission wavelength is directly determined by the bandgap energy of the active layers. The emission wavelength can therefore, at least in principle, be selected within the range 3 μm–40 μm by adding Eu, Sr, or Sn as a second group-IV element [90].

4.7
Conclusions

The Sb-based semiconductor disk lasers covering the wavelength range of 1.9–2.3 μm have already reached a considerable level of maturity. Low threshold pump powers, high optical-to-optical efficiencies, and multiwatt output powers at diffraction-limited beam quality have been demonstrated by several groups.

To realize such high-performance SDLs, several challenges have to be faced. First, specific structural layouts based on a sound knowledge of all relevant optical, electronic, and thermal constants in the III-Sb material system have to be developed. As shown in Section 4.4.2, the precisely controlled alignment of the modal gain curve with respect to the microcavity resonance is a key issue to achieve low threshold pump powers at room temperature and beyond. Second, the MBE growth process has to be optimized to realize layer stacks having a total thickness of 10 μm and more, while simultaneously achieving the highest accuracy regarding the thickness and composition of each individual layer. To this end, the MBE growth has to be accompanied by a comprehensive epi-layer characterization and modeling procedure that can provide detailed information for accurate readjustment of the MBE growth parameters. In addition to these challenges, efficient heat extraction in laser operation has to be ensured by an appropriate thermal management scheme. For the Sb-based SDLs, the best device performance has been achieved using intracavity heat spreaders bonded to the chip surface.

Current R&D focuses increasingly on the optimization of MIR SDLs for specific applications. The spectroscopic applications mostly require tunable, single longitudinal mode laser sources. As shown in Section 4.5, the external cavity configuration of SDLs allows the introduction of frequency-selective elements such as birefringent filters to promote single longitudinal mode operation. Using a birefringent filter, the emission wavelength of MIR SDLs can also be tuned within a range of typically 100–150 nm. When targeting seeding applications, wavelength-stabilized, narrow-linewidth SDLs are required. Such SDLs can be realized by locking the resonator to a Fabry–Pérot reference cavity. When actively stabilizing MIR SDLs in this way, linewidths in the MHz-regime can be easily achieved.

Future work will also focus on further improvements of Sb-based SDLs in terms of power efficiency. In this context, direct pumping of the quantum wells is possibly one of the most promising concepts. This approach also benefits from the advent of efficient Sb-based diode lasers that can serve as a pump source for in-well pumping in future MIR SDL modules.

Another seminal topic in the field of MIR SDLs is the optimization of Sb-based SDLs on GaAs substrates. Such SDLs can benefit from reduced substrate costs, increased growth rates, and an established processing technology. Results from the first Sb-based SDLs on GaAs substrates have demonstrated that these laser structures might prove themselves as a serious alternative to conventional MIR SDLs on GaSb substrates.

Considering the fact that Sb-based diode lasers incorporating GaInAsSb type I quantum wells with an emission wavelength of >3.3 μm has recently been demonstrated, the Sb-based SDLs for the wavelength region above 3 μm are within reach. Compared to their electrically pumped counterparts, the long-wavelength SDLs benefit from the absence of any intentionally doped layers and thus have reduced free-carrier absorption. Since this loss mechanism becomes more important with increasing wavelength, the Sb-based SDLs are promising candidates for high-brightness laser sources emitting at wavelength around 3 μm and beyond.

Acknowledgments

Financial support by the European Commission through the project VERTIGO is gratefully acknowledged. The authors gratefully acknowledge the contributions to this book of Rüdiger Moser, Christian Manz, and Klaus Köhler at Fraunhofer IAF.

References

1 Garnache, A., Liu, A., Cerutti, L., and Campargue, A. (2005) Intracavity laser absorption spectroscopy with a vertical external cavity surface emitting laser at 2.3 μm: application to water and carbon dioxide. *Chem. Phys. Lett.*, **416**, 22–27.

2 Titterton, D.H. (2006) The development of infrared countermeasure technology & systems, in *Mid-Infrared Semiconductor Optoelectronics* (ed. A. Krier), Springer-Verlag, Berlin.

3 Vurgaftman, I., Meyer, J.R., and Ram-Mohan, L.R. (2001) Band parameters for III–V compound semiconductors and their alloys. *J. Appl. Phys.*, **89**, 5815–5875.

4 Shim, K., Rabitz, H., and Dutta, P. (2000) Band gap and lattice constant of $Ga_xIn_{1-x}As_ySb_{1-y}$. *J. Appl. Phys.*, **88**, 7157–7161.

5 Cerutti, L., Garnache, A., Genty, F., Ouvrard, A., and Alibert, C. (2003) Low threshold, room temperature laser diode pumped Sb-based VECSEL emitting around 2.1 μm. *Electron. Lett.*, **39**, 290–292.

6 Schulz, N., Hopkins, J.-M., Rattunde, M., Burns, D., and Wagner, J. (2008) High-brightness long-wavelength semiconductor disk lasers. *Laser Photon. Rev.*, **2**, 160–181.

7 Li, W., Héroux, J.B., Shao, H., and Wang, W.I. (2004) Strain-compensated InGaAsSb/AlGaAsSb mid-infrared quantum-well lasers. *Appl. Phys. Lett.*, **84**, 2016–2018.

8 Bückers, C., Tränhardt, A., Koch, S., Rattunde, M., Schulz, N., Wagner, J., Hader, J., and Moloney, J. (2008) Microscopic calculation and measurement of the gain in a (GaIn)Sb quantum well structure. *Appl. Phys. Lett.*, **92**, 71107-1–71107-3.

9 Corzine, S.W., Geels, R.S., Scott, J.W., Yan, R.H., and Coldren, L.A. (1989) Design of Fabry–Perot surface-emitting lasers with a periodic gain structure. *IEEE J. Quantum Electron.*, **25**, 1513–1524.

10 Shterengas, L., Belenky, G.L., Kim, J.G., and Martinelli, R.U. (2004) Design of high-power room-temperature continuous-wave GaSb-based type-I quantum-well lasers with $\lambda > 2.5$ μm. *Semicond. Sci. Technol.*, **19**, 655–658.

11 Schulz, N., Rösener, B., Moser, R., Rattunde, M., Manz, C., Köhler, K., and Wagner, J. (2008) An improved active region concept for highly efficient GaSb-based optically in-well pumped vertical-external-cavity surface-emitting lasers. *Appl. Phys. Lett.*, **93**, 181113.

12 Tourrenc, J.P., Bouchoule, S., Khadour, A., Harmand, J.C., Miard, A., Decobert, J., Lagay, N., Lafosse, X., Sagnes, I., Leroy, L., and Oudar, J.L. (2008) Thermal optimization of 1.55 μm OP-VECSEL with hybrid metal-metamorphic mirror for single-mode high power operation. *Opt. Quantum Electron.*, **40**, 155–165.

13 Cho, A.Y. (1983) Growth of III–V semiconductors by molecular beam epitaxy and their properties. *Thin Solid Films*, **100**, 291–317.

14 Biefeld, R.M. (2002) The metal-organic chemical vapor deposition and properties of III–V antimony-based semiconductor materials. *Mat. Sci. Eng.*, **R36**, 105–142.

15 Rouillard, Y., Lambert, B., Toudic, Y., Baudet, M., and Gauneau, M. (1995) On the use of dimeric antimony in molecular-beam epitaxy. *J. Cryst. Growth*, **156**, 30–38.

16 Schulz, N., Rattunde, M., Ritzenthaler, C., Rösener, B., Manz, C., Kohler, K., and Wagner, J. (2007) Effect of the cavity resonance-gain offset on the output power

characteristics of GaSb-based VECSELs. *Photon. Technol. Lett.*, **19**, 1741–1743.

17 Schulz, N., Rattunde, M., Ritzenthaler, C., Rösener, B., Manz, C., Kohler, K., Wagner, J., and Brauch, U. (2007) Resonant optical in-well pumping of an (AlGaIn)(AsSb)-based vertical-external-cavity surface-emitting laser emitting at 2.35 µm. *Appl. Phys. Lett.*, **91**, 091113.

18 Schmid, M., Benchabane, S., Torabi-Goudarzi, F., Abram, R., Ferguson, A.I., and Riis, E. (2004) Optical in-well pumping of a vertical-external-cavity surface-emitting laser. *Appl. Phys. Lett.*, **84**, 4860–4862.

19 Beyertt, S.-S., Brauch, U., Demaria, F., Dhidah, N., Giesen, A., Kubler, T., Lorch, S., Rinaldi, F., and Unger, P. (2007) Efficient gallium–arsenide disk laser. *IEEE J. Quantum Electron.*, **43**, 869–875.

20 Beyertt, S.-S., Zorn, M., Kübler, T., Wenzel, H., Weyers, M., Giesen, A., Tränkle, G., and Brauch, U. (2005) Optical in-well pumping of a semiconductor disk laser with high optical efficiency. *IEEE J. Quantum Electron.*, **41**, 1439–1449.

21 Tatebayashi, J., Jallipalli, A., Kutty, M.N., Huang, S.H., Balakrishnan, G., Dawson, L.R., and Huffaker, D.L. (2007) Room-temperature lasing at 1.82 µm of GaInSb/AlGaSb quantum wells grown on GaAs substrates using an interfacial misfit array. *Appl. Phys. Lett.*, **91**, 141102.

22 Rodriguez, J.B., Cerutti, L., and Tournie, E. (2009) GaSb-based, 2.2 µm type-I laser fabricated on GaAs substrate operating continuous wave at room temperature. *Appl. Phys. Lett.*, **94**, 023506-1–023506-2.

23 Rodriguez, J.B., Cerutti, L., Grech, P., and Tournie, E. (2009) Room-temperature operation of a 2.25 µm electrically pumped laser fabricated on a silicon substrate. *Appl. Phys. Lett.*, **94**, 061124-1–061124-2.

24 Hoke, W.E., Lemonias, P.J., Mosca, J.J., Lyman, P.S., Torabi, A., Marsh, P.F., McTaggart, R.A., Lardizabal, S.M., and Hetzler, K. (1998) Molecular beam epitaxial growth and device performance of metamorphic high electron mobility transistor structures fabricated on GaAs substrates. *J. Vac. Sci. Technol. B*, **17**, 1131–1135.

25 Balakrishnan, G., Huang, S., Rotter, T.J., Stintz, A., Dawson, L.R., Malloy, K.J., Xu, H., and Huffaker, D.L. (2004) 2.0 µm wavelength InAs quantum dashes grown on a GaAs substrate using a metamorphic buffer layer. *Appl. Phys. Lett.*, **84**, 2058–2060.

26 Huang, S.H., Balakrishnan, G., Khoshakhlagh, A., Jallipalli, A., Dawson, L.R., and Huffakera, D.L. (2006) Strain relief by periodic misfit arrays for low defect density GaSb on GaAs. *Appl. Phys. Lett.*, **88**, 131911-1–1319911-3.

27 Jallipalli, A., Balakrishnan, G., Huang, S.H., Khoshakhlagh, A., Dawson, L.R., and Huffaker, D.L. (2007) Atomistic modeling of strain distribution in self-assembled interfacial misfit dislocation (IMF) arrays in highly mismatched III–V semiconductor materials. *J. Cryst. Growth*, **303**, 449–455.

28 Balakrishnan, G., Rotter, T.J., Jallipalli, A., Dawson, L.R., and Huffaker, D.L. (2008) Interfacial misfit dislocation array based growth of III-Sb active regions on GaAs/AlGaAs DBRs for high-power 2 µm VECSELs. *Proc. SPIE*, **6871**, 687111-1–687111-7.

29 Kemp, A.J., Valentine, G.J., Hopkins, J.M., Hastie, J.E., Smith, S.A., Calvez, S., Dawson, M.D., and Burns, D. (2005) Thermal management in vertical-external-cavity surface-emitting lasers: finite-element analysis of a heatspreader approach. *IEEE J. Quantum Electron.*, **41**, 148–155.

30 Alford, W.J., Raymond, T.D., and Allerman, A.A. (2002) High power and good beam quality at 980 nm from a vertical external-cavity surface-emitting laser. *J. Opt. Soc. Am. B*, **19**, 663–666.

31 Cerutti, L., Garnache, A., Ouvrard, A., and Genty, F. (2004) High temperature continuous wave operation of Sb-based vertical external cavity surface emitting laser near 2.3 µm. *J. Cryst. Growth*, **268**, 128–134.

32 Schulz, N., Rattunde, M., Manz, C., Kohler, K., Wild, C., Wagner, J., Beyertt, S.-S., Brauch, U., Kubler, T., and Giesen, A. (2006) Optically pumped GaSb-based VECSEL emitting 0.6 W at 2.3 µm. *Photon. Technol. Lett.*, **18**, 1070–1072.

33 Liau, Z.L. (2000) Semiconductor wafer bonding via liquid capillarity. *Appl. Phys. Lett.*, **77**, 651–653.

34 Härkönen, A., Guina, M., Okhotnikov, O., Rössner, K., Hümmer, M., Lehnhardt, T., Müller, M., Forchel, A., and Fischer, M. (2006) 1-W antimonide-based vertical external cavity surface emitting laser operating at 2-μm. *Opt. Express*, **14**, 6479–6484.

35 Härkönen, A., Rautiainen, J., Orsila, L., Guina, M., Rössner, K., Hümmer, M., Lehnhardt, T., Müller, M., Forchel, A., Fischer, M., Koeth, J., and Okhotnikov, O.G. (2008) 2-μm mode-locked semiconductor disk laser synchronously pumped using an amplified diode laser. *Photon. Technol. Lett.*, **20**, 1332–1334.

36 Hopkins, J.M., Hempler, N., Rösener, B., Schulz, N., Rattunde, M., Manz, C., Kohler, K., Wagner, J., and Burns, D. (2008) High-power, (AlGaIn)(AsSb) semiconductor disk laser at 2.0 μm. *Opt. Lett.*, **33**, 201–203.

37 Rösener, B., Schulz, N., Rattunde, M., Manz, C., Kohler, K., and Wagner, J. (2008) High-power high-brightness operation of a 2.25-μm (AlGaIn)(AsSb)-based barrier-pumped vertical-external-cavity surface-emitting laser. *Photon. Technol. Lett.*, **20**, 502–504.

38 Piprek, J., Akulova, Y.A., Babic, D.I., Coldren, L.A., and Bowers, J.E. (1998) Minimum temperature sensitivity of 1.55 μm vertical-cavity lasers at -30 nm gain offset. *Appl. Phys. Lett.*, **72**, 1814–1816.

39 Burns, D., Hopkins, J.M., Kemp, A.J., Rösener, B., Schulz, N., Manz, C., Kohler, K., Rattunde, M., and Wagner, J. (2009) Recent developments in high-power short-wave mid-infrared semiconductor disk lasers. *Proc. SPIE*, **7193**, 719311-1–719311-13.

40 Paajaste, J., Suomalainen, S., Koskinen, R., Härkönen, A., Guina, M., and Pessa, M. (2009) High-power and broadly tunable GaSb-based optically pumped VECSELs emitting near 2 μm. *J. Cryst. Growth*, **311**, 1917–1919.

41 Tropper, A.C. and Hoogland, S. (2006) Extended cavity surface-emitting semiconductor lasers. *Prog. Quantum Electron.*, **30**, 1–43.

42 Hempler, N., Hopkins, J.M., Kemp, A.J., Schulz, N., Rattunde, M., Wagner, J., Dawson, M.D., and Burns, D. (2007) Pulsed pumping of semiconductor disk lasers. *Opt. Express*, **15**, 3247–3256.

43 Hempler, N., Hopkins, J.M., Rösener, B., Rattunde, M., Wagner, J., Moskalev, I., Fedorov, V., Mirov, S., and Burns, D. (2009) Pulsed-pumped 1.9 μm and 2.0 μm semiconductor disk lasers and their use as pump source for Cr2+:ZnSe. Middle Infrared Coherent Sources (MICS'2009), Trouville, France, June 8–12, Paper Tu5.

44 Rösener, B., Rattunde, M., Moser, R., Manz, C., Köhler, K., and Wagner, J. (2009) GaSb-based optically pumped semiconductor laser using multiple gain elements. *Photon. Technol. Lett.*, **21**, 848–850.

45 Rattunde, M., Rösener, B., Moser, R., Hempler, N., Hopkins, J.-M., Burns, D., Manz, C., Köhler, K., and Wagner, J. (2009) Power scaling of GaSb-based semiconductor disk lasers for the 2.X micron wavelength range. Proceedings of CLEO Europe 2009, Paper CB7.5.

46 Moloney, J.V., Hader, J., Li, H., Kaneda, Y., Wang, T.S., Yarborough, M., Koch, S.W., Stolz, W., Kunert, B., Bueckers, C., Chaterjee, S., and Hardesty, G. (2009) OPS laser EPI design for different wavelengths. *Proc. SPIE*, **7193**, 719313-1–719313-15.

47 Rotter, T.J., Huffaker, D.L., Rattunde, M., Wagner, J., Moloney, J., Hader, J., and Yarborough, M. (2009) AlGaInSb-VECSEL grown on GaAs based DBRs for high-power emission at 2 μm. 39th Infrared Colloquium, Freiburg, Germany, February 17–18.

48 Holm, M.A., Burns, D., Ferguson, A.I., and Dawson, M.D. (1999) Actively stabilized single-frequency VECSEL AlGaAs laser. *IEEE Photon. Technol. Lett.*, **11**, 1551–1553.

49 Tittel, F.K., Richter, D., and Fried, A. (2003) Mid-infrared laser applications in spectroscopy, in *Solid-State Mid-Infrared Sources* (eds I.T. Sorokina and K.L. Vodopyanov), Topics in Applied Physics, vol. **89**, Springer, Berlin.

50 Ouvrard, A., Garnache, A., Cerutti, L., Genty, F., and Romanini, D. (2005) Single frequency tunable Sb-based VCSEL

emitting at 2.3. *IEEE Photon. Technol. Lett.*, **17**, 2020–2022.

51 Bertseva, E., Campargue, A., Ding, Y., Fayt, A., Garnache, A., Roberts, J.S., and Romanini, D. (2003) The overtone spectrum of carbonyl sulfide in the region of the $v_1 + 4v_3$ and $5v_3$ bands by ICLAS-VECSEL. *J. Mol. Spectrosc.*, **219**, 81–87.

52 Rothman, L.S., Jacquemart, D., Barbe, A., Benner, D.C., Birk, M., Brown, L.R., Carleer, M.R., Chackerian, C., Chance, K., Coudert, L.H., Dana, V., Devi, V.M., Flaud, J.M., Gamache, R.R., Goldman, A., Hartmann, J.M., Jucks, K.W., Maki, A.G., Mandin, J.Y., Massie, S.T., Orphal, J., Perrin, A., Rinsland, C.P., Smith, M.A.H., Tennyson, J., Tolchenov, R.N., Toth, R.A., Vander Auwera, J., Varanasi, P., and Wagner, G. (2005) The HITRAN 2004 molecular spectroscopic database. *J. Quantum Spectrosc. Ra.*, **96**, 139–204.

53 Garnache, A., Kachanov, A.A., Stoeckel, F., and Planel, R. (1999) High-sensitivity intracavity laser absorption spectroscopy with VECSELs. *Opt. Lett.*, **24**, 826–828.

54 Garnache, A., Kachanov, A.A., Stoeckel, F., and Houdré, R. (2000) Diode-pumped broadband vertical-external-cavity surface-emitting semiconductor laser applied to high-sensitivity intracavity absorption spectroscopy. *J. Opt. Soc. Am. B*, **17**, 1589–1598.

55 Wieman, C.E. and Hollberg, L. (1991) Using diode lasers for atomic physics. *Rev. Sci. Instrum.*, **62**, 1–20.

56 Heim, P.J.S., Fan, Z.F., Cho, S.-H., Nam, K., Dagenais, M., Johnson, F.G., and Leavitt, R. (1997) Single-angled-facet laser diode for widely tunable external cavity semiconductor lasers with high spectral purity. *Electron. Lett.*, **33**, 1387–1389.

57 Holtom, G. and Teschke, O. (1974) Design of a birefringent filter for high-power dye lasers. *IEEE J. Quantum Electron.*, **QE-10**, 577–579.

58 Collins, S.A. and White, G.R. (1963) Interferometer laser mode selector. *Appl. Opt.*, **2**, 448–449.

59 Jacquemet, M., Domenech, M., Dion, J., Strassner, M., Lucas-Leclin, G., Georges, P., Sagnes, I., and Garnache, A. (2006) Single-frequency high-power continuous-wave oscillation at 1003 nm of an optically pumped semiconductor laser. *Proc. SPIE*, **6184**, 61841X-1–6184-11.

60 Garnache, A., Ouvrard, A., and Romanini, D. (2007) Single-frequency operation of external–cavity VCSELs non-linear multimode temporal dynamics and quantum limit. *Opt. Express*, **15**, 9403–9417.

61 van Loon, F., Kemp, A.J., Maclean, A.J., Calvez, S., Hopkins, J.-M., Hastie, J.E., Dawson, M.D., and Burns, D. (2006) Intracavity diamond heatspreaders in lasers: the effects of birefringence. *Opt. Express*, **14**, 9250–9260.

62 Vassen, W., Zimmerman, C., Kallenbach, R., and Hänsch, T.W. (1990) A frequency-stabilized titanium sapphire laser for high-resolution spectroscopy. *Opt. Commun.*, **75**, 435–440.

63 Adams, C.S. and Ferguson, A.I. (1992) Tunable narrow linewidth ultra-violet light generation by frequency doubling of a ring Ti:sapphire laser using lithium tri-borate in an external enhancement cavity. *Opt. Commun.*, **90**, 89–94.

64 Gardner, K.S., Abram, R.H., and Riis, E. (2004) A birefringent etalon as single-mode selector in a laser cavity. *Opt. Express*, **12**, 2365–2370.

65 Abram, R.H., Gardner, K.S., Riis, E., and Ferguson, A.I. (2004) Narrow linewidth operation of a tunable optically pumped semiconductor lase. *Opt. Express*, **12**, 5434–5439.

66 Giet, S., Sun, H.D., Calvez, S., Dawson, M.D., Suomalainen, S., Härkönen, A., Guina, M., Okhotnikov, O., and Pessa, M. (2006) Spectral narrowing and locking of a vertical-external-cavity surface-emitting laser using an intracavity volume Bragg grating. *IEEE Photon. Technol. Lett.*, **18**, 1786–1788.

67 Scholle, K., Lamrini, S., Fuhrberg, P., Rattunde, M., and Wagner, J. (2009) Wavelength stabilization and mode selection of a GaSb-based semiconductor disk laser at 2 μm by using a volume Bragg grating. Proceedings of IEEE CLEO Europe 2009, Paper CB7.4.

68 Giet, S., Lee, C.-L., Calvez, S., Dawson, M.D., Destouches, N., Pommier, J.-C., and Parriaux, O. (2007) Stabilization of a semiconductor disk laser using an intra-cavity high reflectivity grating. *Opt. Express*, **15**, 16520–16526.

69 Garnache, A., Ouvrard, A., and Romanini, D. (2005) Spectro-temporal dynamics of external-cavity VCSELs: single-frequency operation. Proceedings of IEEE CLEO Europe 2005, Paper CB19.

70 Lindberg, H., Larsson, A., and Strassner, M. (2005) Single-frequency operation of a high-power, long-wavelength semiconductor disk laser. *Opt. Lett.*, **30**, 2260–2262.

71 Triki, M., Cermak, P., Cerutti, L., Garnache, A., and Romanini, D. (2008) Extended continuous tuning of a single-frequency diode-pumped VECSEL at 2.3 μm. *IEEE Photon. Technol. Lett.*, **20**, 1947–1949.

72 Cocquelin, B., Holleville, D., Lucas-Leclin, G., Sagnes, I., Garnache, A., Myara, M., and Georges, P. (2009) Tunable single-frequency operation of a diode-pumped vertical external-cavity laser at the cesium D2 line. *Appl. Phys. B*, **95**, 315–321.

73 Hopkins, J.-M., Maclean, A.J., Riis, E., Burns, D., Schulz, N., Rattunde, M., Manz, C., Köhler, K., and Wagner, J. (2007) Tunable, single-frequency, diode-pumped 2.3 μm VECSEL. *Opt. Express.*, **15**, 8212–8217.

74 Maclean, A., Kemp, A.J., Calvez, S., Kim, J.-Y., Kim, T., Dawson, M.D., and Burns, D. (2008) Continuous tuning and efficient intra-cavity generation in a semiconductor disk laser with intra-cavity heatspreader. *IEEE J. Quantum Electron.*, **44**, 216–225.

75 DeVoe, R.G. and Brewer, R.G. (1984) Laser frequency division and stabilization. *Phys. Rev. A*, **30**, 2827–2829.

76 Melngailis, I. (1963) Maser action in InAs diodes. *Appl. Phys. Lett.*, **2**, 176–178.

77 van der Ziel, J.P., Chiu, T.H., and Tsang, W.T. (1985) Optically pumped laser oscillation at 3.83 μm from $InAs_{1-x}Sb_x$ grown by molecular beam epitaxy on GaSb. *Appl. Phys. Lett.*, **47**, 1139–1141.

78 Choi, H.K., Turner, G.W., and Liau, Z.L. (1994) 3.9-μm InAsSb/AlAsSb double-heterostructure diode lasers with high output power and improved temperature characteristics. *Appl. Phys. Lett.*, **65**, 2251–2253.

79 Baranov, A.N., Imenkov, A.N., Sherstnev, V.V., and Yakovlev, Y.P. (1994) 2.7–3.9 μm InAsSb(P)/InAsSbP low threshold diode lasers. *Appl. Phys. Lett.*, **64**, 2480–2482.

80 Shterengas, L., Belenky, G., Hosada, T., Kipshide, G., and Suchalkin, S. (2008) Continuous wave operation of diode lasers at 3.36 μm. *Appl. Phys. Lett.*, **93**, 011103-1–011103-3.

81 Meyer, J.R., Hoffman, C.A., and Bartoli, F.J. (1995) Type-II quantum-well lasers for the mid-wavelength infrared. *Appl. Phys. Lett.*, **67**, 757–759.

82 Kaspi, R., Ongstad, A., Dente, G., Chavez, J., Tilton, M., and Gianardi, D. (2006) High-performance optically pumped antimonide lasers operating in the 2.4–9.3 μm wavelength range. *Appl. Phys. Lett.*, **88**, 41122-1–41122-3.

83 Felix, C., Bewley, W., Vurgaftman, I., Olafson, L., Stokes, D., Meyer, J., and Yang, M. (1999) High-efficiency midinfrared "W" laser with optical pumping injection cavity. *Appl. Phys. Lett.*, **75**, 2876–2878.

84 Rattunde, M., Hopkins, J.-M., Schulz, N., Rösener, B., Manz, C., Köhler, K., Burns, D., and Wagner, J. (2008) High-power GaSb-based optically pumped semiconductor disk laser for the 2.X μm wavelength regime. 9th International Conference on Mid-Infrared Optoelectronics: Materials and Devices (MIOMD-IX), Freiburg, Germany, September 7–11.

85 Rattunde, M., Schmitz, J., Kiefer, R., and Wagner, J. (2004) Comprehensive analysis of the internal losses in 2.0 μm (AlGaIn)(AsSb) quantum-well diode lasers. *Appl. Phys. Lett.*, **84**, 4750–4752.

86 Pankove, J. (1991) *Optical Processes in Semiconductors*, Dover Publications Inc., New York.

87 Rahim, M., Arnold, M., Felder, F., Behfar, K., and Zogg, H. (2007)

Mid-infrared lead–chalcogenide vertical external cavity surface emitting laser with 5 μm wavelength. *Appl. Phys. Lett.*, **91**, 151102-1–151102-3.

88 Rahim, M., Khiar, A., Felder, F., Fill, M., and Zogg, H. (2009) 4.5 μm wavelength vertical external cavity surface emitting laser operating above room temperature. *Appl. Phys. Lett.*, **94**, 201112-1–201112-3.

89 Khokhlov, D.(volume editor) (2003) Lead chalcogenides: physics and applications, in *Optoelectronic Properties of Semiconductors and Superlattices*, vol. **18** (ed. M.O. Manasreh), Taylor & Francis Books Inc., New York.

5
Semiconductor Disk Lasers Based on Quantum Dots
Udo W. Pohl and Dieter Bimberg

5.1
Introduction

Semiconductor disk lasers (SDLs), also termed vertical-external-cavity surface-emitting lasers (VECSELs), were fabricated until recently using exclusively quantum wells (QWs) in the gain section. In the field of semiconductor injection lasers, quantum dots (QDs) were introduced in the 1990s as an alternative to quantum wells. Quantum dots are nanosized semiconductors coherently inserted into a semiconductor material with a larger bandgap [1]. Their size is in the range of the de Broglie wavelength of charge carriers in all spatial directions, leading to atom-like, fully quantized states of confined electrons and holes. The discrete nature of the energy states causes the emergence of unique properties of QDs, not observed with higher dimensionality systems. Device applications of carrier confinement in all three dimensions promised significant advantages over the one-dimensional confinement of quantum wells [2]. Since the early proof of concept, many materials-related obstacles have been overcome and milestones have been achieved with both edge-emitting and surface-emitting devices based on quantum dots.

This chapter provides an introduction to the properties of QDs and highlights their use in diode lasers and semiconductor disk lasers. From the features of quantum dot injection lasers, requirements for the implementation of QDs into semiconductor disk lasers are obtained, and some first QD-SDL devices are presented. Our results indicate advantages of QD gain media in disk lasers and advert to approaches for future improvements.

5.2
Size Quantization in Optical Gain Media

The idea to exploit quantum effects in gain media was introduced in the 1970s [3]. The approach aimed at modifying the electronic density of states in active layers of a heterostructure semiconductor laser to achieve a spectral tuning ability without change of chemical composition. It was also pointed out in Ref. [3] that quantum effects may lower the lasing threshold.

Size quantization leads to an increase in the density of states near the band edges. The effect increases with the progressive reduction in dimensionality: from the square root dependence of a bulk solid, through a step-like dependence of quantum wells and an inverse square root dependence of quantum wires to the discrete delta function of quantum dots. Consequently, nonequilibrium charge carriers injected into a size-quantized semiconductor gradually concentrate at the band edges. The impact on optical gain was analyzed in the 1980s for laser structures based on GaAs and InP materials [4]. Modeling assumed a single active layer and, in the case of QDs, a cubic shape of uniform size in the dot ensemble, and hence an equal localization of carriers in the dots. The result given in Figure 5.1 for active GaAs layers embedded in $Al_{0.2}Ga_{0.8}As$ shows a clear increase in the maximum gain as the dimensionality is reduced. The increase in maximum gain is also reflected in a decrease in the lasing threshold current density J_{th} marked by dashed lines at the curves.

The lasing threshold is generally reached, if the modal gain g_{mod} just equilibrates the internal losses α_{int} and those of the cavity mirrors α_{mirr}.

$$g_{mod} = \Gamma g_{mat} = \alpha_{int} + \alpha_{mirr}.$$

Γ is the optical confinement factor that is given by the overlap of the total light intensity in the cavity with the active material. Values are in the order of the ratio of the total volume of the quantum confinement material and the total volume of

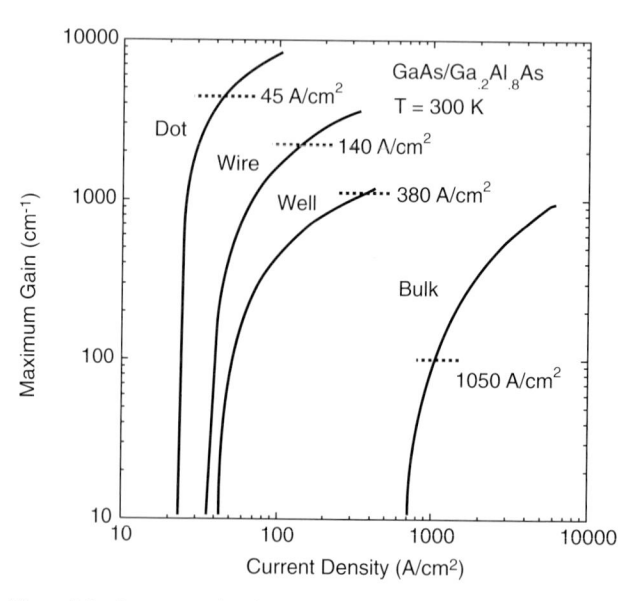

Figure 5.1 Dimensionality dependence of the maximum gain of a double-heterostructure laser diode, calculated for $GaAs/Al_{0.2}Ga_{0.8}As$ quantum dots, quantum wires, quantum wells, and bulk as active media. Data reproduced from Ref. [4] by permission of the IEEE.

the waveguide. In a single quantum well of thickness t_{QW}, the confinement factor is about the ratio of t_{QW} and half the lasing wavelength in the material, that is, $\Gamma_{QW} \approx t_{QW}/(\lambda/2n)$, n being the refractive index. Typical values are some 10^{-2}. The confinement factor Γ_{QD} of a QD array is composed of an in-plane part Γ_{xy} and a vertical part Γ_z. Γ_{xy} depends on the coverage with QDs, Γ_z is given by the average height. Typical numbers result from an areal QD density of mid $10^{10}\,\mathrm{cm}^{-2}$, a lateral extension of 10 nm, and a height of 5 nm, yielding for $\Gamma_{xy} \times \Gamma_z$ some 10^{-3}. The optical confinement factor Γ of a QD layer is hence significantly smaller than that of a QW. On the contrary, the material gain of QD media g_{mat} exceeds that of QWs by far [5, 6]. Two reasons contribute to this finding. First, the increase in the density of states described above. Second, quantum confinement leads to an increased overlap of confined electron and hole wave functions and to an increase in the exciton binding energy. The oscillator strength of radiative recombination is thereby significantly enhanced. As a result of the counteracting effects of confinement factor and material gain, a largely comparable modal gain is found in QD and QW lasers.

Since the number of states in a QD layer is small compared to those of a QW, a low lasing threshold and also a low gain saturation level are easily reached in single active layers. To increase both confinement factor and gain saturation level, actual QD lasers generally comprise a *stack* of QD layers.

5.2.1
Quantum Dots in Lasers

Work on low dimensional structures first focused on quantum wells because no method for fabricating quantum dots of sufficient quality existed at that time. Still a proof of concept was provided in the early 1980s by placing a quantum well laser in a strong magnetic field [7]: in-plane quantization originating from the magnetic field lead to some decrease in the temperature sensitivity of the threshold current density. Such behavior is expected as a result of reduction in dimensionality, leading in the ultimate limit of a zero-dimensional structure to a lasing threshold current density J_{th} being independent of temperature. The feature is expressed in terms of an infinite characteristic temperature T_0 in the relation $J_{th}(T) = J_0 \times \exp(T/T_0)$.

Low temperature sensitivity of zero-dimensional structures is a direct consequence of the modified electronic density of states. In bulk material, charge carriers are redistributed within the continuous distribution function to higher energy states as the temperature is increased. These carriers do not contribute to the inversion at lasing energy. Additional injection current is required to maintain a constant inversion condition. On the other hand, the occupation in a quantum dot is described by a delta function. The charge carriers cannot be redistributed if excited states lie sufficiently above the ground state ($\Delta E \gg k_B T$). Therefore, $T_0 = \infty$ was predicted for an ideal quantum dot laser.

In real QD lasers, two effects must be taken into account. First, confinement energies of charge carriers localized in a QD are finite. The carriers captured in a dot may consequently be re-emitted. Second, the real QDs show a significant variation of localization energy within an ensemble. The QD ensemble therefore exhibits an inhomogeneously broadened density of states, which is composed of numerous single dot states. Both effects were addressed in a numeric model, considering the microstates of a QD ensemble [8]. The charge carriers are assumed to be captured from quantum well states of the wetting layer (cf. Section 5.2.2) in a time τ_c and recombine in the QD in a time τ_s or are emitted back to the quantum well in a time τ_e. At high temperature, τ_e is small and the QD ensemble may equilibrate via the quantum well. Consequently, a Fermi distribution is approached at high temperature. In the low-temperature range, τ_e gets quite large (order 10 ms compared to τ_s of order 1 ns). Re-emission is then suppressed, and the average occupation in each dot is made equal within the entire inhomogeneously broadened distribution. An equal occupation of QDs with different ground-state energies represents a highly nonequilibrium distribution.

Modeling results for a realistic QD ensemble referring to data of InAs dots in GaAs matrix are given in Figure 5.2 [9]. In addition to previous work, an excited state is taken into account: Two inhomogeneously broadened QD levels are assumed at −100 and −40 meV with respect to the barrier energy set to 0, cf. gray DOS function. Figure 5.2 illustrates the strong deviation from thermal equilibrium at cryogenic temperatures and the approach to a Fermi function at room temperature under the assumed conditions. The occupation distribution is essentially controlled by $\tau_e/\tau_s \times \exp((E_c - E_f)/(k_B T))$ [8]. The transition between the two regimes, therefore, depends strongly on the localization energy of the confined charge carriers.

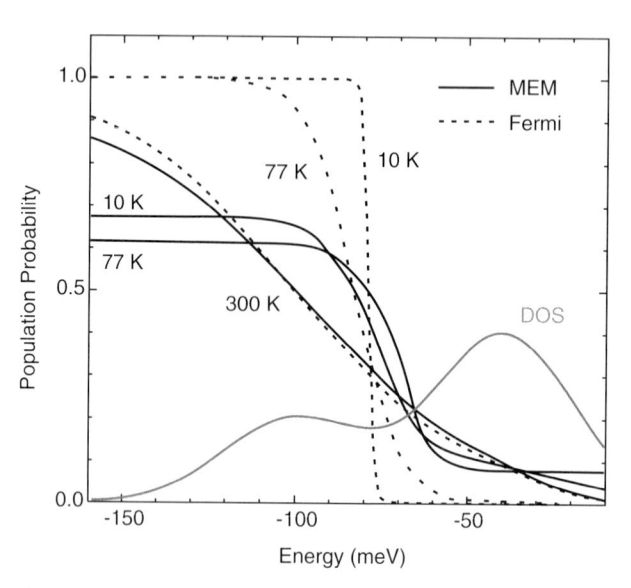

Figure 5.2 Carrier distribution function calculated using either master equations for the microstates (MEM, black lines) or a Fermi function (dashed). DOS denotes the assumed density of states of the quantum dot ensemble. Data reproduced from Ref. [9] by permission of Elsevier B.V.

5.2.2
Species of Quantum Dots

Early attempts to fabricate QDs by patterning of quantum wells resulted in dots with poor optical efficiency. In the 1990s, an effective approach for fabricating coherent, defect-free semiconductor QDs was developed by employing strained-layer epitaxy in the Stranski–Krastanow growth mode [10, 11]. InAs quantum dots in GaAs matrix represent the best studied model system for such self-organized transformation from two-dimensional layer-by-layer growth to a three-dimensional mode. Driving force is a reduction of the elastic strain energy. The material in the spontaneously formed islands is not constrained by the surrounding material and may relax laterally, thereby forming typically mid 10^{10} dots cm^{-2}. The action of entropy at growth temperature leads to ensembles of dots that vary in size, shape, and composition [12]. Consequently, the energies of bound states vary within the ensemble. Some part of the InAs material does not redistribute to form islands but remains as a two-dimensional layer because the surface free energy of InAs is lower than that of the GaAs substrate [13]. This wetting layer forms a quantum well after covering the structure by a cap layer.

The existence of a wetting layer connected to the Stranski–Krastanow QDs turned out to give rise to limitations in the performance of QD devices. Capture and re-emission of charge carriers into and from the dots proceed essentially via the wetting layer. The localization energy of carriers is hence significantly lowered because wetting layer states lie well below barrier states. Moreover, carrier transfer from the wetting layer into the dots is considered to be a rate-limiting step in the dynamics of relaxation [14].

The approach of submonolayer quantum dots [15] was developed to build quantum dot ensembles *without* a wetting layer. InAs/GaAs submonolayer QDs are fabricated by a cycled alternate deposition of InAs with an average thickness below a single monolayer (i.e., below half a lattice constant of InAs) and a thin GaAs spacer layer (typically few monolayers). The amount of InAs in each individual deposition is well below the critical thickness for three-dimensional QD formation in the Stranski–Krastanow mode. The first InAs submonolayer deposition causes initial island formation on the surface [16]. After rapid coverage of these islands by a thin GaAs layer, locations of islands nucleating in the next InAs submonolayer are controlled by the nonuniform strain distribution of the buried InAs islands [17, 18]. The described procedure should not be confused with seemingly similar preparations. Cycled depositions of submonolayers for *both* InAs *and* GaAs depositions lead to dots comparable to the Stranski–Krastanow QDs [19]. Cycled depositions applied on top of Stranski–Krastanow QDs result in dots termed columnar QDs [20], which are also connected to a wetting layer.

5.2.3
Energies of Confined Charge Carriers

A general feature of size-quantized media is the ability to adjust the energy of confined carriers by a variation of size at constant material composition. In a

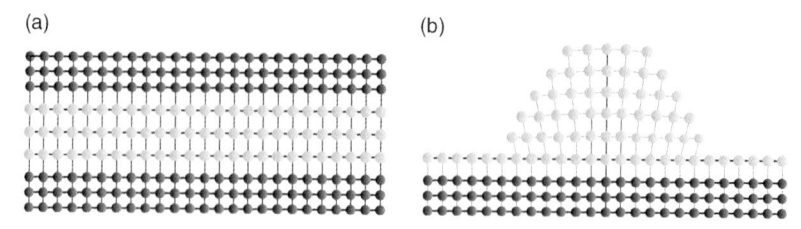

Figure 5.3 Scheme illustrating elastic relaxation in a quantum well (a) and a quantum dot (b). Light and dark circles represent atoms of size-quantized materials and barrier materials, respectively.

quantum well, quantization is given in a single spatial direction. For infinite barriers, the energy eigenvalues with quantum number n are simply given by $E_n = \hbar^2 \pi^2 n^2 / (2m^* t^2)$, t being the thickness of the well. Energies of bound states in a quantum well with *finite* barriers, normalized to E_1, have comparable, somewhat decreased values. A quantum well has a homogeneous thickness and chemical composition. The strain in the well is then solely a function of composition x, and consequently, also homogenous in the well. Therefore, the energy of an exciton confined in a well with given barrier materials is just determined by the two parameters, thickness t and composition x, of the well. The emission wavelength of exciton recombination may therefore be readily calculated.

In a quantum dot, the strain is inhomogeneous, as illustrated in Figure 5.3. The resulting shear deformations of a dot made of zincblende material then give rise to piezoelectric fields. The piezoelectric potential has a large impact on the electronic properties of a QD [21–23]. Such contribution does not exist in biaxially strained quantum wells. Moreover, usually also the composition of the QD material is not homogeneous. The eigenenergies of carriers confined in a QD are therefore a complex function of size, strain, and composition of the dot. Determination of the emission wavelength of a QD and a targeted adjustment by changing the dot geometry requires advanced modeling using, for example, **k·p** theory [24, 25] or density functional theory [22].

Interesting target wavelengths for frequency-doubled semiconductor disk lasers in full color display applications are in the range 940–1220 nm. Realistic 8-band **k·p** calculations of QD structures were performed to provide a roadmap for a directed and controlled growth with defined QD size, composition, spacers, and barriers [25]. In addition to emission wavelength, the objective of QD development must consider high-density stacked QD arrays to accomplish large modal gain and thermal stability at operation temperatures.

Simulations of electronic and optical properties must naturally assume the size, shape, and composition of the quantum dots. The structure of a typical InGaAs/GaAs Stranski–Krastanow QD designed for laser applications is shown in the cross-sectional STM image (Figure 5.4). Bright areas indicate In-rich material. We note that the capped, alloyed InGaAs QD has an inhomogeneous distribution of the In/Ga ratio within the dot. Strain-induced In migration during growth leads to an inverse truncated cone composition [26, 27]. Moreover, the shape is found to be quite flat. This is in contrast to uncovered dots like that depicted in Figure 5.3. Quantum dots

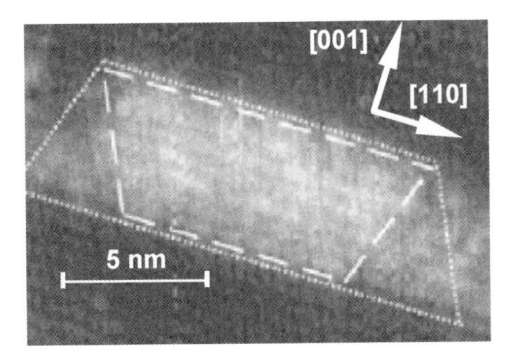

Figure 5.4 Cross-sectional scanning tunneling micrograph of an $In_{0.8}Ga_{0.2}As$ quantum dot, overgrown with a 3 nm thick $In_{0.1}Ga_{0.9}As$ quantum well for wavelength tuning. The contours of the dot (bright area) are indicated by a dotted line, and those of its In-rich core by dashed lines. Data reproduced from Ref. [26] by permission of the American Institute of Physics.

undergo strong structural changes upon deposition of the cap layer. The cap introduces strain that creates a general tendency to build a flat top facet as studied in detail for InAs/GaAs QDs [28, 29].

Based on the structural data illustrated in Figure 5.4 basic QD structures with an emphasis on a realistic composition profile have been varied systematically to assess their importance in view of electronic and optical properties and to create guidelines for targeted wavelength tuning. Figures 5.5 and 5.6 summarize results obtained from **k·p** modeling on relevant QD structures [23, 25] (A. Schliwa, personal communication). Calculations include single- and multiple particle energies because excited states are known to contribute significantly to lasing transitions due to state filling. The total amount of InAs in each QD has usually been kept fix to separate volume effects from the effect of the studied parameter. A fixed InAs amount is important because the nominal amount of InAs deposited during epitaxy and the QD density are well known. Therefore, an average InAs content per QD can be estimated from growth conditions. The total amount of InAs needs to be constant to discriminate between the different manifestations of a QD by comparing to calculated spectra.

The quantum size effect known from quantum wells may also be clearly observed in quantum dots of varied size for fixed shape and composition. Figure 5.5a exemplifies for an InAs/GaAs dot of truncated pyramidal shape that both electron and hole localization increase as the QD size increases. The effect is particularly pronounced for the ground-state electrons because most of the wave function lies in the high-energy barrier material if the dot is small [25]. In practice, it is difficult to preserve a constant, size-independent shape during QD formation. Usually the vertical aspect ratio a given by height over base length changes. Consequences of such shape change are shown in Figure 5.5b. The assumed shape of an InAs/GaAs dot varies from a very flat truncated pyramid with $a = 0.04$ to a full pyramid with $\{110\}$ side facets, a base length of 17 nm, and $a = 0.5$ [25]. The initial strong decrease of electron and hole energies for enlarged aspect ratio originates from the dominant size quantization in growth direction and a connected delocalization in flat structures.

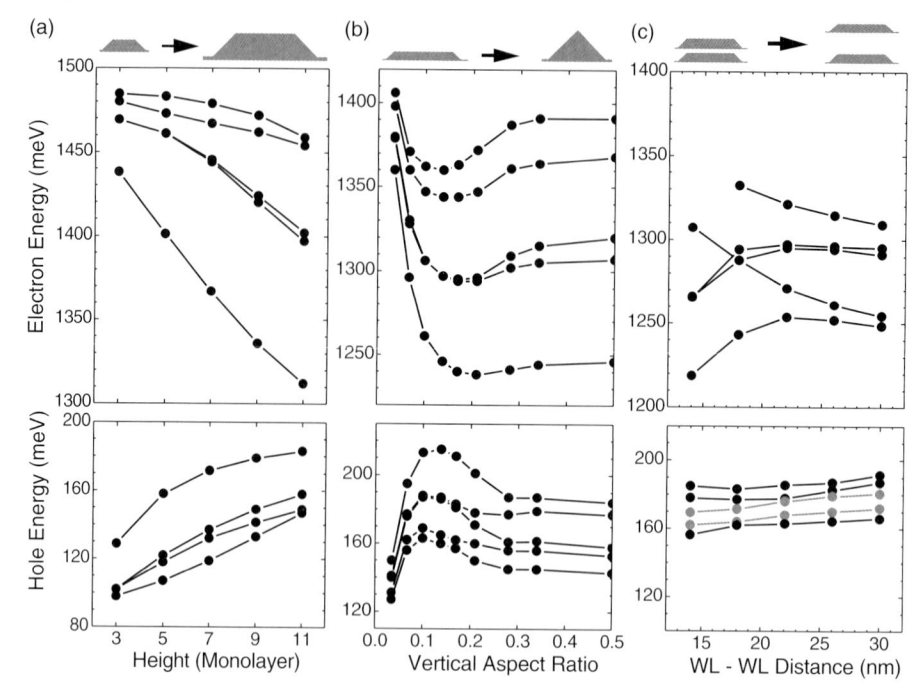

Figure 5.5 Simulated energies of electrons (top) and holes (bottom) confined in InAs/GaAs quantum dots with varied structural parameters: effect of (a) size and (b) vertical aspect ratio. (c) Two identical InAs QDs, which are separated by a GaAs spacer layer of varied thickness (A. Schliwa, personal communication). Energy values refer to the valence–band edge of GaAs, a large hole energy corresponds to a strongly confined hole. (a) Reproduced from Ref. [25] and (b) reproduced from Ref. [23] by permission of the American Physical Society.

Above $a = 0.1$, we note a reversal of the trend, resulting from a counteracting effect induced by strain: the strain character changes from biaxial in flat dots to more hydrostatic. This effect increases the local bandgap and thereby also the quantization energy. The increased splitting of the first two excited electron states results from an increased piezoelectric field in a full pyramid.

The effect of a *homogeneous* change in the composition x of an $In_xGa_{1-x}As/GaAs$ dot is largely comparable to the effect of size. An increase in the In fraction x on the cation sublattice reduces the bandgap and increases localization (not shown). Figure 5.4 shows that the In distribution in a QD may be quite *inhomogeneous*. The effect of a changed aspect ratio a on energy was calculated assuming a circular lens-shaped dot with a reverse cone composition. Such composition profile was found in structural investigations [26, 27]. The In-rich core concentrates in the apex, leading to a reduced total strain in the dot. Figure 5.6a shows that the aspect ratio has an only minor impact on electron and hole energies [23]. Similar results were also obtained when the inhomogeneous distribution of the In-rich cone was assumed to have another angle.

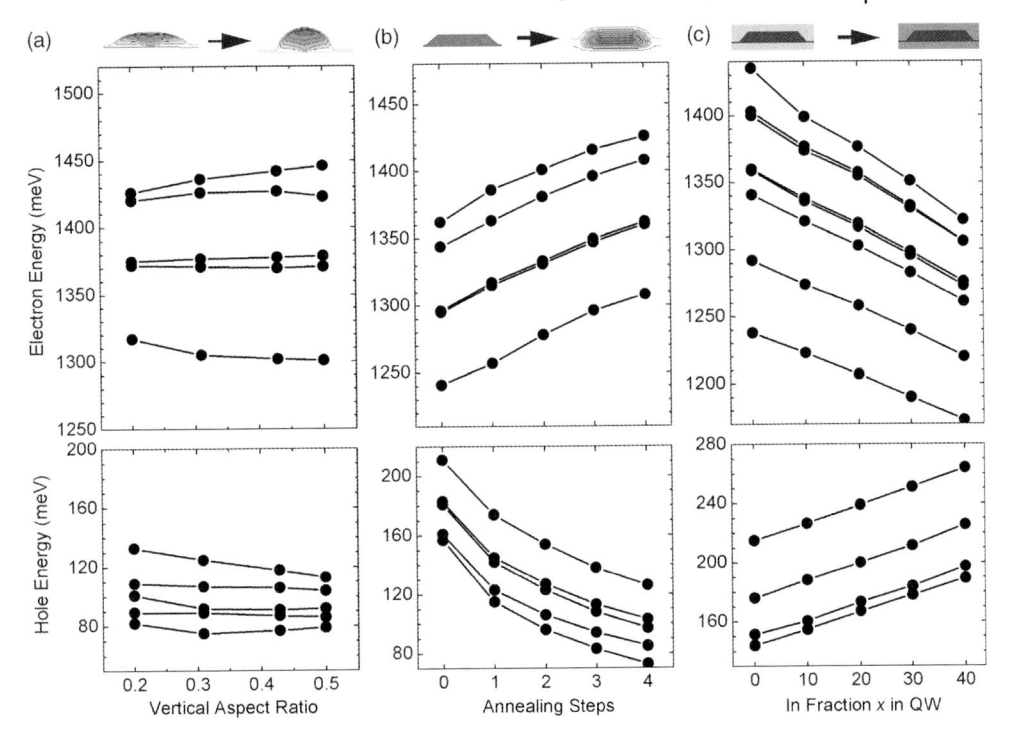

Figure 5.6 Calculated electron and hole energies for In$_x$Ga$_{1-x}$As/GaAs dots with inhomogeneous composition (a and b) and an InAs/GaAs dot of fixed size in an In$_x$Ga$_{1-x}$As quantum well (c). In (a) the vertical aspect ratio of a lens-shaped dot is considered, and in (b) the successive interface softening of an InAs dot in GaAs is assumed. (a) Reproduced from Ref. [23] by permission of the American Institute of Physics.

An inhomogeneous Ga/In distribution may also originate from intermixed interfaces between the InAs QD and the surrounding GaAs matrix. Such intermixing occurs during annealing or due to growth at elevated temperatures above 500 °C. The intermixed interfaces have a strong impact on the energies, as demonstrated in Figure 5.6b (A. Schliwa, personal communication). The interdiffusion is simulated by applying a smoothing algorithm with a variable number of smoothing steps. The intermixing leads to a larger Ga composition and hence a larger bandgap within the volume of the wave function. The effect is largely comparable to an increase of homogeneous Ga composition in the dot.

A widely applied means to adjust the emission wavelength of a quantum dot is the overgrowth by a QW or, additionally, to place a QW underneath the dot to form a dot-in-a-well (DWELL) structure [30, 31]. The QW reduces the hydrostatic stress exerted on the QD by the matrix, and consequently, decreases the bandgap energy. The impact on confined carrier energies is assessed using a QD shape of a truncated pyramid embedded in a QW of lower In content. The results for a QD with 20 nm base length and 3.5 nm height, embedded in a 4.5 nm thick QW of varied In composition, are given in Figure 5.6c (Schliwa, A., personal communication). With increasing In content of the QW, a strong increase in localization is found, in agreement with the experimentally observed red shift of respective structures. Calculated absorption spectra (not shown here) indicate a very strong contribution of excited states to the emission in the case of population.

In a VECSEL structure, the dot layers are stacked with separating spacer layers. The effect of a vertical stacking of two identical InAs/GaAs QDs on the electronic properties is shown in Figure 5.5c. The mutual separation is given in terms of the distance of one wetting layer (WL) to the other WL. The general conclusion from such calculations is that no coupling of the usually well-confined holes occurs. Also, no coupling occurs for electrons, provided the WL–WL distance exceeds ∼35 nm. This is a typical range for closely stacked QD layers. Note that energy shifts also occur for much thicker spacers due to long-range strain fields.

The simulations provide useful guidelines to tune the emission wavelength of actual InGaAs/GaAs QD structures to targeted wavelengths.

5.2.4
Quantum Dot Lasers

Early attempts to fabricate quantum dots by employing dry etching of quantum well structures lead to QDs with low luminescence efficiency. The implementation of such dots into a laser resulted in devices with poor performance [32]. A breakthrough occurred when the approach of self-organized Stranski–Krastanow growth was applied to realize ensembles of defect-free dots with a high areal density. The technique is basically very simple and does not require any significant modification of the growth equipment or any processing technology. It should be noted that early studies were accompanied by pessimistic statements about inherent properties of quantum dot ensembles. The relaxation of captured carriers was assumed to be seriously hampered by a phonon bottleneck effect, and the orthogonality of electron and hole wave functions was supposed to strongly limit the radiative recombination [33]. Moreover, the inhomogeneous broadening within a QD ensemble and the assumed small total number of dots participating in lasing operation were considered a major drawback to achieve sufficient output power [34].

5.2.4.1 Edge-Emitting Quantum Dot Lasers
The first injection lasers based on self-organized quantum dots grown using molecular beam epitaxy (MBE) were edge-emitting ridge-waveguide devices with a single layer of $In_{0.5}Ga_{0.5}As/GaAs$ Stranski–Krastanow dots [35]. The devices demonstrated the predicted advantages [4, 7] of zero-dimensional structures over those of higher dimensionality. A low threshold current density of $120\,A\,cm^{-2}$ with a low temperature sensitivity expressed by a high characteristic temperature of 350 K was observed at low-temperature operation. The excellent performance was, however, maintained only up to 120 K. A number of processes were claimed to effect the depart from the characteristics of an idealized structure: thermal redistribution of carriers to nonlasing states within the dots [36], thermally induced escape of carriers

out of the dots [37], and nonradiative mechanisms such as Auger [38] and defect mechanisms.

Substantial improvements of device characteristics were achieved by various approaches. An excitation of carriers to nonlasing states is particularly critical for holes because their energy levels are more closely spaced due to their larger effective mass. A permanent occupation of low-lying hole states by p-type modulation doping largely suppressed the effect and lead to a T_0 of 213 K up to 80 °C [39]. An excitation of carriers to the wetting layer and barrier states may occur essentially for electrons and remains presently a critical issue, particularly for long-wavelength DWELL structures. A significant progress was achieved in materials improvement. The steps noted here were achieved by metal-organic vapor-phase epitaxy (MOVPE). The first MOVPE-grown QD lasers used either a 3-fold InAs QD stack [40] or a 10-fold $In_{0.5}Ga_{0.5}As$ QD stack [41]. Both approaches achieved ground-state lasing at room temperature, featuring a T_0 of 385 K up to 50 °C for the 10-fold stack laser. The stacks were grown at the same low temperature for QD layers and GaAs spacers. Substantially improved performance was obtained by introducing temperature cycling to grow the spacer layers at increased temperature and to smoothen the spacer–dot interfaces [42]. For threefold stacked dot layers emitting at 1.16 μm, threshold and transparency current densities of 110 and 18 A cm^{-2}, respectively, with internal quantum efficiency exceeding 90% were achieved [43]. An internal loss below 1.5 cm^{-1} substantiates the benefit of high-temperature spacer growth between the low-temperature QD depositions. All these devices were grown using arsine as group-V precursor. An appreciating data of 30 A cm^{-2} transparency current and 91% internal quantum efficiency were likewise reported for lasers grown using the less-hazardous replacement tertiarybutylarsine [44].

Quantum dot lasers benefit from the reduced lateral charge carrier diffusion [45] due to trapping in the dots. The effect was shown to suppress beam filamentation observed in quantum well lasers [46]. Furthermore, nonradiative surface recombination decreases. Consequently, the robustness against facet degradation in high-power operation is enhanced, and deep etch through the active region in narrow-stripe ridges is possible without surface passivation [47].

Much work was devoted to exploit the ability of InGaAs/GaAs Stranski–Krastanow dots to emit at wavelength near and beyond 1.3 μm. The widely applied DWELL approach modeled in Figure 5.6c or the application of a strain-reducing layer (SRL) basically lowers the emission energy by reducing the strain exerted on the buried QD layer with respect to a GaAs cap. It must be noted that the *overall* strain in the QD stack is *increased* by the well. Furthermore, the *local* strain at large In-rich QDs required for long wavelength emission is rather high. This easily leads to formation of In-rich large dislocated clusters, which degrade the device characteristics. Using the DWELL approach and carefully adjusting the growth parameters for each individual QD layer in the stack to avoid defect formation, lasing at 1250 nm with a very low threshold current density of 66 A cm^{-2} and 94% internal quantum efficiency was obtained [48]. The target of lasing beyond 1.3 μm was accomplished with DWELL devices grown using MBE. By applying a temperature cycling comparable to that described above for MOVPE, operation at 1.307 μm with an extraordinary low threshold of 33 A cm^{-2}

($17\,\text{A cm}^{-2}$ with facet coating) was achieved [49]. InGaAs-based devices grown using MOVPE with addition of antimony demonstrated lasing at 1.35 μm with a high modal gain of $19.3\,\text{cm}^{-1}$ [50].

The submonolayer approach was also employed to fabricate quantum dot lasers. A modal gain as high as $44\,\text{cm}^{-1}$ was measured for a 10-fold cycled SML dot layer [51], and a high-power operation of 3.9 W with good conversion efficiency was reported [52].

5.2.4.2 Surface-Emitting Quantum Dot Lasers

Vertical-cavity surface-emitting lasers (VCSELs) make high demands on the gain medium due to a short cavity length in the range of the emission wavelength λ. The key issues are similar to those required for VECSELs, where the upper distributed Bragg mirror (DBR) is replaced by a window layer and an output-coupling mirror in an external cavity: a high modal gain must be combined with minimal internal losses to achieve sufficient amplification per round-trip. Devices based on quantum dots, therefore, need a high volume density of dots, precisely placed in the antinodes of the optical field. The issue implies both a high areal density and a close stacking of dot layers.

The density of the dot ensemble and the inhomogeneous broadening are related to the maximum modal gain g_{max}. The saturation gain of a single dot layer is given by [53]

$$g_{max} \propto \frac{s\lambda^2}{\tau} \frac{\Gamma \varrho_{QD}}{(\Delta\varepsilon)_{inhom}},$$

where s, τ, Γ, and ϱ_{QD} are the degeneracy of the involved states, the spontaneous radiative lifetime, the modal optical confinement factor at transition energy, and the QD area density, respectively. $(\Delta\varepsilon)_{inhom}$ is the inhomogeneous broadening of the ensemble emission.

The lasing wavelength of a VCSEL is not defined by the spectral position of the peak gain but by the cavity band pass. Transitions from the excited states at the cavity wavelength may therefore significantly contribute to lasing. The impact of excited states on the gain spectrum is illustrated in Figure 5.7. A model assuming master equations for the microstates (MEM) was employed to evaluate the change of gain for increased injection [54], assuming typical values for an ensemble of InGaAs dots in GaAs matrix, that is, $(\Delta\varepsilon)_{inhom} = 30\,\text{meV}$ for ground- and excited-state transitions, $E_1 - E_0 = 60\,\text{meV}$, $\varrho_{QD} = 5 \times 10^{10}\,\text{cm}^{-2}$, and $\tau = 1\,\text{ns}$. The gain given in Figure 5.7 is scaled in units of the maximum gain of the ground-state transition. This value results from an excitation of two excitons per dot and per lifetime.

The gain spectrum of the dot ensemble in Figure 5.7a indicates saturation of the ground-state gain near 1.12 eV already at moderate injection. The contributions of excited-state transitions near 1.18 eV gradually lead to a broadening and a blueshift of the gain spectrum as the injection is increased. They also enhance the modal gain at the ground-state transition energy. Transitions from the excited states involve a number of closely spaced levels. They have in total a higher degeneracy and oscillator

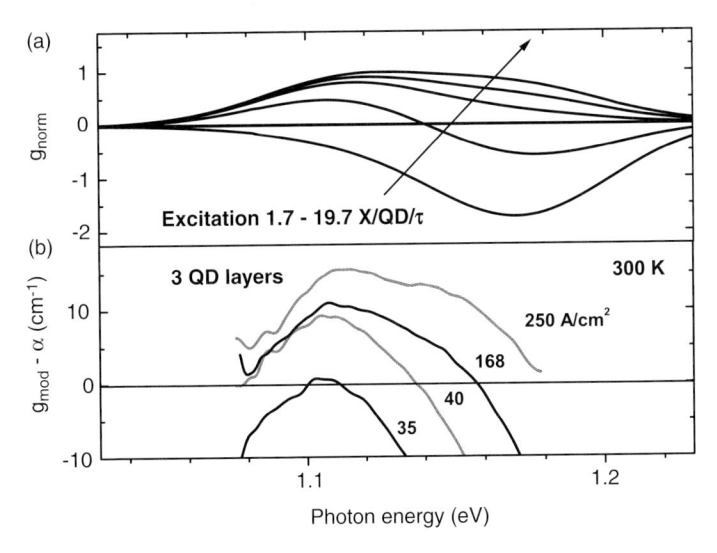

Figure 5.7 (a) Calculated normalized gain of a dot ensemble for gradually increased occupation with excitons ranging from 1.7 to 19.7 excitons per dot and per lifetime τ. (b) Measured modal gain of threefold stacked InGaAs/GaAs dots for gradually increased injection current. Reproduced from Ref. [56] by permission of the American Institute of Physics.

strength than the ground state, shifting the gain maximum eventually to the energy of excited-state transitions (not shown here). Figure 5.7b plots measured gain spectra of a threefold dot stack for similar values of inhomogeneous broadening and $E_1 - E_0$. We note the saturation of the ground-state gain near 1.1 eV and good qualitative agreement with the MEM calculation. Modeling demonstrates that dots used for VCSELs must have a high volume density and small inhomogeneous broadening.

GaAs-based VCSELs are advantageous with respect to InP-based devices, particularly due to large refractive index steps in lattice-matched GaAs/AlAs DBR mirrors and a higher thermal conductivity obtained to remove dissipated heat. The feasibility of fabricating VCSELs based on quantum dots was demonstrated for GaAs substrates applying both growth techniques, MBE [55] and MOVPE [56]. The implemented $In_{0.5}Ga_{0.5}As$ dots were grown in the Stranski–Krastanow mode either 10-fold or 3 × 3-fold stacked, respectively. Operation requires a good matching of the QD emission energy to the cavity dip, which defines the lasing wavelength. Optical *in situ* reflectance applied during MOVPE allowed the efficient optimization of the growth procedure [57, 58]. By placing 3 × 3 groups of InGaAs/GaAs dot layers in each of the three central field antinodes of a 4λ cavity, ground-state lasing at 20 °C with 1.45 mW cw (6 mW pulsed) output power at 1.1 μm was demonstrated. Limitations were basically given by nonradiative recombination in the barrier or wetting layers at high carrier densities.

VCSEL operation was also demonstrated with submonolayer quantum dots fabricated as described in Section 5.2.2 [59, 60]. The SML dots are typically formed by cycled depositions of 0.5–0.8 monolayer (ML) InAs and ∼2.5 ML GaAs. The absence of a wetting layer promises a fast dynamics of carrier relaxation. In fact, an extraordinary large modulation bandwidth allowing a high-speed 20 Gbit s^{-1} operation even at elevated temperature up to 120 °C was recently reported [61]. The submonolayer approach hence provides an interesting alternative to the conventional Stranski– Krastanow growth in quantum dot-based gain media.

5.3
Development of Disk Lasers Based on Quantum Dots

The unique properties of quantum dots stimulated studies to implement them into semiconductor disk lasers. Initial work focused on their applications in semiconductor saturable absorber mirrors (SESAMs) for passive mode locking [62]. QD SESAMs provide an independent control of saturation fluence and modulation depth, which may be adjusted by the dot density [63]. Thus, low saturation fluence at moderate modulation depth can be achieved. Due to a fast recovery dynamics, pulses as short as 114 fs were demonstrated [64].

Advances in growth control eventually allowed the implementing of quantum dots also in gain media of semiconductor disk lasers. VECSEL operation was demonstrated with both the Stranski–Krastanow QDs [65, 66] and submonolayer QDs [67]. The stepwise development of these devices is summarized in the following.

5.3.1
Concepts of Gain Structures

Gain considerations pointed out in Section 5.2.4.2 indicate the crucial requirement of a high dot density in the volume of the gain medium, realized by a high areal density and a dense stacking of QD layers. The issue is particularly challenging for dots emitting at long wavelength because these dots are comparably large and have consequently large strain fields. The design and growth procedure must avoid a strain-induced formation of defects (e.g., Ref. [68]), which degrade the performance of the device. Moreover, the strain-induced structural coupling of dots in adjacent layers may lead to a vertical alignment with an increased inhomogeneous broadening due to a gradually increasing QD size [69]. A significant progress in the development of edge emitters was achieved by growing the dots at low temperature, but the separating spacer layers at increased temperature [42]. Based on the thermal stability of covered dots below 600 °C [70], a procedure to flatten the growth front prior to dot-layer deposition and to overgrow the dots was established that maintains a high radiative recombination efficiency of the QDs. The procedure was also applied in metal-organic vapor-phase epitaxy of the first QD-based SDLs.

The basic design considerations for the gain section of a semiconductor disk laser are similar for quantum dots and quantum wells. The first realized QD gain chips implemented a layout of resonant periodic gain to selectively enhance gain at the operating wavelength. The chips consist of a semiconductor Bragg mirror, a GaAs pump light absorber containing stacks of active QD layers, and a confinement window that prevents diffusion of photoexcited charge carriers toward the

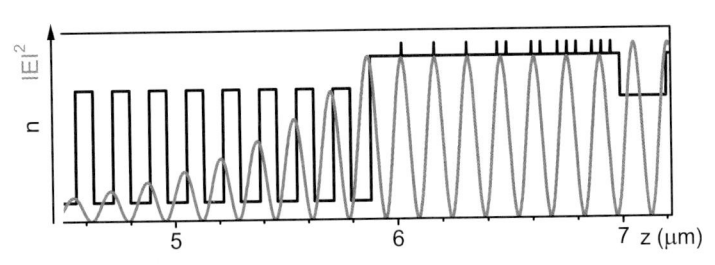

Figure 5.8 Vertical refractive index profile (black lines) and calculated optical field (gray curve) of a semiconductor disk laser with implemented quantum dots (design 1).

surface [65–67]. The dot layers are located at the antinodes of the optical field, and the window layer thickness is adjusted to match the subcavity resonance of the lasing wavelength. The structures were modeled using the transfer matrix method for normal incidence of the light. The nominal reflectivities of the DBR mirrors were 99.92 and 99.96%, achieved by using either 35.5 or 37.5 $Al_{0.98}Ga_{0.02}As/Al_{0.2}Ga_{0.8}As$ pairs of $\lambda/4$ layers, respectively. An Al composition of 0.2 was chosen in the high index part of the mirror pairs so as to avoid light absorption at operation wavelength. The structure of the device is illustrated in Figure 5.8. Quantum dot layers represent small grouped peaks in the gain region, which is clad by the DBR and a window layer on the left- and right-hand sides, respectively.

Two different approaches were used for the gain region. The first design intends to maximize the modal gain per dot layer, while the second design aims for a maximum total modal gain by increasing the number of dot layers. Quantum dots for the three target wavelengths 940, 1040, and 1220 nm were implemented into the resonant gain structures. The submonolayer QDs were employed for 940 [71] and 1040 nm emission [67]. Stranski–Krastanow QDs emitting from the ground state were used for 1220 nm [66], and Stranski-Krastanow QDs operating at the first excited state were used for 1040 nm emission [65]. VECSEL cw operation was demonstrated for all approaches.

The first design of maximum gain per dot layer was used for 1040 and 940 nm targets. The resonant gain region shown in Figure 5.8 comprises 13 dot layers divided into two groups of three (2 × 3), two groups of two (2 × 2), and three single layers (3 × 1). The three single layers were omitted in the 940 nm device. Each group is centered at the position of an antinode of the standing optical wave. The nonuniform arrangement accounts for the exponentially decreasing pump light intensity within the absorbing GaAs matrix and assures a more homogenous pumping of all QD layers. The gain structure is clad by a DBR comprising 35.5 $\lambda/4$ pairs and an $Al_{0.3}Ga_{0.7}As$ window capped by a 10 nm thick GaAs layer to prevent oxidation.

The second design of maximum total gain was employed for wavelengths exceeding 1200 nm. The larger dots cannot be fabricated with as high areal density and cannot be stacked as dense as short wavelength dots. The number of QD layers is therefore increased to 7 threefold groups, which are symmetrically positioned at optical antinode positions. The reflectance of the DBR is slightly increased by adding 2 mirror-layer pairs. From the characteristics of QD edge emitters based on comparable dots, a transparency condition of about 10 A cm^{-2} can be deduced [72]. Using 21 QD layers, this condition is already fulfilled at an incident pump power density below 1 kW cm^{-2}. Given a modal gain value of $\sim$5 cm^{-1} per QD layer for the ground-state transition [30], modeling shows that lasing can be achieved with 21 QD layers using a highly reflecting output coupling mirror ($R > 99.8\%$) in the external cavity. The topmost window is realized by a lattice-matched 65 nm thick In$_{0.48}$Ga$_{0.53}$P top layer, which is transparent for the pump light and does not require an (absorbing) GaAs oxidation protection.

5.3.2
Adjustment of Quantum Dot Emission Wavelength

The Stranski–Krastanow and the submonolayer growth modes represent two alternatives for high-density QDs. In the InGaAs material system, SML depositions are particularly well suited for short wavelengths due to the large amount of GaAs within the range of confined charge carriers. Long wavelengths beyond 1200 nm are rather accessible using the ground-state emission of Stranski–Krastanow dots. Both kinds of quantum dots have quite different optical properties. The photoluminescence (PL) of Stranski–Krastanow dots and submonolayer dots is compared in Figure 5.9.

The Stranski–Krastanow dots show a very broad spectrum. This leads to a virtually constant gain, which hardly depends on the spectral position of the subcavity resonance formed by the DBR and the window layer. Hence, a change from room temperature to operating conditions (366 K) does not cause an appreciable resonance-gain detuning. On the other hand, the submonolayer dots exhibit a narrow

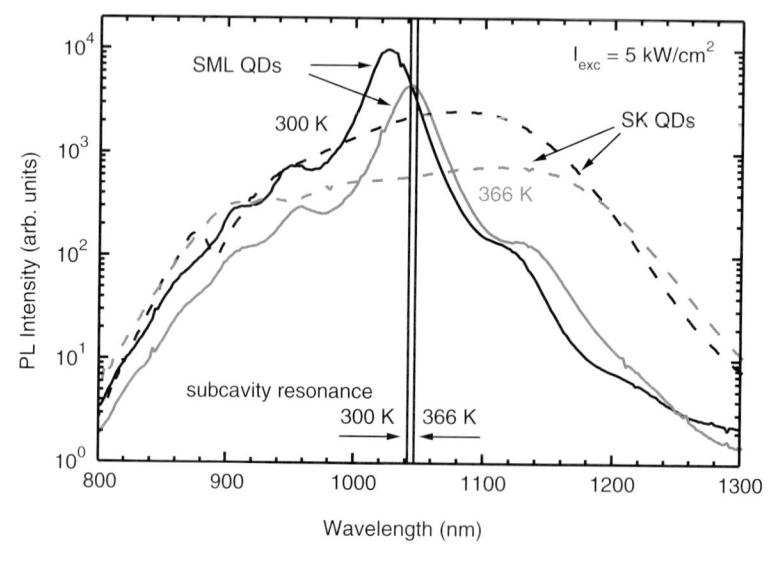

Figure 5.9 Photoluminescence of the threefold stacks of submonolayer dots (solid lines) and Stranski–Krastanow dots (dashed lines), measured at 300 K (black) and 366 K (gray). The two vertical lines signify the thermal shift of a subcavity resonance at these temperatures. Data reproduced from Ref. [71] by permission of Elsevier B.V.

luminescence with a significant thermal shift. Careful adjustment of emission wavelength at operation temperature is therefore required.

5.3.2.1 Tuning of Stranski–Krastanow Quantum Dots

The Stranski–Krastanow QDs were fabricated for 1040 and 1220 nm emission. Dots with ground-state emission at 1040 nm showed comparatively poor emission intensities at operation temperatures yielding low gain. Therefore, the QD ground-state emission was tuned to 1100 nm to match *the first excited state* of the ensemble emission to 1040 nm. The gain generated per QD layer is then theoretically doubled with respect to ground-state emission due to the larger degeneracy of the excited states. The quantum dots were grown from $\sim$2.7 ML $In_{0.65}Ga_{0.35}As$ layers without a quantum well overgrowth. A high areal dot density of $10^{11}\,cm^{-1}$ is obtained for the given growth conditions. Stacking studies demonstrated that a minimum spacer thickness of 35 nm was necessary to maintain high optical quality. Low V/III ratios (<5) along with moderate growth rates ($\sim$1 $\mu m\,h^{-1}$) were found to enhance PL emission intensities and to narrow PL linewidth. Figure 5.10a demonstrates very good optical characteristics of threefold QD stacks.

Dots for 1220 nm operation wavelength range require ground-state emission of the Stranski–Krastanow QDs. The dots were formed from 2.7 ML $In_{0.68}Ga_{0.32}As$ overgrown with a strain-reducing 3.9 nm thick $In_{0.12}Ga_{0.88}As$ well. Details of the growth procedure are given in Ref. [44]. Stacking was investigated to reduce the spacer width and to maximize the total stacking number. The higher total strain per QD layer compared to the 1040 nm QDs required a larger spacing distance of 45 nm. In addition, the duration of growth interruption after QD deposition, the thickness and composition of the InGaAs strain-reducing layer on top of the dots, and the V/III ratio of the strain-reducing layer were optimized. Figure 5.10b shows the response of the QD emission on parameter variation for one example. A reduction in the InGaAs layer thickness for QD growth by only 5% results in a significant change in inhomogeneous broadening.

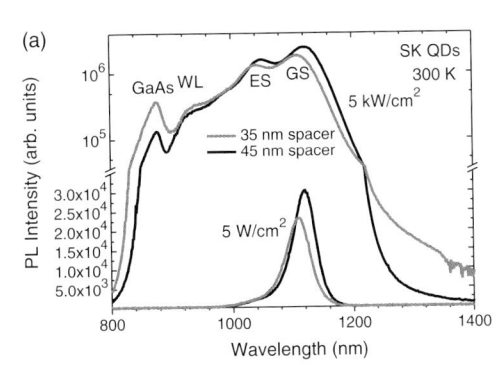

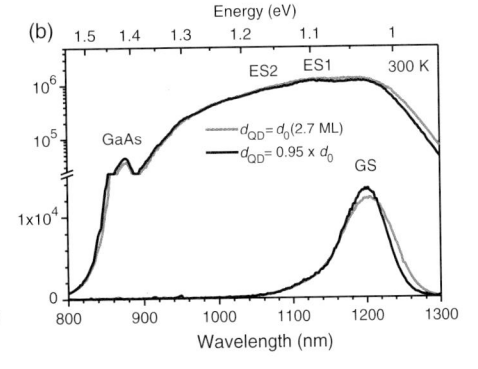

Figure 5.10 (a) PL of threefold stacked S–K QDs with varied spacer thickness. The emission wavelength is adjusted for the first excited state (ES) being resonant to the subcavity of the VECSEL. GS and WL denote emissions from the ground state and the wetting layer, respectively. (b) PL of threefold stacked S–K QDs with slightly varied InGaAs deposition thickness.

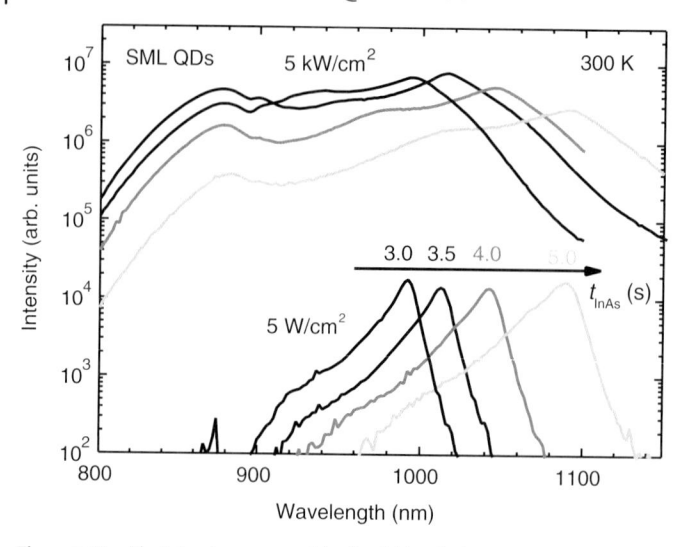

Figure 5.11 Photoluminescence of the fivefold cycled InA/GaAs submonolayer QD samples grown with varied duration of the InAs deposition. Bottom and top spectra are excited using a low and high excitation density, respectively.

5.3.2.2 Tuning of Submonolayer Quantum Dots

Submonolayer QDs were fabricated for 940 nm and 1040 nm emission. The SML QDs were grown by cycled depositions of pure binaries InAs and GaAs. The tuning of the emission wavelength may be performed by adjusting either the InAs deposition duration or the number of InAs/GaAs cycles. The effect of InAs thickness variation is demonstrated in Figure 5.11. The wavelength control is somewhat facilitated by adjusting the number of cycles, yielding robust reproducibility with similar optical performance.

The wavelength tuning of SML QDs must consider proper alignment of the PL peak wavelength with the cavity resonance of the VECSEL structure. Since the actual operating temperature is determined by extrinsic parameters like pump power and heat dissipation, a thermal shift of 80 K was assumed. Hence, a detuning of 35 nm of the SML-QD emission at room temperature was applied. The inhomogeneous broadening of SML-QD emission could be varied over a comparatively small range of 25 meV.

The submonolayer QDs for 940 nm emission were grown using a fivefold cycle of 0.5 ML InAs and 2.3 ML GaAs. Studies of stacking three such SML-QD layers using GaAs spacers from 60 to 10 nm showed a constant high PL intensity down to 20 nm thickness and a drop of intensity for thinner spacers. A thickness of 20 nm for three layers in an antinode was therefore chosen for device applications to obtain a large overlap to the optical field in the gain region.

SML QDs for 1040 nm operation were grown in the same way as those for 940 nm emission. Wavelength tuning was performed by increasing the number of submonolayer cycles per SML-QD layer. Limits for the given conditions were found to be 1070 and 960 nm for the long and short wavelength side, respectively. An emission of 1040 nm was obtained with 10 InAs/GaAs cycles of nominally 0.5 ML InAs and 2.3 ML GaAs. A minimum spacer thickness of 20 nm could be achieved without structural degradation.

5.3.3
Characteristics of Quantum Dot Disk Lasers

First results on SDLs based on quantum dots have been reported recently [65–67]. Data presented here will certainly soon be improved.

For the demonstration of VECSEL operation, the gain chips were capillary water bonded to a 300 µm thick natural diamond heat spreader and mounted on a water-cooled copper heat sink. The QD gain mirrors were placed into a V-shaped cavity comprising a curved mirror with 200 mm radius of curvature and a plane output coupler (OC) mirror. An 808 nm fiber-coupled diode laser was used for pumping the SDL structure at an incident angle of 35° to the surface normal. The pump light was focused on a 180 µm diameter spot that accurately matched the size of the cavity mode.

5.3.3.1 Disk Lasers with Stranski–Krastanow Quantum Dots

The first implementation of dots in semiconductor disk lasers employed S–K dots for 1040 nm emission from the first excited state [65]. The QD layers were integrated into a resonant gain structure according to design 1 given in Figure 5.8. The spectral characteristics of the gain chip are shown in Figure 5.12a. Surface PL occurs at the wavelength of the small dip in the reflectivity of the DBR stopband that is attributed to the subcavity resonance. The small thermal shift of the PL peak position by 0.075 nm K^{-1} is determined by the shift of the subcavity dip.

The lasing spectrum of the S–K-QD SDL given in Figure 5.13 shows Fabry–Perot fringes due to the etalon effect of the diamond intra cavity heat spreader. Using 1% of output coupling and a heat sink kept at 15 °C, a cw output power of 0.28 W is obtained. The threshold pump power is 6.5 W and the slope efficiency is 6.7%. At high pump power, an onset of thermal rollover occurs. Data of this QD-SDL represent an encouraging first step.

A potential benefit of employing quantum dots instead of quantum wells in a gain medium is the achievement of a low lasing threshold with a small temperature dependence. Moreover, the ability of elastic strain relaxation in dots allows the extending of the operation wavelength of GaAs-based devices to the near-infrared range. These issues were addressed in a S–K-QD device designed for 1220 nm operation [66]. Design 2 (Section 5.3.1) with a uniform arrangement of 7 × 3 dot layers was used.

The device showed a threshold pump power of 0.48 W at 15 °C for a high-reflectivity (99.8%) output coupler mirror (cf. Figure 5.14). The results demonstrate temperature-stable operation in the measured range. The dependence on pump power density is only 0.027 nm (kW cm^{-2})$^{-1}$. The center wavelength of the emission shifts only by 0.06 nm K^{-1}, leaving the output power largely unaffected. Such shift is almost an order of magnitude smaller than typical values of 0.3 nm K^{-1} observed for SDLs based on GaInNAs quantum wells [73]. The temperature-independent differ-

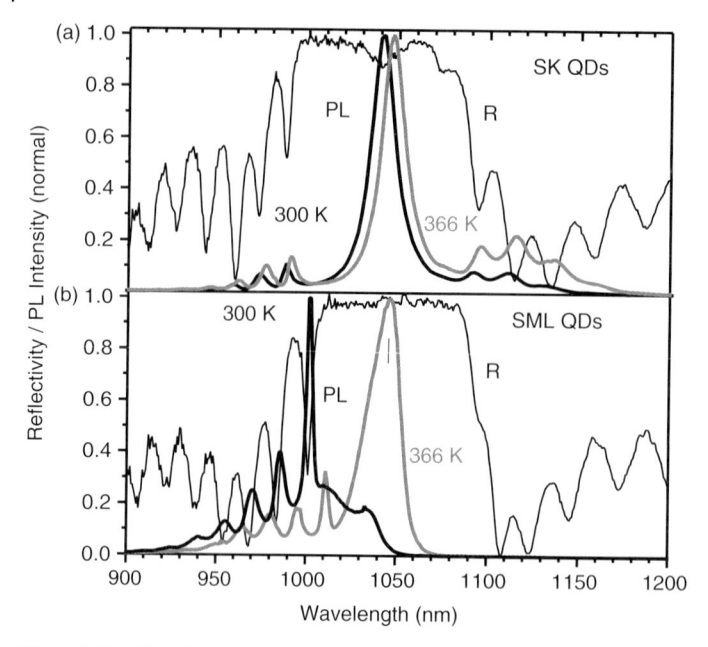

Figure 5.12 Photoluminescence (PL) from the surface of (a) an S–K-QD SDL and (b) a SML-QD SDL, recorded at 300 (black) and at 366 K (gray). The thin curves are reflectivities (R) of the devices measured at 300 K. Data reproduced from Ref. [71] by permission of Elsevier B.V.

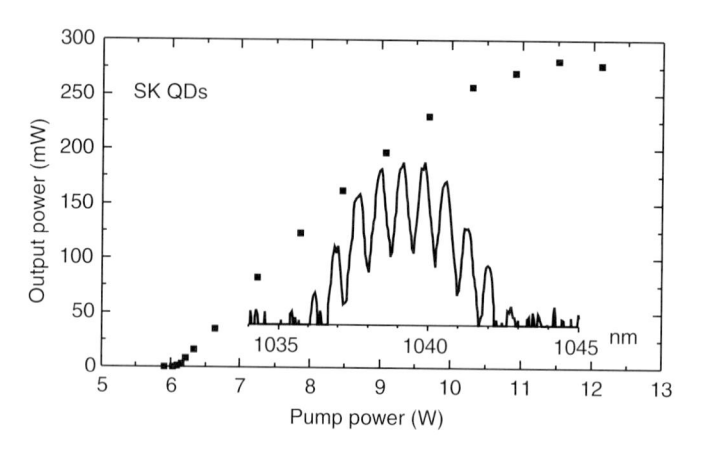

Figure 5.13 Output power from the first S–K-QD SDL. The operation wavelength of 1040 nm is matched to the emission of the first excited state of the dots. *Inset*: Laser spectra with fringes from the diamond heat spreader. Sample TU Berlin, data courtesy of O.G. Okhotnikov (University of Tampere).

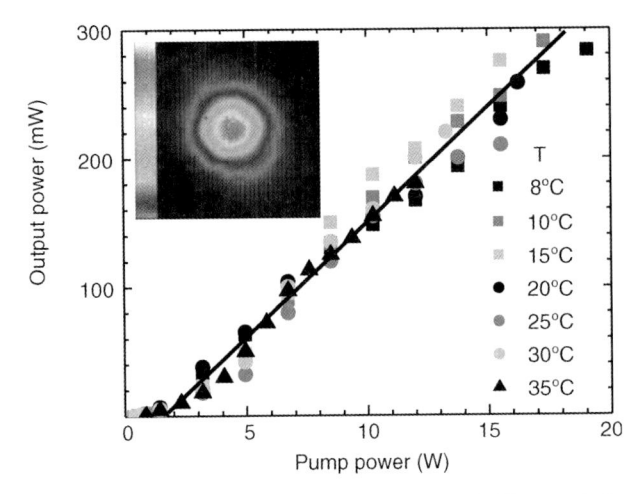

Figure 5.14 Output power from a S–K-QD SDL at 1210 nm emission wavelength for various temperature settings T of the heat sink. The inset gives the spatial intensity profile of the beam recorded at 170 mW. Reproduced from Ref. [66] by permission of the American Institute of Physics.

ential efficiency of the device is 2%. The low value originates from the limited modal gain of the structure and is an issue of future improvements.

5.3.3.2 Disk Lasers with Submonolayer Quantum Dots

Results on SDLs based on quantum dots obtained so far yield a higher gain achieved with submonolayer dots, accompanied by a smaller wavelength range of gain. Using design 1 (Figure 5.8) for 1035 nm emission, 13 SML-QD layers, each comprising 10-fold cycled depositions of nominally 0.5 ML InAs and 2.3 ML GaAs, were integrated into a gain structure [67]. The larger gain of SML-QDs compared to SK-QDs translates to an increased slope efficiency. Characteristics of the device given in Figure 5.15 yield a slope efficiency of 12.4% (15% if optical interfaces are taken into account). Due to a larger efficiency, no significant thermal rollover is observed. A maximum cw output power of 1.4 W is obtained for 1% outcoupling.

Good characteristics were likewise obtained from an SML-QD SDL fabricated for 940 nm emission [71]. An output power of 0.5 W at 1% outcoupling was achieved using 10 layers of fivefold cycled SML-QDs, limited by the available pump source.

5.4 Conclusions

The performance of semiconductor disk lasers based on quantum wells benefits from the quite mature fabrication technology of classical heterostructures. The field of quantum dots is comparably young and still rapidly developing. First results on SDLs based on quantum dots proved advantages of using zero-dimensional structures in gain media, comparable to those previously found in QD-based injection laser

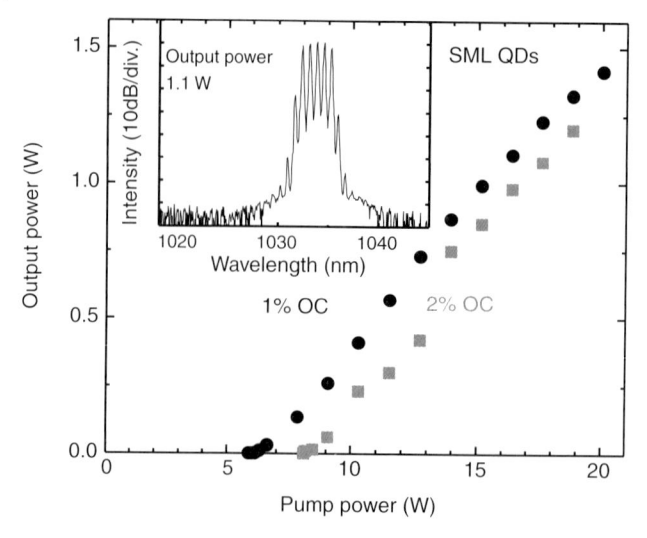

Figure 5.15 Output power from a SML-QD SDL for two output couplers (OC). *Inset*: Laser spectra recorded at 1.1 W. Reproduced from Ref. [67] by permission of the American Institute of Physics.

diodes. SDLs with Stranksi–Krastanow QDs demonstrated a low lasing threshold of 0.48 W with a remarkable small temperature dependence of 0.06 nm K^{-1}.

An important issue for the further development of QD-based SDLs is the dense stacking of QD layers. A significant progress was very recently reported for QD-SDLs emitting at 1060 nm grown using MBE [74]. The achievement of employing GaAs spacers as thin as 10 nm in sevenfold InGaAs dot stacks in a device comprising 5 × 7 QD layers leads to the demonstration of 27 mW output power in mode-locked 18 ps pulses at 2.57 GHz repetition rate. Very recently an SDL with such InAs/GaAs dot stacks produced a CW output power of 4.35 W with 22% slope efficiency at 1032 nm [75]. The reported maximum output was limited by the available pump power. Data presented here will be significantly improved in the near future.

The reviewed recent proof of concept introduced quantum dots to the field of semiconductor disk lasers. Implementation of dots instead of wells will enrich applications of vertical-external-cavity surface-emitting lasers.

References

1 Bimberg, D., Grundmann, M., and Ledentsov, N.N. (1998) *Quantum Dot Heterostructures*, John Wiley & Sons, Ltd, Chichester, UK.
2 Bimberg, D., Kirstaedter, N., Ledentsov, N.N., Alferov, Zh.I., Kop'ev, P.S., and Ustinov, V.M. (1997) *IEEE J. Sel. Top. Quantum Electron.*, **3**, 196.
3 Dingle, R. and Henry, C.H. (1976) Quantum effects in heterostructure lasers. US Patent 3,982,207.
4 Asada, M., Miyamoto, Y., and Suematsu, Y. (1986) *IEEE J. Quantum Electron.*, **QE-22**, 1915.
5 Kirstaedter, N., Schmidt, O.G., Ledentsov, N.N., Bimberg, D., Ustinov, V.M.,

Egorov, A.Yu., Zhukov, A.E., Maximov, M.V., Kop'ev, P.S., and Alferov, Zh.I. (1996) *Appl. Phys. Lett.*, **69**, 1226.

6 Schmidt, O.G., Kirstaedter, N., Ledentsov, N.N., Mao, M.-H., Bimberg, D., Ustinov, V.M., Egorov, A.Yu., Zhukov, A.E., Maximov, M.V., Kop'ev, P.S., and Alferov, Zh.I. (1996) *Electron. Lett.*, **32**, 1302.

7 Arakawa, Y. and Sakaki, H. (1982) *Appl. Phys. Lett.*, **40**, 939.

8 Jiang, H. and Singh, J. (1999) *J. Appl. Phys.*, **85**, 7438.

9 Grundmann, M. (2000) *Physica E*, **5**, 167.

10 Leonard, D., Pond, K., and Petroff, P.M. (1994) *Phys. Rev. B*, **50**, 11687.

11 Grundmann, M. (ed.) (2002) *Nano-Optoelectronics*, Springer, Berlin.

12 Shchukin, V.A., Ledentsov, N.N., and Bimberg, D. (2002) Entropy effects in self-organized formation of nanostructures, in *NATO Advanced Workshop on Atomistic Aspects of Epitaxial Growth* (eds M. Kortla *et al.*) Kluwer, Dordrecht.

13 Wang, L.G., Kratzer, P., Scheffler, M., and Moll, N. (1999) *Phys. Rev. Lett.*, **82**, 4042.

14 Vallaitis, T., Koos, C., Bonk, R., Freude, W., Laemmlin, M., Meuer, C., Bimberg, D., and Leuthold, J. (2008) *Opt. Express*, **16**, 170.

15 Straßburg, M., Heitz, R., Türck, V., Rodt, S., Pohl, U.W., Hoffmann, A., Bimberg, D., Krestnikov, I.L., Shchukin, V.A., Ledentsov, N.N., Alferov, Zh.I., Litvinov, D., Rosenauer, A., and Gerthsen, D. (1999) *J. Electron. Mater.*, **28**, 506.

16 Bressler-Hill, V., Lorke, A., Varma, S., Petroff, P.M., Pond, K., and Weinberg, W.H. (1994) *Phys. Rev. B*, **50**, 8479.

17 Xie, Q., Madhukar, A., Chen, P., and Kobayashi, N.P. (1994) *Phys. Rev. Lett.*, **75**, 2542.

18 Shchukin, V.A., Bimberg, D., Malyshkin, V.G., and Ledentsov, N.N. (1998) *Phys. Rev. B*, **57**, 12262.

19 Huffaker, D.L. and Deppe, D.G. (1998) *Appl. Phys. Lett.*, **73**, 520.

20 Kita, T., Tamura, N., Wada, O., Sugawara, M., Nakata, Y., Ebe, H., and Arakawa, Y. (2006) *Appl. Phys. Lett.*, **88**, 211106.

21 Grundmann, M., Stier, O., and Bimberg, D. (1995) *Phys. Rev. B*, **52**, 11969.

22 Bester, G., Wu, X., Vanderbilt, D., and Zunger, A. (2006) *Phys. Rev. Lett.*, **96**, 187602.

23 Schliwa, A., Winkelnkemper, M., and Bimberg, D. (2007) *Phys. Rev. B*, **76**, 205324.

24 Stier, O., Grundmann, M., and Bimberg, D. (1999) *Phys. Rev. B*, **59**, 5688.

25 Schliwa, A., Winkelnkemper, M., and Bimberg, D. (2009) *Phys. Rev. B*, **79**, 075443.

26 Lenz, A., Timm, R., Eisele, H., Hennig, Ch., Becker, S.K., Sellin, R.L., Pohl, U.W., Bimberg, D., and Dähne, M. (2002) *Appl. Phys. Lett.*, **81**, 5150.

27 Liu, N., Tersoff, J., Baklenov, O., Holmes, A.L., and Shih, C.K. (2000) *Phys. Rev. Lett.*, **84**, 334.

28 Pohl, U.W., Pötschke, K., Schliwa, A., Guffarth, F., Bimberg, D., Zakharov, N.D., Werner, P., Lifshits, M.B., Shchukin, V.A., and Jesson, D.E. (2005) *Phys. Rev. B*, **72**, 245332.

29 Costantini, G., Rastelli, A., Manzano, C., Acosta-Diaz, P., Songmuang, R., Katsaros, G., Schmidt, O.G., and Kern, K. (2006) *Phys. Rev. Lett.*, **96**, 226106.

30 Lester, L.F., Stintz, A., Li, H., Newell, T.C., Pease, E.A., Fuchs, B.A., and Malloy, K.J. (1999) *IEEE Photon. Technol. Lett.*, **11**, 931.

31 Ustinov, V.M., Maleev, N.A., Zhukov, A.E., Kovsh, A.R., Egorov, A.Yu., Lunev, A.V., Volovik, B.V., Krestnikov, I.L., Musikhin, Yu.G., Bert, N.A., Kop'ev, P.S., Alferov, Zh.I., Ledentsov, N.N., and Bimberg, D. (1999) *Appl. Phys. Lett.*, **74**, 2815.

32 Forchel, A., Leier, H., Maile, B.E., and Germann, R. (1988) Fabrication and optical spectroscopy of ultra-small III–V compound semiconductor structures, in *Advances in Solid State Physics*, vol. **28** (ed. U. Rössler), Pergamon/Vieweg, Braunschweig, pp. 99.

33 Benisty, H., Sotomayor-Torrès, C.M., and Weisbuch, C. (1991) *Phys. Rev. B*, **44**, 10945.

34 Vahala, K.J. (1988) *IEEE J. Quantum Electron.*, **24**, 523.

35 Kirstaedter, N., Ledentsov, N.N., Grundmann, M., Bimberg, D., Ustinov, V.M., Ruvimov, S.S., Maximov, M.V., Kop'ev, P.S., Alferov, Zh.I., Richter, U.,

Werner, P., Gösele, U., and Heydenreich, J. (1994) *Electron. Lett.*, **30**, 1416.

36 Kapteyn, C.M.A., Lion, M., Heitz, R., Bimberg, D., Brunkov, P.N., Volovik, B.V., Konnikov, S.G., Kovsh, A.R., and Ustinov, V.M. (2000) *Appl. Phys. Lett.*, **76**, 1573.

37 Le Ru, E.C., Fack, J., and Murray, R. (2003) *Phys. Rev. B*, **67**, 245318.

38 Marko, I.P., Andreev, A.D., Adams, A.R., Krebs, R., Reithmeier, J., and Forchel, A. (2003) *Electron. Lett.*, **39**, 58.

39 Shchekin, O.B. and Deppe, D.G. (2002) *IEEE Photon. Technol. Lett.*, **14**, 1231.

40 Heinrichsdorff, F., Mao, M.-H., Kirstaedter, N., Krost, A., Bimberg, D., Kosogov, A.O., and Werner, P. (1997) *Appl. Phys. Lett.*, **71**, 22.

41 Maximov, M.V., Kochnev, I.V., Shernyakov, Y.M., Zaitsev, S.V., Gordeev, N.Yu., Tsatsul'nikov, A.F., Sakharov, A.V., Krestnikov, I.L., Kop'ev, P.S., Alferov, Zh.I., Ledentsov, N.N., Bimberg, D., Kosogov, A.O., Werner, P., and Gösele, U. (1997) *Jpn. J. Appl. Phys.*, **36**, 4221.

42 Sellin, R.L., Heinrichsdorff, F., Ribbat, Ch., Grundmann, M., Pohl, U.W., and Bimberg, D. (2000) *J. Cryst. Growth*, **221**, 581.

43 Sellin, R.L., Ribbat, Ch., Grundmann, M., Ledentsov, N.N., and Bimberg, D. (2001) *Appl. Phys. Lett.*, **78**, 1207.

44 Sellin, R.L., Kaiander, I., Ouyang, D., Kettler, T., Pohl, U.W., Bimberg, D., Zakharov, N.D., and Werner, P. (2003) *Appl. Phys. Lett.*, **82**, 841.

45 Kim, J.K., Strand, T.A., Naone, R.L., and Coldren, L.A. (1999) *Appl. Phys. Lett.*, **74**, 2752.

46 Ribbat, Ch., Sellin, R.L., Kaiander, I., Hopfer, F., Ledentsov, N.N., Bimberg, D., Kovsh, A.R., Ustinov, V.M., Zhukov, A.E., and Maximov, M.V. (2003) *Appl. Phys. Lett.*, **82**, 952.

47 Ouyang, D., Ledentsov, N.N., Bognár, S., Hopfer, F., Sellin, R.L., Kaiander, I., and Bimberg, D. (2004) *Semicond. Sci. Technol.*, **19**, L43.

48 Strittmatter, A., Germann, T.D., Kettler, Th., Posilovic, K., Pohl, U.W., and Bimberg, D. (2006) *Appl. Phys. Lett.*, **88**, 262104.

49 Sellers, I.R., Liu, H.Y., Groom, K.M., Childs, D.T., Robbins, D., Badcock, T.J., Hopkinson, M., Mowbray, D.J., and Skolnick, M.S. (2004) *Electron. Lett.*, **40**, 1412.

50 Guimard, D., Ishida, M., Hatori, N., Nakata, Y., Sudo, H., Yamamoto, T., Sugawara, M., and Arakawa, Y. (2008) *IEEE Photon. Technol. Lett.*, **20**, 827.

51 Xu, Z., Birkedal, D., Juhl, M., and Hvam, J. (2004) *Appl. Phys. Lett.*, **85**, 3259.

52 Mikhrin, S.S., Zhukov, A.E., Kovsh, A.R., Maleev, N.A., Ustinov, V.M., Shernyakov, Yu.M., Soshnikov, I.P., Livshits, D.L., Tarasov, I.S., Bedarev, D.A., Volovik, B.V., Maximov, V.M., Tsatsul'nikov, A.F., Ledentsov, N.N., Kop'ev, P.S., Bimberg, D., and Alferov, Zh.I. (2000) *Semicond. Sci. Technol.*, **15**, 1061.

53 Asryan, L.V. and Suris, R.A. (1997) *IEEE J. Sel. Top. Quantum Electron.*, **3**, 148.

54 Grundmann, M., Stier, O., Bognar, S., Ribbat, C., Heinrichsdorff, F., and Bimberg, D. (2000) *Phys. Status Solidi A*, **178**, 255.

55 Saito, H., Nishi, K., Ogura, I., Sugou, S., and Sugimoto, Y. (1996) *Appl. Phys. Lett.*, **69**, 3140.

56 Hopfer, F., Kaiander, I., Lochmann, A., Mutig, A., Bognar, S., Kuntz, M., Pohl, U.W., Haisler, V.A., and Bimberg, D. (2006) *Appl. Phys. Lett.*, **89**, 061105.

57 Pohl, U.W., Pötschke, K., Kaiander, I., Zettler, J.-T., and Bimberg, D. (2004) *J. Cryst. Growth*, **272**, 143.

58 Kaiander, I., Hopfer, F., Kettler, T., Pohl, U.W., and Bimberg, D. (2004) *J. Cryst. Growth.*, **272**, 154.

59 Hopfer, F., Mutig, A., Kuntz, M., Fiol, G., Bimberg, D., Ledentsov, N.N., Shchukin, V.A., Mikhrin, S.S., Livshits, D.L., Krestnikov, I.L., Kovsh, A.R., Zakharov, N.D., and Werner, P. (2006) *Appl. Phys. Lett.*, **89**, 141106.

60 Blokhin, S.A., Maleev, N.A., Kuzmenkov, A.G., Sakharov, A.V. Kulagina, M.M., Shernyakov, Yu.M., Novikov, I.I., Maximov, M.V., Ustinov, V.M., Kovsh, A.R., Mikhrin, S.S., Ledentsov, N.N., Lin, G., and Chi, J.Y. (2006) *IEEE J. Quantum Electron.*, **42**, 851.

61 Mutig, A., Fiol, G., Moser, P., Arsenijevic, D., Shchukin, V.A., Ledentsov, N.N., Mikhrin, S.S., Krestnikov, I.L., Livshits, D.L., Kovsh, A.R., Hopfer, F., and

Bimberg, D. (2008) *Electron. Lett.*, **44**, 1303.

62 Garnache, A., Hoogland, S., Tropper, A.C., Gerard, J.M., Thierry-Mieg, V., and Roberts, J.S. (2001) CLEO Europe 2001.

63 Maas, D.J.H.C., Bellancourt, A.-R., Hoffmann, M., Rudin, B., Barbarin, Y., Golling, M., Südmeyer, T., and Keller, U. (2008) *Opt. Express*, **16**, 18646.

64 Lagatsy, A.A., Bain, F.M., Brown, C.T.A., Sibbett, W., Livshits, D.A., Erbert, G., and Rafailov, E.U. (2007) *Appl. Phys. Lett.*, **91**, 231111.

65 Strittmatter, A., Germann, T.D., Pohl, J., Pohl, U.W., Bimberg, D., Rautiainen, J., Guina, M., and Okhotnikov, O.G. (2008) *Electron. Lett.*, **44**, 290.

66 Germann, T.D., Strittmatter, A., Pohl, J., Pohl, U.W., Bimberg, D., Rautiainen, J., Guina, M., and Okhotnikov, O.G. (2008) *Appl. Phys. Lett.*, **93**, 051104.

67 Germann, T.D., Strittmatter, A., Pohl, J., Pohl, U.W., Bimberg, D., Rautiainen, J., Guina, M., and Okhotnikov, O.G. (2008) *Appl. Phys. Lett.*, **92**, 101123.

68 Lenz, A., Eisele, H., Timm, R., Hennig, Ch., Becker, S.K., Sellin, R.L., Pohl, U.W., Bimberg, D., and Dähne, M. (2004) *Appl. Phys. Lett.*, **85**, 3848.

69 Lita, B., Goldman, R.S., Philips, J.D., and Bhattacharya, P.K. (1999) *Appl. Phys. Lett.*, **74**, 2824.

70 Heinrichsdorff, F., Grundmann, M., Stier, O., Krost, A., and Bimberg, D. (1998) *J. Cryst. Growth*, **195**, 540.

71 Germann, T.D., Strittmatter, A., Pohl, U.W., Bimberg, D., Rautiainen, J., Guina, M., and Okhotnikov, O.G. (2008) *J. Cryst. Growth*, **310**, 5182.

72 Germann, T.D., Strittmatter, A., Kettler, T., Posilovic, K., Pohl, U.W., and Bimberg, D. (2007) *J. Cryst. Growth*, **298**, 591.

73 Konttinen, J., Harkonen, A., Tuomisto, P., Guina, M., Rautiainen, J., Pessa, M., and Okhotnikov, O. (2007) *New J. Phys.*, **9**, 140.

74 Hoffmann, M., Barbarin, Y., Maas, D.J.H.C., Golling, M., Krestnikov, I.L., Mikhrin, S.S., Kovsh, A.R., Südmeyer, T., and Keller, U. (2008) *Appl. Phys. B*, **93**, 733.

75 Butkus, M., Wilcox, K.G., Rautiainen, J., Okhotnikov, O., Mikhrin, S.S., Krestnikov, I.L., Kovsh, A.R., Hoffmann, M., Südmeyer, T., Keller, U., and Rafailov, E.U. (2009) *Optics Letters*, **34**, 1672.

6
Mode-Locked Semiconductor Disk Lasers
Thomas Südmeyer, Deran J.H.C. Maas, and Ursula Keller

6.1
Introduction

6.1.1
Ultrafast Lasers

Lasers generating short pulses with picosecond or femtosecond duration (which we refer to as ultrafast lasers) enabled numerous breakthroughs since the early 1990s in both fundamental science and industrial applications [1]. Their great impact on scientific research is reflected by two Nobel Prizes. In 1999, A. Zewail received the Chemistry Nobel Prize for providing a deeper understanding of the dynamics of chemical reactions [2], thanks to the extremely short pulse duration. But ultrafast lasers have shown unique properties in the spectral domain also. High repetition rate pulse trains provide a comb-shaped optical spectrum [3–5]. The excellent comb stability enables precision optical frequency metrology for which J.L. Hall and T.W. Hänsch received the 2005 Nobel Prize in Physics. The impact of ultrafast lasers on technology is equally impressive. Many experiments clearly demonstrate that ultrafast lasers enable new applications and improve the existing technologies in fields as diverse as manufacturing, biology, medicine, communications, and instrumentation.

The ultrashort light pulses are in most cases generated with passively mode-locked lasers. In this regime of operation, usually a single pulse propagates inside a laser cavity and generates an output pulse each time it hits the output coupler (Figure 6.1). The pulse-forming mechanism relies on a saturable absorber, a device that exhibits lower losses for higher intensities, thus favoring pulsed operation against other regimes such as continuous-wave (cw) operation. The emitted beam consists of a sequence of pulses separated by the round-trip time T_R of the resonator. The term "mode locking" for this operation regime refers to the situation in frequency space. In Fourier space, a period pulse train corresponds to a frequency comb with a spacing that is equal to the pulse repetition rate $f_{rep} = 1/T_R$. The individual frequency lines have a fixed phase relationship, they are "locked" to each other (note that the

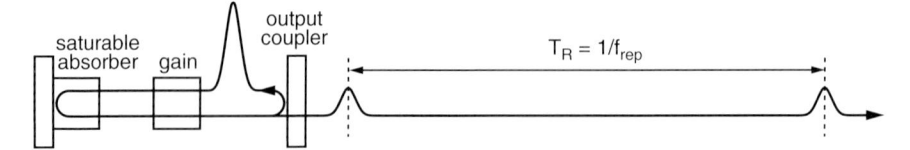

Figure 6.1 Schematic of a passively mode-locked laser oscillator. A single pulse oscillating in the cavity with a round-trip time T_R produces an output pulse each time it hits a partially transmitting output coupler, resulting in a periodic pulse train with a repetition rate $f_{rep} = 1/T_R$.

frequency lines usually deviate slightly from the original frequency of the longitudinal resonator modes, which are not exactly equidistant due to dispersion and cavity nonlinearities). The peak power of the output pulse is higher than the average power; the enhancement depends on the ratio between the round-trip time and the pulse duration. If the pulse duration is short compared to the round-trip time, the peak power of the pulses is substantially higher than the average power. High peak powers are particularly interesting for applications involving nonlinear effects, such as frequency conversion or multiphoton absorption. Reducing the cavity length allows to access repetition rates of several tens of gigahertz, which is important in optical communication or for new applications such as optical clocking of multicore microprocessors.

At present, most ultrafast lasers are not semiconductor lasers, but they use an ion-doped glass or crystal as gain material that is optically pumped. Depending on their gain geometry, these lasers are usually referred to as fiber lasers, waveguide lasers, or bulk solid-state lasers. The term solid-state laser most commonly refers only to lasers relying on a bulk laser crystal or glass, despite the fact that semiconductor, fiber, and waveguide lasers are also solid-state systems. In many aspects, their performance is at present superior to that of ultrafast semiconductor lasers. The shortest pulses directly generated from laser oscillators are achieved by Ti:sapphire lasers. This gain crystal provides a very broad amplification spectrum and enables the generation of pulses with only 5 fs [6–8]. Ultrafast Ti:sapphire lasers are also often used for applications that require wavelength tuning, a tuning range from below 800 nm to above 1000 nm can easily be achieved for longer pulse durations (typically around 100 fs). However, the Ti:sapphire gain material requires complex and expensive green pump lasers. A major advancement was the development of directly diode-pumped solid-state lasers (DPSSL), which enabled the realization of simpler and more reliable ultrafast lasers [9]. The dominant technology for achieving stable and reliable pulse formation in DPSSL are semiconductor saturable absorber mirrors (SESAMs) [10, 11] that are also used for passive mode locking of semiconductor disk lasers (SDLs). SESAMs can be custom designed over a wide range, allowing the precise control of the absorption parameters by the design of the semiconductor layers, the growth parameters, and optional postprocessing. This flexibility makes them ideally suited for the mode locking of various types of lasers. A SESAM is easy to use, usually it simply replaces one end mirror of the laser. SESAM mode-locked DPSSL enabled record high average powers of up to 80 W in femtosecond pulses [12, 13]. Pulse energies above 10 µJ can be directly generated from the ultrafast femtosecond oscillators [14, 15], which allows high-speed machining or high-field science experiments [16]. A very impressive progress was also demonstrated in the area of ultrafast fiber lasers [17]. The performance of ultrafast lasers in terms of repetition rate is equally impressive. The passively mode-locked solid-state lasers have achieved up to 2 W at 10 GHz and up to 110 mW at 160 GHz repetition rate in the 1 μm spectral region [18], and up to 35 mW at 100 GHz in the 1.5 μm telecom spectral region [19]. Fiber lasers also achieve very high repetition rates [20], however, at several tens of gigahertz, they cannot be fundamentally mode locked anymore. Instead, harmonic mode locking (H-ML) is used, where multiple pulses simultaneously circulate in the cavity at once, which is susceptible to pulse drop out and higher noise levels.

6.1.2
Ultrafast Semiconductor Lasers

Despite the numerous demonstrations of the benefit for ultrashort pulses in various applications and excellent performance of ultrafast lasers based on ion-doped laser crystals or glasses, their current market penetration is still relatively low. One reason for this is the high cost. Typical ultrafast solid-state or fiber lasers have still a high degree of complexity, such lasers typically consist of many different optical components that have to be adjusted precisely. Direct electrical pumping is not feasible, and cost-effective manufacturing of the gain element is challenging (often high-quality crystal growth and expensive postprocessing is needed). A crucial task for bringing ultrafast lasers to an increasing number of widespread applications is reducing their complexity and price, while at the same time improving their reliability. Semiconductor lasers are ideally suited for cost-effective mass production. The continuous-wave semiconductor lasers are widely spread over a vast range of applications; they can even be found in everyday devices such as audio CD players, laser printers, optical mice, or barcode readers in supermarkets. Similarly, ultrafast semiconductor lasers allow a high level of integration, which can result in compact and simple devices.

During the past few years, the performance of ultrafast edge-emitting semiconductor lasers substantially improved, the highest output power to date is 250 mW at a pulse repetition rate of 4.3 GHz [21]. Unfortunately, it is very challenging to increase the power levels into the watt regime because the long interaction length in the device introduces significant dispersion and nonlinearities that can lead to strongly chirped pulses and increased timing jitter. Furthermore, end facet damage can occur at high peak powers. It appears difficult to achieve several hundreds of milliwatts average power at pulse repetition rates well above 10 GHz because gain guiding at higher current densities gives rise to higher order transverse modes that destabilize the mode locking. In addition, edge-emitting semiconductor lasers have strongly asymmetric beam profiles that often need to be corrected with precisely mounted lenses. Typically, the same epitaxial layer forms both the gain (with a forward-biased section) and the saturable absorber (with a reverse-biased section), and can therefore not be optimized independently.

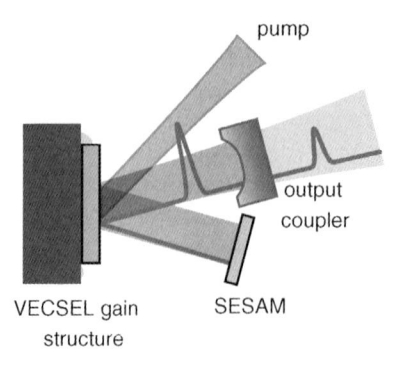

Figure 6.2 Typical setup of a SESAM-mode-locked VECSEL.

Semiconductor disk lasers, also called vertical-external-cavity surface-emitting lasers (VECSELs),[1] bring together the advantages of ultrafast DPSSLs and semiconductor lasers (see Ch. 1, Section 1.2). They combine high average power levels in the multiwatt regime with diffraction-limited, round output beams. Moreover, semiconductor bandgap engineering allows a large design freedom for the laser wavelength and its gain bandwidth. Similar to DPSSLs, introducing a SESAM into the semiconductor disk laser cavity enables self-starting and reliable formation of ultrashort pulses (Figure 6.2) [22]. So far, all passively mode-locked semiconductor disk lasers have relied on this technique.

The first passively mode-locked VECSEL was demonstrated in 2000 by Hoogland et al. [22]. It generated 22 mW average output power in 22 ps pulses at a repetition rate of 4 GHz. During the following years, a large progress in terms of spectral coverage, output power, pulse duration, and repetition rate has been made. Today, the ultrafast VECSELs generate the shortest pulses than generated by any passively mode-locked semiconductor laser without external pulse compression [23, 24], recently even sub-100 fs pulse duration was achieved [25]. They also generate substantially higher average power levels than mode-locked edge emitters, up to 2.1 W was achieved in 4.7 ps pulses [26]. The pulse repetition rate was increased to up to 50 GHz [27] at a power level of 100 mW. All these results had diffraction-limited, circular output beams. With such performance, ultrafast semiconductor disk lasers start to reach a performance level that was previously restricted to ultrafast DPSSLs. Especially, the combination of high average output power and several tens of gigahertz of repetition rate (Figure 6.3) makes these lasers well suited for numerous novel applications.

6.1.3
Application Areas

There are numerous applications that rely on cost-effective, compact ultrafast lasers with high output power and repetition rate. The most promising area in terms of total

---

1) Please note that in the following we will use VECSEL and SDL synonymously.

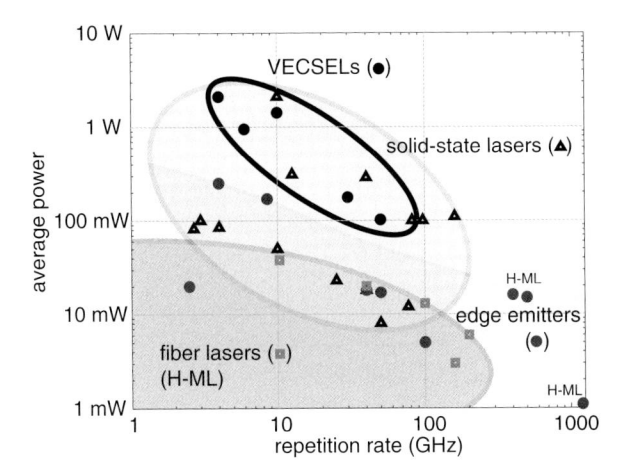

Figure 6.3 Output power for different types of ultrafast lasers operating in the gigahertz regime. To date, the available laser sources in this regime are based on edge-emitting semiconductor lasers, harmonically mode-locked (H-ML) fiber lasers, diode-pumped ion-doped solid-state lasers, or semiconductor disk lasers. Above 100 GHz, the high average-power edge-emitting lasers are also harmonically mode locked ([28], updated in Ref. [29]).

market size is in computer technology. Optical links have better data transmission capabilities than electrical systems, which have made them the standard for long haul telecom networks. Today, optical links become more and more attractive for other transmission systems like rack-to-rack connections having a distance range of 1–100 m. With Moore's law saying that the number of transistors on an integrated circuit is increasing exponentially [30], doubling approximately every 2 years, the required communication bandwidth has also been increasing exponentially. Electrical interconnects have limited scaling properties, and optical interconnects can increase the short-distance communication data rates of the processor with other parts in a computer (such as memory or graphics processing unit). High-speed optical connections have been demonstrated to achieve 30 Gbit per second by direct modulation of a continuous-wave laser [31]. With the combination of multiple lasers, more than 1 Tbit per second was demonstrated. The next step is the use of the pulses from a mode-locked laser, rather than shaping the pulses with a modulator. The short pulse durations, high peak power, wide spectral bandwidth, and low timing jitter lead to simplified synchronization and improved receiver sensitivity [32]. Clock distribution can also profit from pulsed laser sources. In an integrated circuit, a clock signal is needed to synchronize the logic. The important parameters are skew, jitter, and power. A large skew means a large time difference between the arrivals of clock pulses on different locations on the chip, while jitter is a measure for the variation of the pulse to pulse time. While for transistors, the performance improves with the scaling down of the size, the copper clock interconnects have an increased delay because of the smaller wire cross sections. An interconnected bottleneck is expected in the near future if the scaling proceeds as planned. The main advantages brought by optics are a reduction in skew and jitter, a lower sensitivity to temperature variations, and a reduction in power consumption [33, 34]. A powerful mode-locked laser would be an ideal source for optical clocking and enable further performance improvement of microprocessors.

Another promising area for ultrafast lasers is nonlinear frequency conversion into the visible spectral region for laser display applications. Projectors operating with red, green, and blue laser sources achieve excellent color saturation with a wide color gamut. The large focal depth allows projection even onto complex surfaces. With the development of powerful laser projectors, the conventional analogue film rolls could be replaced by digital film with numerous benefits resulting from end-to-end digital handling; for example, higher image quality to loss-free copying, editing, and playing of movie material, as well as faster and cheaper distribution over digital channels, for example, via the Internet. Laser projection displays are also very promising for the consumer market, for example, for efficient back projection televisions or microprojectors that can be directly integrated into notebooks or cell phones. The required projection technology is readily available, and the largest challenge for this application is the availability of suitable visible laser sources. As explained in section (see Ch. 3, Section 3.2), it is challenging to provide compact and powerful lasers directly operating in the visible spectral region, and frequency doubling of an efficient infrared laser is an excellent alternative. Instead of intracavity doubling of continuous-wave lasers, ultrafast high-power lasers can efficiently be frequency converted in a simple single pass through a nonlinear crystal due to their high peak power, which simplifies the conversion stage. Furthermore, such lasers may operate with very short pulses, and their large optical bandwidth could reduce image-disturbing speckle patterns.

Ultrafast lasers are also being used in an increasing number of applications in biology, medicine, and metrology. Focusing pulses with high peak power onto highly nonlinear fibers creates a broad optical spectrum, which can cover a huge spectral region of several hundreds of nanometers or more. Such supercontinuum sources can be used in optical coherence tomography (OCT) [35], which provides high-resolution two- and three-dimensional images of biological tissue in a fast, noninvasive way. OCT is mostly used in ophthalmology [36, 37], but its medical impact has also been proven for gastroenterology, dermatology, intra-arterial imaging, and dentistry. The nonlinear optical effects in biological structures enable high-resolution nonlinear multiphoton imaging. Fluorescence imaging techniques like two-photon excitation microscopy allow to image living tissue up to a depth of 1 mm. Frequency combs generated by femtosecond pulses enable high-precision metrology, which can significantly improve the accuracy of clocks. Such lasers may even be used in space, for example, for future global positioning systems with higher accuracy.

The ultrafast semiconductor disk lasers are promising for all these applications because they have a large design freedom in terms of operation wavelengths and achieve high power levels and short pulse durations. In the following, we describe the most relevant aspects of mode-locked VECSELs. In Section 6.2, we will review passive mode locking of VECSELs with SESAMs. We introduce the main parameters of the

SESAM, discuss the optimization of the key parameters for achieving stable and reliable pulse formation in different operation regimes, and present different designs using quantum well (QW) and quantum dot (QD) absorbers. Section 6.3 gives a detailed overview on the current state-of-the-art ultrafast semiconductor disk lasers. A clear advantage of optically pumped semiconductor disk lasers compared to DPSSLs is that both the gain section and the saturable absorber are realized with semiconductor materials. This allows integrating both functionalities into one single element, which was realized for the first time in 2007 [38]. In Section 6.4, we describe such semiconductor disk lasers with integrated saturable absorbers that are referred to as mode-locked integrated external cavity surface-emitting lasers (MIXSEL).

6.2
SESAM Mode Locking of Semiconductor Disk Lasers

There are various techniques that can force a laser into mode-locked operation. These are classified into two main categories: active mode locking and passive mode locking. In active mode locking, the resonator losses, gain, or phase changes are periodically modulated. In most cases, an electric signal at the round-trip frequency drives an intracavity optical modulator. For electrically pumped semiconductor lasers, it is also possible to directly modulate the driving current. In passive mode locking, the loss modulator is a saturable absorber, a nonlinear passive element that has lower loss for laser pulses with higher energy. As the loss modulation is controlled by the oscillating pulse itself, no synchronization to the pulse repetition frequency is required; therefore, it is called passive mode locking. The shorter the circulating intracavity pulse, the faster the modulation becomes (if the recovery time of the saturable absorber is sufficiently fast). Because the dynamics in saturable absorbers can be faster than the switching speeds of typical active loss modulators, substantially shorter pulses can be generated with passive mode locking compared to active mode locking. There are various techniques to achieve a saturable absorption mechanism, for example, using the saturable absorption of semiconductors or carbon nanotubes [39–41], employing nonlinear frequency conversion [42] or nonlinear phase shifts such as polarization rotation [43] or the Kerr-effect [44]. An overview of the various techniques is given in Refs [1, 9, 45].

Semiconductor saturable absorber mirrors [10, 11, 46] have become key devices for mode locking of various laser types, including diode-pumped solid-state fiber and semiconductor lasers. The invention of the SESAM resolved the long-standing Q-switching instability problems for DPSSLs [47]. The SESAM technology enabled fundamentally mode-locked laser oscillators with record high repetition rates (up to 160 GHz in the 1 μm spectral region [18] and 100 GHz in the 1.5 μm telecom spectral region [19]) and pulse energies of more than 10 μJ [14, 15]. A semiconductor absorbs light when the photon energy is sufficient to excite carriers from the valence band to the conduction band. At sufficiently high pulse energies, the absorber is saturated because possible initial states of the pump transition are depleted while the final states are partially occupied; as a result, the absorption decreases. Semiconductors

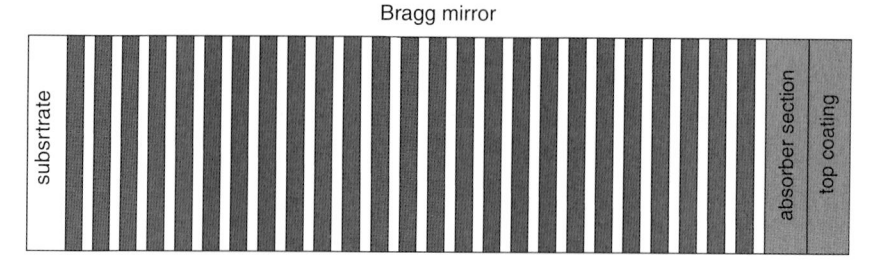

Figure 6.4 A typical SESAM design. The SESAM contains an absorber section integrated into a Bragg mirror. The design of the absorber section and the top coating section enables precise control of the nonlinear reflectivity.

are ideally suited as saturable absorbers because they can cover a broad wavelength range from the visible to the infrared and can access short recovery times, supporting the generation of picosecond to femtosecond pulse durations. A SESAM is a mirror structure with an integrated semiconductor saturable absorber (Figure 6.4), which can be bulk semiconductor, quantum wells, or layers consisting of quantum dots. All these absorbers have the same nonlinear absorption property – the saturation is bleached at high pulse fluences, which increases the reflectivity of the device. The macroscopic nonlinear optical parameters for mode locking can be optimized over a wide range by the properties of the absorber layer and the design of the field propagation inside the absorber section. This flexibility enables stable and self-starting ultrafast operation of many different laser types. So far, all passively mode-locked semiconductor disk lasers have relied on SESAM mode locking.

6.2.1
Macroscopic Key Parameters of a SESAM

Following are the main macroscopic optical parameters of a SESAM determining its suitability for mode locking:

1) The nonlinear optical reflectivity that depends on the saturation behavior of the saturable absorber and the mirror structure.
2) The temporal recovery of the saturable absorption, which is mainly determined by the saturable absorber.

6.2.1.1 Nonlinear Optical Reflectivity
The measurement principle for the nonlinear optical reflectivity is simple: a pulse from an ultrafast laser is reflected by a SESAM, thereby saturating the absorber to a certain degree and experiencing an absorption, which depends on the pulse energy and the transverse mode area on the SESAM. Both the incident pulse energy and the reflected pulse energy are measured as function of the incident pulse energy, which is varied over several orders of magnitude. In Figure 6.5, a simple measurement setup is shown, which uses a beam splitter (BS) and two separate photodiodes to record incident and reflected pulse energies.

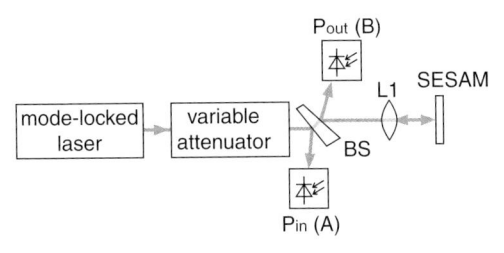

Figure 6.5 Simple measurement scheme for the optical reflectivity of a SESAM.

The recorded macroscopic nonlinear reflectivity of a SESAM is shown as a function of incident pulse fluence F_p on the absorber given by

$$F_p = E_p/A_{abs},$$

where E_p is the incident pulse energy and A_{abs} is the $1/e^2$ intensity mode area of a Gaussian beam on the SESAM. It usually follows a characteristic curve as shown as in Figure 6.6, which is described by the following parameters:

- The *modulation depth* ΔR, that is, the difference in reflectivity between a fully saturated and an unsaturated SESAM (in the absence of induced absorption, see below).
- The *nonsaturable losses* ΔR_{ns}, which are caused by absorption or scattering.
- The *saturation fluence* F_{sat}, which is the pulse fluence for which the SESAM is saturated.

For high fluences, additional nonlinear optical effects can reduce reflectivity, such as two-photon absorption (TPA), or other effects [48, 49]. This behavior is described by the *induced absorption* F_2, which is the fluence where the reflectivity has dropped to $1/e$ due to induced absorption [48, 49].

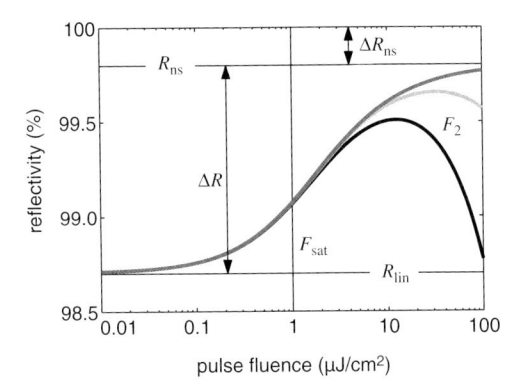

Figure 6.6 Reflectivity as function of the incident pulse fluence. Initially, the reflectivity increases with pulse fluence. Inverse saturable absorption can lead to a rollover and decrease the reflectivity for high fluences. The light gray curve has a small induced absorption parameter F_2, the black curve has a large F_2, and for the dark gray curve, $F_2 = \infty$.

It can be influenced by the SESAM design and depends on the semiconductor material composition, pulse duration, and wavelength of the laser. It is to be noted that in the presence of induced absorption, the modulation depth does not correspond anymore to the maximum achievable difference in reflectivity (see Figure 6.6).

To understand the characteristic shape of the nonlinear reflectivity, it is helpful to first consider a simple model that describes the nonlinear absorption of a traveling-wave ultrashort pulse by a thin semiconductor slab with thickness d.

$$I(z=0, t) = I_{in}(t) \qquad I(z=d, t) = I_{out}(t)$$

We start with the rate equation for a semiconductor close to the quasi-equilibrium [50], neglect carrier diffusion, and assume a linear dependence between the amplitude gain and the carrier density N:

$$\frac{dN(z,t)}{dt} = \Lambda - \gamma_{nr} N - BN^2 - a(N-N_0)\frac{2I}{\hbar\omega}. \tag{6.1}$$

The optical pumping rate Λ is zero in the case of a saturable absorber. The second term represents the nonradiative recombination due to capture by defects in the semiconductor with a decay constant γ_{nr}. The third term is the spontaneous emission with the parameter B. Ultrafast VECSELs usually operate in the slow saturable absorber regime, in which the pulse duration is substantially shorter than the recovery time. Therefore, we can neglect both recombination terms during the pulse interval. The third term represents the absorption ($N < N_0$) or stimulated emission ($N > N_0$), with a being the differential gain coefficient, N_0 the transparency carrier density, and I the time-dependent pulse intensity. The absorber is completely saturated when $N = N_0$. We can rewrite Equation 6.1 in terms of the intensity absorption $\alpha = -2g = -2a(N - N_0)$, obtaining

$$\frac{d\alpha(z,t)}{dt} = -\frac{2a}{\hbar\omega}\alpha(z,t)I = -\frac{1}{F_{sat}}\alpha(z,t)I,$$

with $F_{sat} = \hbar\omega/(2a)$ being the saturation fluence expressed in μJ cm^{-2}. To calculate the transmitted pulse energy through the saturable absorber, we have to solve the following system of differential equations:

$$\frac{\partial I(z,t)}{\partial z} = -\alpha(z,t)I(z,t),$$

$$\frac{\partial \alpha(z,t)}{\partial t} = -\frac{\alpha(z,t)}{F_{sat}}I(z,t).$$

With the initial conditions $\alpha(z, 0) = \alpha_{lin}$ and $I(0, t) = I_{in}(t)$, this can be solved analytically (see Refs [51, 52]). The saturated intensity transmission after the pulse can then be expressed as function of the input pulse fluence F_{in}:

$$T_\infty(F_{in}) = \frac{T_{lin}}{T_{lin} - (T_{lin} - 1)e^{-F_{in}/F_{sat}}},$$

with $T_{\text{lin}} = \exp(-\alpha_{\text{lin}} d)$ being the linear transmission. This expression is the optical transmission that a small probe pulse would see after a strong pulse and can be used, for example, in pump probe experiments to obtain the saturation fluence F_{sat}. Calculating the transmission, the pulse experiences itself from the equations above results in

$$T(F_{\text{in}}) = \frac{F_{\text{out}}}{F_{\text{in}}} = \frac{\ln[1 + T_{\text{lin}}(e^{F_{\text{in}}/F_{\text{sat}}} - 1)]}{F_{\text{in}}/F_{\text{sat}}}.$$

Note that the solution does not depend on the absorption parameters α and the thickness of the saturable absorber, but only on the linear transmission T_{lin} and the saturation fluence F_{sat}. Also, the specific temporal shape of the intensity is not relevant, which is due to the assumption of no internal relaxation during the pulse. The temporal pulse shape of the reflected pulse differs slightly from the incident pulse shape because the leading edge experiences more absorption than the trailing edge. This results in a temporal shift of the center of energy. Another point to mention is that the absorption changes cause instantaneous changes in the refractive index according to the Kramers–Krönig relations.

In a SESAM, there are usually one or several absorber layers with a smaller thickness compared to the wavelength (the $\sim$5–10 nm thickness of a typical InGaAs quantum well is more than an order of magnitude smaller than the wavelength). The absorber layers are integrated into a mirror structure, leading to standing wave effects. Nevertheless, the saturation shows a similar behavior, if the enhancement of the field strength in the absorber is taken into account (for a detailed discussion, see Ref. [52]).

To fully model the nonlinear reflectivity of a SESAM, three other corrections have to be taken into account:

Nonsaturable losses: A real absorber always has some nonsaturable losses, which means that the reflectivity will never be 100%, independent of the incident fluence. Among the mechanisms, those causing these losses are transmission losses through the mirror, nonsaturable defect absorption, free-carrier absorption (FCA), Auger recombination, and scattering losses from rough surfaces. The nonsaturable losses are taken into account by multiplying them with a maximum reflectivity R_{ns}.

Induced absorption: Taking into account the induced absorption at higher fluences, the model function is multiplied with a correction factor $\exp(-F/F_2)$, where F_2 is the induced absorption coefficient:

$$R(F) = R_{\text{ns}} \frac{\ln[1 + R_{\text{lin}}/R_{\text{ns}}(e^{F/F_{\text{sat}}} - 1)]}{F/F_{\text{sat}}} e^{-\frac{F}{F_2}}.$$

One source of induced absorption is two-photon absorption [49, 53], in this case the induced absorption coefficient can be computed using

$$F_2 = \frac{\tau_p}{0.585 \int \beta_{\text{TPA}}(z) n^2(z) |\varepsilon(z)|^4 dz},$$

with $\varepsilon(z)$ being the normalized electric field in the structure, n the refractive index, and β_{TPA} the TPA coefficient expressed in cm GW^{-1}. The induced absorption also depends on the pulse length τ_p; for fs pulses, the induced absorption is very strong and can be approximated by TPA only, whereas for ps pulses, the induced absorption is weaker, however, stronger than that expected from TPA alone [49]. The physical origin of this additional induced nonlinear absorption in the ps regime remains to be clarified.

Correction for the Gaussian transverse mode profile of the laser beam: The transverse mode profile of the pulses does not have a top-hat shape but is usually Gaussian. To obtain the reflectivity for this beam profile and the fluence defined as $F_p = E_p/(\pi w^2)$, we have to integrate over the entire beam profile and obtain [54]

$$R^{Gauss}(F_p) = \frac{1}{2F_p} \int_0^{2F_p} R(F) dF,$$

which can only be computed numerically.

The influence of these three corrections is graphically illustrated in Figure 6.7. The dashed lines indicate the reflectivity without the induced absorption, and the solid lines show the effect of the induced absorption at higher pulse fluences. By integrating the model function according to the Gaussian beam profile, the measured reflectivity is more averaged, resulting in less steep slopes as indicated by the black lines.

The experimentally observed nonlinear reflectivity agrees very well to the fitting function of this model, as is shown in Figure 6.7. The shown SESAM has a nonlinear reflectivity that is well suited for VECSEL mode locking. The parameters of the SESAM have to be matched to the laser type for which mode locking is targeted. For

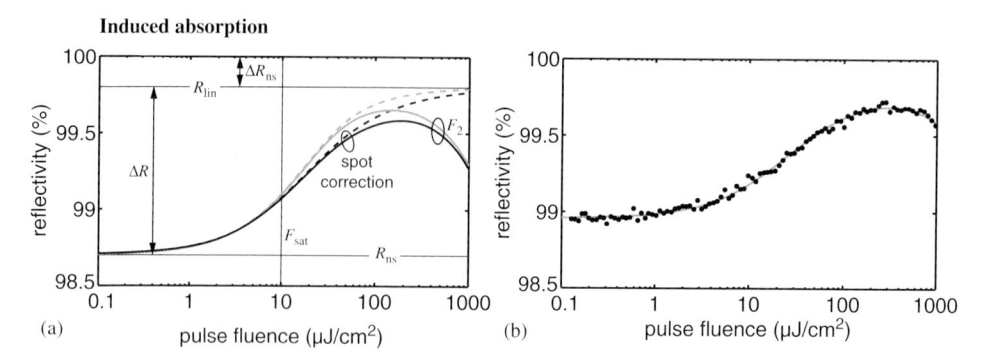

Figure 6.7 Corrections on the model function needed for better agreement with real SESAMs. (a) An absorber always has some nonsaturable losses and induced absorption at higher fluences (solid lines versus dashed lines). Measured with a Gaussian beam, the slope becomes less (black lines versus gray lines). (b) These corrections result in good agreement with measurements of the nonlinear reflectivity, here shown for a typical QD-SESAM (see Section 6.2.3.2).

diode-pumped solid-state lasers and surface-emitting semiconductor lasers, a small modulation depth in the range of 0.5–2% is usually sufficient for self-starting passive mode locking. The amount of acceptable nonsaturable loss strongly depends on the available gain and the output coupling rate, high-power DPSSLs often operate at <0.1% to reduce excessive thermal heating, which otherwise becomes severe at several hundred watts of intracavity average power.

For designing a mode-locked laser, the precise knowledge of the SESAM absorber parameters is crucial. Despite the simple measurement principle, such measurements are challenging: the total change of reflectivity for the above shown sample is below 1%, and an absolute accuracy of better 0.1% is normally needed for a variation of the input pulse fluence by four orders of magnitude. Using the measurement setup shown in Figure 6.5, great care has to be taken to obtain a linear response of the two photodetector (PD) arms. A simpler measurement method was presented in Ref. [55]. Instead of detecting the incident and reflected power levels simultaneously by two different detectors, these signals are separated in time and measured with the same detector system. In this way, it is possible to improve the accuracy and linearity, while at the same time reducing the requirements on the measurement system. The measurement part consists of a nonpolarizing beam splitter cube, a lens (L1), a chopper wheel, and a photodetector. The separation in time is achieved by a chopper wheel that successively blocks the reference and the SESAM arms (Figure 6.8). Despite the simple and cost-effective approach, an accuracy of <0.05% was achieved over a dynamic range of four orders of magnitude.

6.2.1.2 Temporal SESAM Response

Another important characteristic of the SESAM is its temporal response. In many mode-locked lasers, the minimum achievable pulse duration is determined by the recovery time of the SESAM (it is to be noted that this minimum duration is often substantially shorter than the recovery time, see also Section 6.2.2). The SESAM recovery dynamics are measured with a pump-probe setup, a time-resolved differential reflection measurement. A beam from a mode-locked laser is first split into

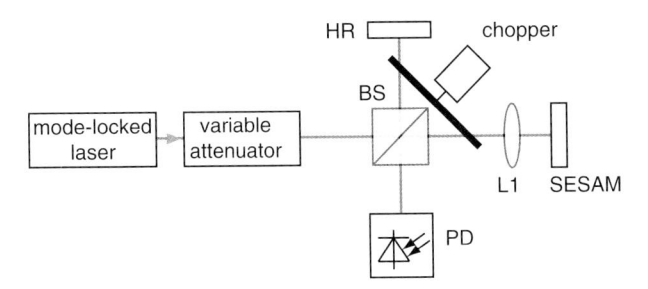

Figure 6.8 Improved measurement scheme for the optical reflectivity of a SESAM. The output of the mode-locked laser is variably attenuated, the reflectivity of the SESAM is obtained by measuring the response from both arms separated in time (by successively opening and closing the two arms with a chopper wheel).

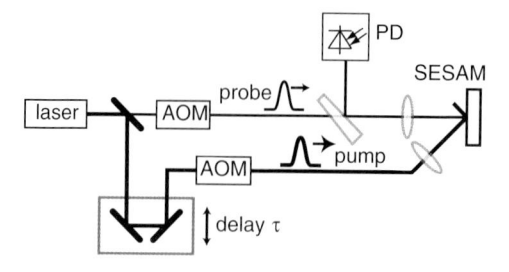

Figure 6.9 Schematic of the pump probe setup for SESAM measurements. By modulating the pump, it is possible to directly measure the reflectivity change of the SESAM. The probe beam is also modulated to suppress stray light from the pump beam. Lock-in detection at the difference frequency makes the setup very sensitive.

a strong pump beam and a weak probe beam. The energy of the probe pulse is sufficiently strong to saturate the SESAM. The power ratio of both beams should be at least 10:1 to make sure that the probe pulses do not influence the measurement. The probe beam is delayed with respect to the pump pulses by an optical delay line on a computer-controlled translation stage. Both beams overlap on the sample. The pump beam hits the sample first under a small angle. A part of the reflected probe beam is separated with a beam splitter and measured with a photodiode. For each time delay, the changes in probe transmission induced by the pump pulse are measured. This sampling technique allows observing phenomena that are much faster (picosecond to femtosecond timescale) than the used photodetector and its detection electronics, which usually operate in the nano- to microsecond range. To achieve high resolution and good signal-to-noise ratio, both pump and probe beams are usually intensity modulated by acousto-optic modulators (AOMs) and lock-in detection is used (Figure 6.9).

Figure 6.10 shows a representative time-resolved response trace for a SESAM with a quantum dot saturable absorber layer (see Section 6.2.3.2) and for a QW-SESAM (operation wavelength 1030 nm, low temperature-grown InGaAs QW) [56]. The temporal resolution of the trace results from the time delay and is only limited by

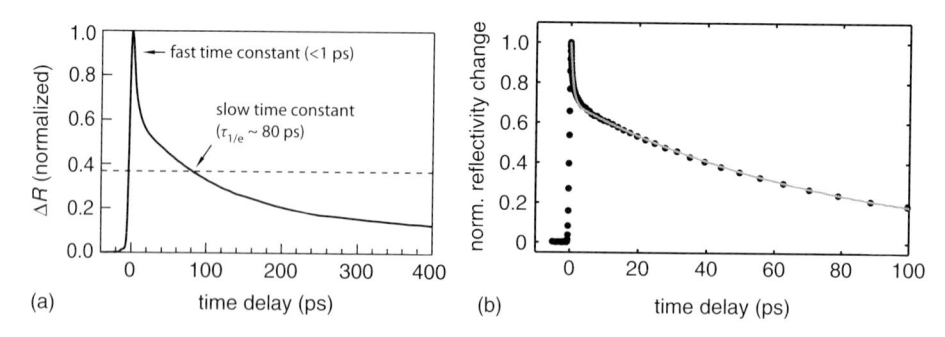

Figure 6.10 Typical temporal impulse response of a QW-SESAM (a) and a QD-SESAM (b). By convention, the time delay is positive when the probe pulse follows the pump pulse.

the cross-correlation of pump and probe pulses. For most SESAMs, one observes a bitemporal impulse response, which usually is fitted with two time constants:

$$\Delta R_{pp}(\tau) = A e^{-\tau/\tau_{slow}} + (1-A) e^{-\tau/\tau_{fast}},$$

where A is the amplitude of the slow component with time constant τ_{slow} and $(1 - A)$ is the amplitude of the fast component with time constant τ_{fast}. In case of the shown QD-SESAM response, $\tau_{fast} = 0.77$ ps, $\tau_{slow} = 74$ ps, and $A = 70\%$. The fast recovery is due to transitions in the dots, whereas the slow recovery is due to recombination. For the low temperature (LT)-grown QW-SESAM, the fast time constant is in the order of several tens to hundreds of femtoseconds. It is determined by carrier–carrier scattering processes that lead to thermalization within the bands. In the measurement, 2.7 ps pulses were used [56], which are not short enough to fully resolve this fast time constant. The slow time constant is due to interband trapping and recombination processes. Depending on the growth parameters, it is in the order of picoseconds to nanoseconds. The low-temperature growth of the absorber results in the incorporation of defects, which can serve as midgap traps for faster recovery. This allows the slow time constant to be reduced to values well below 100 ps. However, increasing the defect density in the absorber typically comes at the cost of increasing the nonsaturable losses of the SESAM. Another option to achieve faster modulation is to employ the AC-Stark effect, as will be discussed in Section 6.3.4.2.

6.2.2
Pulse Formation

The mechanism of passive mode locking depends on the recovery time of the saturable absorber and the level of gain saturation. One distinguishes fast and slow saturable absorbers, depending on the comparison of pulse duration with the recovery time. For a fast absorber, the recovery time is shorter than the pulse and the duration of the loss modulation is similar to the duration of the circulating pulse. But stable pulse formation can even be achieved with pulses that are much shorter than the recovery time of the absorber, in which the absorber is referred to as slow saturable absorber.

Figure 6.11 shows three simplified models that are useful for an initial discussion on the pulse formation. It compares the saturable loss and gain dynamics for three cases, a fast saturable absorber in combination with weak gain saturation (which corresponds, for example, to a typical solid-state gain medium), a slow saturable absorber with weak gain saturation, and a slow saturable absorber with strong gain saturation (e.g., dye or semiconductor gain medium). In all shown cases, the net gain is positive during a short time window when the pulse passes.

In the case of the *fast saturable absorber*, the loss modulation follows the pulse shape, and the positive net gain window is closed immediately after the pulse has passed. This can be achieved, for example, by Kerr lens mode locking (KLM) [57–59]. Here, the self-focusing due to the Kerr effect in combination with an aperture causes higher losses for lower intensities. The very fast Kerr response enabled the shortest

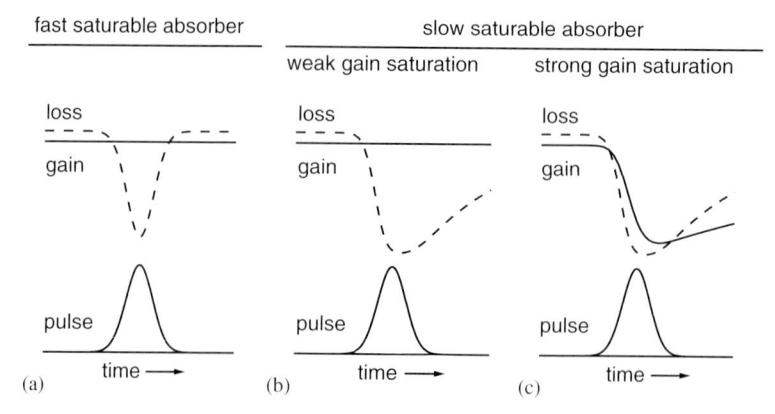

Figure 6.11 Mode locking mechanisms for fast and slow saturable absorbers. (a) For a fast saturable absorber, the recovery time is shorter than the pulse duration. In case of slow recovery, one distinguishes (b) weak gain saturation, which occurs usually in typical solid-state gain media or (c) strong gain saturation, which is usually present in semiconductor gain structures.

pulses directly generated by a laser oscillator; pulse durations around 5 fs were obtained using the Ti:sapphire gain material [6–8]. One disadvantage of KLM is that a significant change in mode size from the Kerr effect is achieved only by operating the laser cavity near a stability limit, making operation more sensitive to mechanical drift and temperature changes. Furthermore, mode-locked operation is often not self-starting but has to be started by introducing a perturbation to the laser (like a mechanical vibration of a cavity element). Intense pulses are usually required to achieve sufficient Kerr lensing, which are currently not within reach of semiconductor lasers. An alternative for this are Stark-SESAMs that enabled the generation of subpulses of 300 fs (see Section 6.3.4.2).

For the *slow saturable absorber*, the recovery time is longer than the pulse duration. One distinguishes two dynamic situations, which depend on the gain medium used in the laser. If the upper state gain lifetime is substantially longer than the cavity round-trip time, the gain can be considered as constant. This is usually the case for ion-doped solid-state gain media that have typical cross sections in the range of 10^{-21}–10^{-18} cm^2. Semiconductors, on the other hand, have gain cross sections that are several orders of magnitude larger, and significant dynamic gain saturation can occur during the amplification of the pulse. As shown in Figure 6.11b, the slow recovery of the absorber results in a net gain after the pulse. At first sight, it is surprising that stable operation can be achieved because any fluctuations behind the pulse are amplified and may introduce instabilities. However, stable operation can be achieved even if the pulse duration is more than a magnitude shorter than the net gain window. One stabilizing method is soliton mode locking [60–62]. Soliton pulses can be formed when self-phase modulation (SPM) and group delay dispersion (GDD) are balanced in the laser cavity. In this case, the pulse duration is determined by the SPM coefficient, the total intracavity GDD, and the pulse energy. The saturable absorber has only a low influence on the pulse shaping; it is mainly responsible for starting and stabilizing the mode locking process. But even without soliton mode locking, stability can be achieved in the slow absorber regime because the absorber absorbs only the leading edge of the pulse, causing the pulse to move backward each round-trip. In this way, the growing noise behind the pulse will be swallowed after some time by the pulse. Numerical simulations and experiments have shown that the pulses can be more than 20 times shorter than the recovery time [63].

In the strong gain saturation regime, for which the upper state gain lifetime is in the order of the cavity round-trip time or below, the pulse can saturate the gain such that the positive net gain window is closed at the end of the pulse. This is typically the case in dye lasers [64, 65] or semiconductor lasers [66, 67]. Achieving stable mode locking requires the absorber to saturate at lower pulse energies than the gain, that is,

$$\frac{E_{sat,abs}}{E_{sat,gain}} \ll 1.$$

In this regime, the absorber losses drop faster than the gain and a short window is formed where the gain is higher than the losses, as is shown in Figure 6.11c. This regime is present when both gain and absorber are based on the same type of material, and the above equation puts an important boundary condition on the operation parameters of the SESAM. Most of the presented results on ultrafast semiconductor disk lasers operate in the regime of slow saturable absorption with dynamic gain saturation.

6.2.2.1 Model for the Pulse Shaping

A simple numerical simulation can provide a first insight [29, 68] into how the important physical effects such as gain saturation, saturable absorption, and group delay dispersion influence the evolution of pulses in the laser. For providing a basic understanding, no exhaustive treatment is required, and neither the transversal mode effects nor the longitudinal effects have to be taken into account. The simulation is based on numerical iteration of a pulse inside the cavity until a stable solution is found (Figure 6.12). A pulse, defined on a time grid $(-T/2, T/2)$, interacts with successive cavity elements. Starting with an initial pulse (which can also be quantum noise), numerous cavity round-trips are simulated until a stable solution is obtained. We use the slowly varying envelope approximation to describe the time evolution of the pulse:

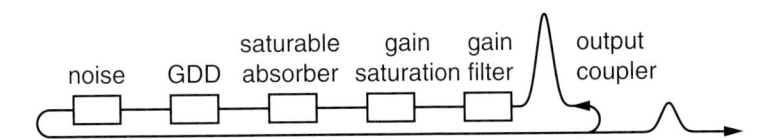

$$\tilde{E}(t) = \text{Re}\{A(t)\exp(-i\omega_0 t)\},$$
$$P(t) = |A(t)|^2,$$

Figure 6.12 Model used to simulate the pulse propagation. The pulse interacts with operators describing the cavity elements: gain, SESAM, and output coupler.

with $\tilde{E}(t)$ being the time-varying electrical field, $P(t)$ the instantaneous power, $A(t)$ the complex envelope, and ω_0 the reference frequency. The discretization time step ΔT is chosen between the optical period and the pulse duration. Suppose that $A(t)$ is almost constant within ΔT, the error is negligible, and the approximation is good. The effect of an optical element is applied to the pulse by operators, which are evaluated either in time or frequency domain, whichever is more suitable. A fast Fourier transform (FFT) converts the pulse into the other domain.

The gain filter represents the limited spectral *behavior* of the gain. It determines the wavelength of the output pulses and also restricts the pulse duration (the smaller the bandwidth, the longer the minimum achievable pulses). In the real devices, the spectral gain is determined by the gain enhancement and the intrinsic gain.

For both gain and absorber saturation, we use the same operator. The dynamic saturation is treated in time domain using a differential equation based on Equation 6.1:

$$\frac{dg(t)}{dt} = \frac{g_0 - g(t)}{\tau} - g(t)\frac{P(t)}{E_{sat}},$$

where $g(t)$ is the dimensionless wavelength-independent gain, g_0 is the small signal gain, τ is the recombination time constant, and E_{sat} is the saturation energy. The small signal gain g_0 depends on the pumping rate Λ, differential gain a, transparency density N_0, and the recombination rate γ_{nr}. For the gain structure, we have a positive gain $g_0 > 0$; for the saturable absorber, the small signal gain can be approximated with $g_0 = -\Delta R/2$. Because of the saturation of gain and absorption, the real part of the refractive index changes causing a nonlinear phase change. The complex amplitude reflectivity of gain and absorber is computed using

$$r(t) = \exp[(1 - i\alpha_{lef})g(t)],$$

with α_{lef} being the linewidth enhancement factor. The phase change affects the frequency spectrum in a similar way as self-phase modulation.

Group delay dispersion is important in a mode-locked laser cavity as it determines the pulse duration and has an important influence on the stability region. Like the gain filter operator, the GDD is applied in the frequency domain. The phase change as function of radial frequency is given by

$$\varphi(\omega) = \frac{1}{2}D(\omega - \omega_0)^2,$$

where D is the GDD coefficient expressed in fs^2.

The noise operator is important for evaluating the initial pulse formation. It adds random, statistically independent complex amplitudes with a variance $\sigma^2 = P/2$ to the amplitudes of the time trace. Because the SESAMs absorb the leading edge of the pulse, the pulse shifts backward; we use a center operator for numerical stability, which keeps the pulse peak centered around $t = 0$. This model is capable of simulating the complete buildup process in a mode-locked VECSEL as shown in Figure 6.13.

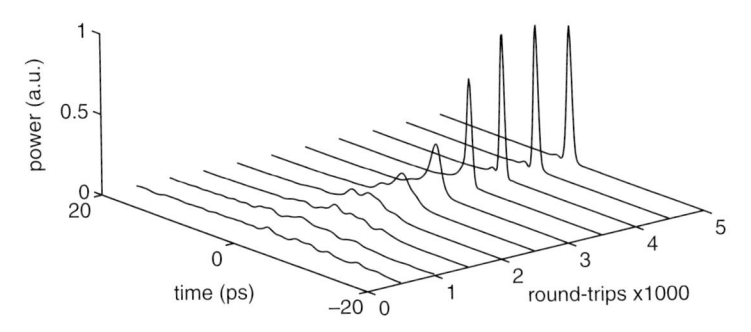

Figure 6.13 Simulated pulse buildup in a passively mode-locked VECSEL, the repetition rate is 5 GHz. The output pulse duration is 0.9 ps. After 5000 round-trips, the average output power is stable.

6.2.2.2 Mode-Locking Stability and the Importance of Gain and SESAM Saturation

Such simulations allow evaluating the required operation parameters for stable and self-starting mode locking. There are several challenges that can prevent stable mode locking. An important instability is the Q-switched mode locking (QML). In this regime of operation, Q-switching and mode locking occur simultaneously, leading to large fluctuations of the pulse energy. Hönninger et al. [47] have studied the QML stability limits. Stable cw mode locking is obtained if the intracavity pulse energy E_p is higher than a threshold value,

$$E_p^2 > E_{\text{sat,gain}} E_{\text{sat,abs}} \Delta R,$$

where $E_{\text{sat,gain}}$ and $E_{\text{sat,abs}}$ are the saturation energies of gain and absorber, respectively, and ΔR is the modulation depth. Scaling to high repetition rates can be challenging because for the same average power, the pulse energy E_p decreases. This is particularly critical for ion-doped solid-state gain materials with a small gain cross section and therefore a large gain saturation energy $E_{\text{sat,gain}}$. But also for these laser materials, QML can be suppressed in many cases if an optimized SESAM with low modulation depth and low saturation fluence is employed [69, 70]. An advantage of the substantially lower saturation fluence and upper state lifetime of semiconductor disk lasers compared to ion-doped bulk or fiber laser is that stable mode locking at very high repetition rates can be more easily achieved. Semiconductor lasers have substantially higher differential gain and smaller $E_{\text{sat,gain}}$, reducing efficiently the tendency toward QML.

Especially for operation at low repetition rates, multiple pulsing can destabilize the laser. Saarinen et al. [71] observed that the number of pulses circulating in a mode-locked VECSEL increases with pump power. At a certain pump level, two pulses with lower energy have a gain advantage over a single pulse with higher energy. Scaling to lower repetition rates in mode-locked VECSELs is limited by harmonic mode locking, as the QWs are not able to store much energy because of the short carrier lifetime and two circulating intracavity pulses will, therefore, have a higher overall gain than a single pulse.

Another type of instability is the tendency toward cw operation. Stable mode locking is possible only when short pulses have a gain advantage over continuous-wave operation. An important role is played by the saturation fluence of the saturable absorber. We studied the mode-locking stability and pulse duration as function of the ratio $E_{sat,abs}/E_{sat,gain}$. For typical operation parameters in the slow saturable absorber regime, our simulations show that the saturation energy of the absorber should be at least 10 times smaller than the saturation energy of the gain:

$$\frac{E_{sat,abs}}{E_{sat,gain}} = \frac{A_{abs} F_{sat,abs}}{A_{gain} F_{sat,gain}} < 0.1,$$

with A_{abs} and A_{gain} being the mode areas on absorber and gain, respectively. A higher modulation depth results in a larger stability region and enables shorter pulse durations; however, if the SESAM is not completely saturated, the average output power drops because of the larger losses. There are two possibilities to satisfy this inequality. If gain and absorber have similar saturation fluences (as is usually the case for QW-SESAMs), the only way to fulfill the requirement is to choose a smaller spot size on the SESAM than on the gain structure. However, this is challenging for high repetition rates (see Section 6.3.3) and impossible for the integration of both elements into a single semiconductor structure (see Section 6.4). For these applications, it is important to reduce the saturation fluence, which can, for example, be achieved by using QD saturable absorbers (see Section 6.2.3.2).

6.2.2.3 Importance of Group Delay Dispersion

Another important simulation result is the explanation of the strong influence that group delay dispersion introduces on the pulse duration of mode-locked VECSEL. Already in 2002, the most relevant parameters for the temporal pulse shaping were identified and a qualitative theory on a quasi-soliton pulse shaping mechanism was developed by Paschotta et al. [68]. Mode-locked VECSELs show a strong saturation within the absorber and also within the gain (Figure 6.14a). Both of these saturations

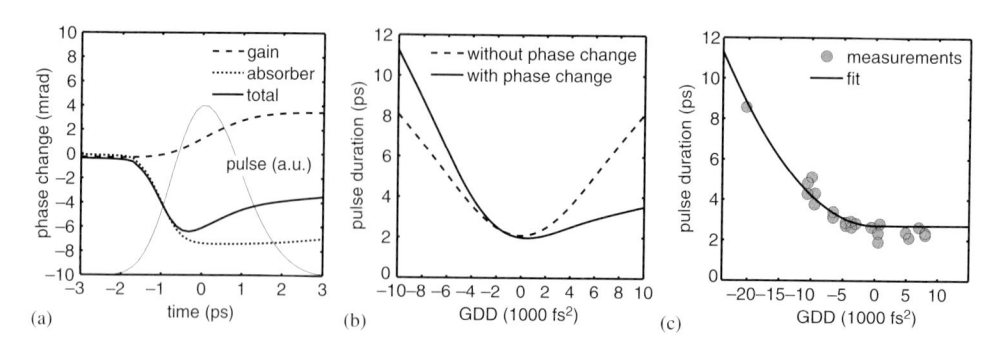

Figure 6.14 (a) Temporal nonlinear phase change experienced by the pulse due to saturation of the gain and absorber. (b) Simulated pulse duration as a function of the intracavity GDD with and without taking gain and absorber saturation into account. (c) Measured pulse duration as a function of the intracavity GDD for a wavelength of 956 nm.

lead to a total temporal phase change, which is similar to that of self-phase modulation, however with an opposite sign. Similar to soliton mode locking [60], this phase change can be matched with the appropriate amount of positive GDD, and quasi-soliton pulses are obtained (for details, refer to Ref. [68]). Recently, this quasi-soliton pulse formation theory was experimentally studied in detail and confirmed [72]. In Figure 6.14b, the pulse formation simulations with and without this temporal phase change are compared. Neglecting it would result in a symmetric dependence of the pulse duration on the GDD. However, simulations including the nonlinear phase change show an asymmetric dependence. Negative GDD leads to longer pulses, while substantially shorter pulses are obtained for positive GDD.

The experimental results show a good agreement with the theory of quasi-solitons. The results also confirm that higher positive GDD, compared to higher negative GDD, does not strongly increase the pulse duration, which releases the VECSEL design constraints in that GDD regime.

6.2.3
SESAM Designs

The main advantage of a SESAM for passive mode locking is its design flexibility, which allows optimizing its macroscopic optical parameters in a wide range for various laser types. The number of saturable absorber layers as well as the field enhancement can be easily designed. The most prominent designs are either based on a resonant or on an antiresonant design, which typically have approximately an order of magnitude difference in field enhancement, but intermediate solutions are also important for many laser types [70]. So far, most SESAMs employ QW absorbers, but recently there has also been an increasing interest in QD-based SESAMs because the strong localization of the wave function leads to an atom-like density of states that enables novel SESAMs with tunable optical properties. In the following section, we will first discuss different SESAM structures for field enhancement control. Afterward, we will discuss the properties of SESAMs using QW and QD absorbers.

6.2.3.1 SESAM Structure for Field Enhancement Control

A SESAM structure usually contains several tens of semiconductor layers with a total thickness in the order of 10 μm (see also Figure 6.4). They are epitaxially grown onto a wafer by standard techniques, such as metal-organic vapor-phase epitaxy (MOVPE) and molecular beam epitaxy (MBE). The bottom section of a SESAM typically consists of an alternating sequence of two materials with different indices of refraction and an optical layer thickness that corresponds to one quarter of the design wavelength, forming a Bragg mirror. In this way, the Fresnel reflections from each interface of the two materials interfere constructively, resulting in a high reflection for the design wavelength. The total reflectivity is determined by the number of layer pairs and the difference in reflective index of the two materials. For a target wavelength in the 1 μm spectral region and AlAs and GaAs layers, a theoretical reflectivity of >99.9% is achieved with 25 pairs. Note that there are also other multilayer bottom mirror designs that use a more complex layer design, for example, by using a varying layer thickness for dispersion control [73].

On top of the bottom mirror structure, an absorber section and often an additional top section are placed, which enables precise control of the nonlinear reflectivity. To analyze the SESAM structure in detail, numerical simulations are required. By assuming an incident plane wave, the electrical field inside and outside the structure can be computed using a well-known transfer matrix algorithm for multilayer structures [74]. For an incident field with intensity I_{inc}, the intensity inside the structure at position z (from the substrate) is given by

$$I(z) = n(z) \cdot |\varepsilon(z)|^2 \cdot I_{inc}, \tag{6.2}$$

where n is the local refractive index and ε is the electric field, normalized to 1 for the incident wave. The quantity $|\varepsilon(z)|^2$ is referred to as the field enhancement; for a 100% reflective mirror, the field enhancement oscillates between 0 and 4 outside the structure due to destructive and constructive interference of incident and reflected fields. The field strength at the absorber position determines the modulation depth and the saturation fluence: for a low field enhancement, the modulation depth is low and the saturation fluence is large. The Fresnel reflection at the semiconductor–air interface of the absorber section in combination with the bottom Bragg reflection leads to a resonance effect. Most commonly, the absorber section is based on a resonant or on an antiresonant design and a single absorber layer. In Figure 6.15, these cases are shown for AlAs and GaAs materials at an operation wavelength of 960 nm. The absorber is embedded between spacer layers. The first spacer layer after the mirror is chosen such that the absorber is in the antinode of the standing wave pattern of the electrical field. The second spacer layer determines the field enhancement. By making the round-trip phase change in the last layer $\pi(2n-1)$, the structure is antiresonant; by making the round-trip phase $2\pi n$, the structure becomes resonant.

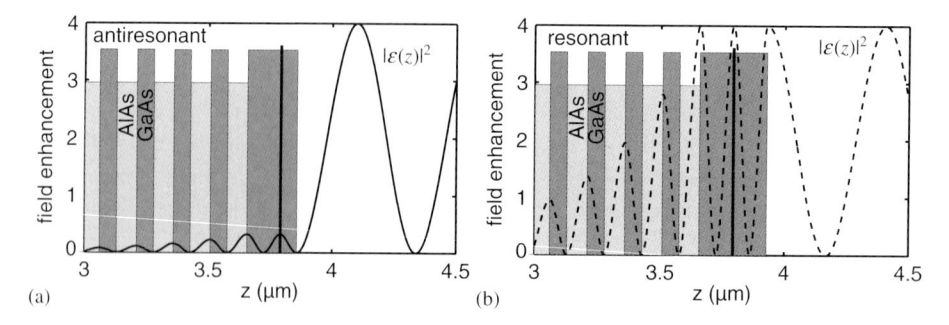

Figure 6.15 Refractive index pattern and field enhancement of an antiresonant SESAM (a) and a resonant SESAM (b).

We define the field enhancement factor ξ_{abs} as the field enhancement inside the absorber

$$\xi_{abs} = |\varepsilon(z_{abs})|^2, \tag{6.3}$$

with z_{abs} being the position of the absorber. In the antiresonant SESAM, the field enhancement is $\xi_{abs} = 4/n^2 \approx 0.32$, while in the resonant case, $\xi_{abs} = 4$. The (extrinsic) modulation depth is proportional to the enhancement factor $\Delta R \propto \xi_{abs}$ and the (extrinsic) saturation fluence is inversely proportional to $F_{sat} \propto 1/\xi_{abs}$. By changing the field enhancement, the product $F_{sat} \cdot \Delta R$ cannot be altered: the absorbed pulse fluence in a SESAM is $F_p[1 - R(F_p)]$, for a transparent SESAM ($F_p \to \infty$), this converges to $F_{sat} \cdot \Delta R$ (assuming $\Delta R_{ns} = 0$ and $F_2 = \infty$). Thus, $F_{sat} \cdot \Delta R$ is the energy needed to completely saturate the SESAM. This energy must be proportional to the number of states in the absorber, since these states have to be filled with carriers for transparency. If one does not change the density of states, the absorbed pulse fluence to obtain complete saturation stays the same and the product $F_{sat} \Delta R$ is maintained. The only solution to reduce both F_{sat} and ΔR is to change the properties of the saturable absorber layer by reducing the density of states.

The antiresonant SESAM design is the most common design for standard DPSSLs because it exhibits a relatively broad operation wavelength range with nearly constant saturable absorber parameters. For lasers with high intracavity pulse energies, the use of an antiresonant design is beneficial because it minimizes the field intensity inside the structure. This reduces the nonsaturable losses and results in a higher damage threshold. Further, antiresonant SESAMs have broadband properties and higher saturation fluence that are crucial for mode-locking high-power lasers with multimicrojoule pulse energy [14]. In Figure 6.16a, the field enhancement factor ξ_{abs} is shown as function of the wavelength. The resonant structure is strongly wavelength sensitive. For some laser designs, this is advantageous because it is possible to tune the modulation depth with the operation wavelength. On the other hand, the operation wavelength range is limited and the structure is sensitive to growth errors,

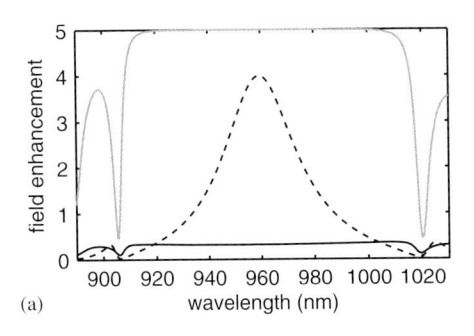

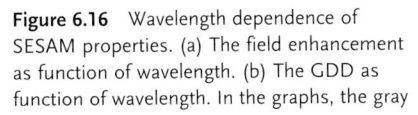

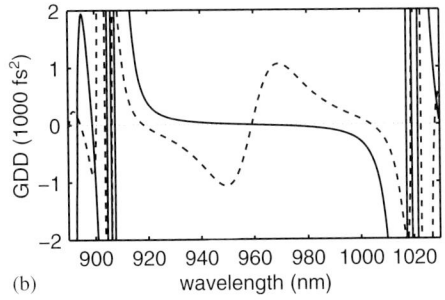

Figure 6.16 Wavelength dependence of SESAM properties. (a) The field enhancement as function of wavelength. (b) The GDD as function of wavelength. In the graphs, the gray curve indicates the DBR reflectivity (arbitrary units), the solid black curves the antiresonant, and the dashed black curves the resonant structure.

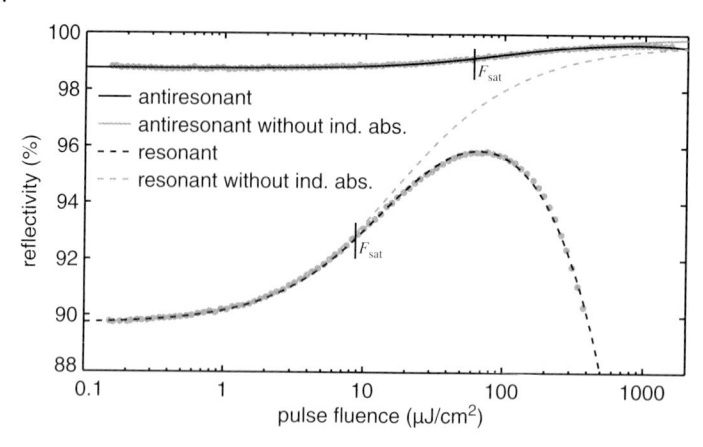

Figure 6.17 Example of an antiresonant (solid curves) and a resonant (dashed curves) SESAM having the same saturable absorber. The black curves show the least-square fit. The gray curves show the fit function without induced absorption, thus the reflectivity obtained by long ps pulses.

a 1% mismatch in the layer thickness shifts the resonance by 10 nm. The gray line indicates the mirror reflectivity that has a bandwidth of over 80 nm, the dashed line represents the resonantly designed structure, and the solid line represents the antiresonantly designed structure. In Figure 6.16b, the group delay dispersion as function of the operation wavelength is shown. In the resonant case, the reflected beam has a strong wavelength-dependent phase and therefore larger GDD fluctuations.

Furthermore, the resonant design has a higher electrical field inside the structure resulting in more two-photon absorption and a smaller F_2 parameter. A nonlinear reflectivity measurement with 100 fs pulses of an antiresonant and resonant SESAM, using the same saturable absorber layer, is shown in Figure 6.17. The difference in modulation depth, saturation fluence, and induced absorption is clearly visible. The gray lines show the SESAM curves without induced absorption, thus the reflectivity for longer pulses in the picosecond regime.

An antiresonant design has clear advantages in terms of wavelength tolerance; however, it also exhibits a larger modulation depth that can be a challenge for VECSEL mode locking.

6.2.3.2 Comparison of Quantum Well and Quantum Dot SESAMs

So far, most SESAMs employ quantum well absorbers, which enabled stable mode locking of various laser materials operating in the femtosecond and picosecond regimes [9]. As described in Section 6.2.2.2, the absorber needs to saturate at lower pulse energies than the gain. The gain structure in VECSELs typically consists of several QWs, with a saturation fluence similar to the QW used in standard SESAMs. Stable pulse formation is usually achieved by strong focusing onto the QW-SESAM (area being 10–40 times smaller than in the gain, see Figure 6.18), which limits the geometrical size and restricts the maximum achievable repetition rate [27, 28].

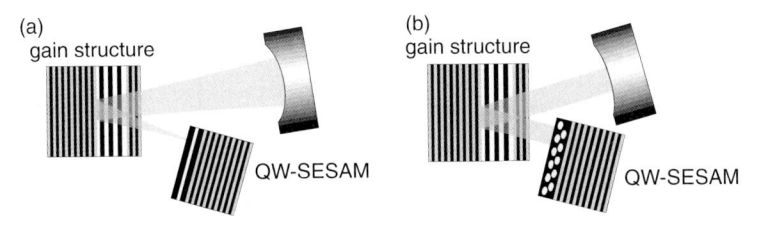

Figure 6.18 Cavity setup of a mode-locked VECSEL with large mode area ratio using a QW-SESAM (a) and with the same mode area using a low-fluence QD-SESAM (b).

The saturation fluence can be reduced by a resonant design (see Section 6.2.3.1), however, as the product $F_{sat} \cdot \Delta R$ is maintained, such resonant QW-SESAMs usually exhibit a fairly large modulation depth, which is similar or higher than the available VECSEL gain. The only solution to reduce both F_{sat} and ΔR is to reduce the joint density of states. A solution to this issue is the use of QD-based SESAMs because the strong localization of the wave function leads to an atom-like density of states. For QDs, the density of states is simply proportional to the dot density, which enables novel SESAMs with tunable optical properties. The QD saturable absorbers were first used to mode lock the semiconductor edge emitters in 1999 [75]; 2 years later in 2001, the first QD-SESAM was reported by Garnache et al. [76]. Rafailov et al. reported fast recovery dynamics, which can be beneficial for achieving shorter pulse durations and pulse durations as short as 114 fs have been shown [78]. The key advantage of QD-SESAMs for VECSEL mode locking is the possibility to decouple modulation depth and saturation fluence [38]. The modulation depth can be controlled by the dot density, while the saturation fluence can be controlled by the field enhancement. In this way, low saturation fluence SESAMs with optimized modulation depth were realized, which enabled mode locking with the same mode areas on SESAM and VECSEL (Figure 6.18). The effect of the QD-growth parameters on the macroscopic optical SESAM parameters was investigated in Ref. [79]. A set of self-assembled InAs QD-SESAMs optimized for an operation wavelength around 960 nm with varying dot density and growth temperature was studied. The InAs QDs are grown by MBE using Stranski–Krastanow growth, which depends on the following parameters: substrate temperature, arsenic pressure, growth rate in number of monolayers (MLs) per second (i.e., $ML\,s^{-1}$), and indium monolayer coverage (ML coverage). The indium coverage is controlled by the opening time of the indium source shutter, a longer opening time results in higher ML coverage, leading to higher dot density [80]. In Figure 6.19, an atomic force microscopy (AFM) image of a QD layer is shown (note that the QD layer in a SESAM is placed inside a cover layer that can change its properties).

In Figure 6.20, the nonlinear reflectivity measurements of five antiresonant QD-SESAMs with different ML coverages are shown (measured at a laser center wavelength of 960 nm). The measurement data (dots) are fitted (solid lines) with the model function taking into account the finite spot sizes (see Section 6.2.1). By increasing the ML coverage (corresponding to an increased dot density), the modulation depth also increases. The modulation depth is proportional to the dot

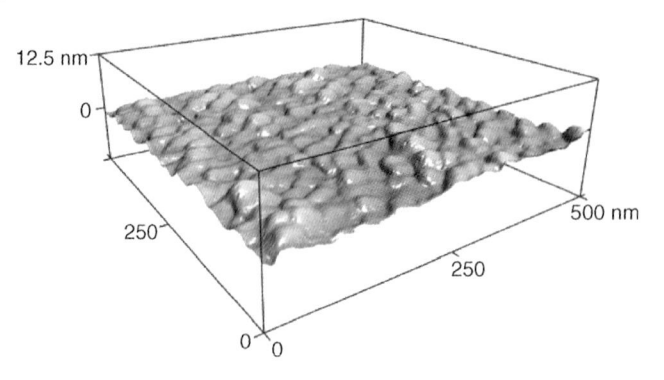

Figure 6.19 Atomic force microscopy (AFM) image of a QD layer.

density, while the saturation fluence remains constant. The rollover due to two-photon absorption also remains constant. The rollover depends on the pulse duration of the laser and the design of the structure, that is, the integrated intensity in the GaAs layers, which is similar for all structures [49] (note that the measurements were done with a Ti:sapphire fs laser, lasers operating in the ps regime have less induced absorption and therefore the rollover would occur at a higher fluence) (Figure 6.21).

The mode-locked VECSELs with similar spot size on gain and absorber require SESAMs with low saturation fluence (typically $<10\,\mu J\,cm^{-2}$) and a modulation depth of around 1%. For the samples shown here, this can easily be achieved by a partially resonant design with an appropriate field enhancement in the absorber layer. However, as discussed in Section 6.2.3.1, such a design exhibits a smaller wavelength tolerance and higher GDD than an antiresonant design. Recently, low-saturation

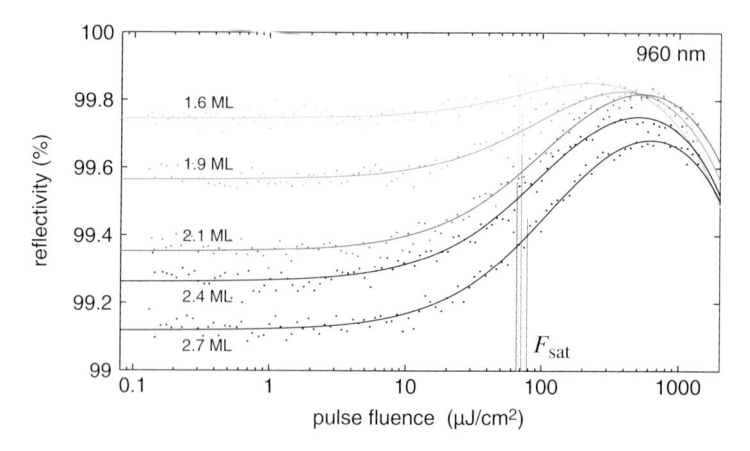

Figure 6.20 Top: Nonlinear reflectivity measurements of QD-SESAMs done at 960 nm with 140 fs pulses, the upper curve has the smallest ML coverage (dot density) and therefore the smallest modulation depth. By increasing the dot density, the modulation depth also increases, while the saturation fluence remains constant.

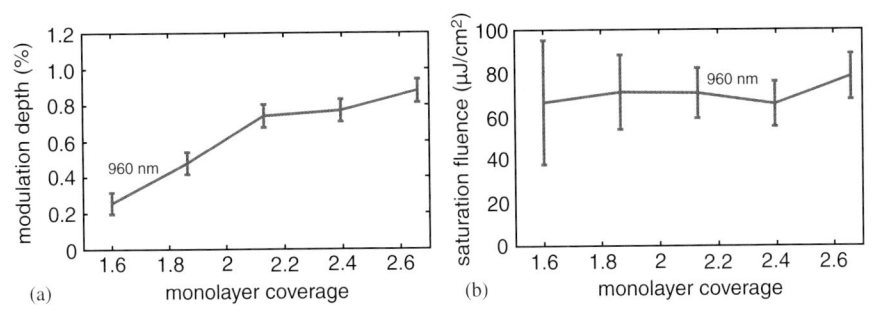

Figure 6.21 SESAM parameters modulation depth (a) and saturation fluence (b) as function of monolayer coverage.

fluence QD-SESAMs were developed, which achieved mode locking with similar spot sizes from an antiresonant design [81]. This was achieved by post-growth annealing, which improved the QD quality and significantly reduced the saturation fluence. The antiresonant design with its reduced GDD is very promising for achieving passively mode-locked VECSELs with higher pulse repetition rates and shorter pulse durations. Furthermore, the antiresonant QD saturable absorber substantially simplifies the MIXSEL design (see Section 6.4) and provides better growth tolerance.

6.3
Mode Locking Results

6.3.1
Introduction

During the past few years, a tremendous progress in the performance of ultrafast VECSELs in terms of wavelength coverage, average output power, repetition rate, and achievable pulse duration was achieved. Ultrafast VECSELs became attractive sources for numerous applications relying on short pulse laser in areas as diverse as biology, medicine, telecommunications, optical clocking and interconnects, or nonlinear frequency conversion.

The initial work on mode-locked VECSELs was based on synchronous pumping with another mode-locked pump laser [82]. The first passively mode-locked VECSEL was demonstrated in 2000 by Hoogland et al. [66] with moderate output powers of 21.6 mW in 22 ps pulses at a repetition rate of 4 GHz and a wavelength of 1030 nm. To date, ultrafast VECSELs [28] have been successfully realized with average output powers of up to 2.1 W [26], pulse repetition rates of up to 50 GHz [27], and pulse durations below 300 fs [23, 24]. All these results had diffraction-limited, circular output beams.

A major difference between semiconductor gain materials and more conventional gain materials such as ion-doped bulk or fiber lasers is their substantially larger absorption and emission cross sections that are several orders of magnitude higher,

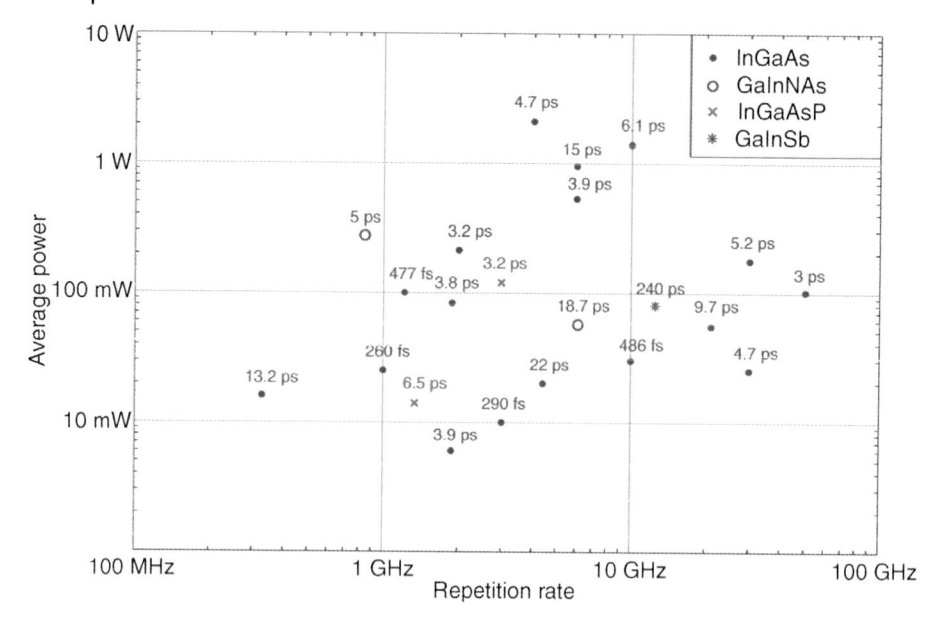

Figure 6.22 Overview of passively mode-locked optically pumped VECSELs in terms of average power and repetition rate (see Table 6.1). The pulse duration is written next to each material symbol.

resulting in low gain saturation fluences and low upper state lifetime. This limits the amount of stored energy in the gain material and the minimum achievable repetition rates. The low repetition rates in the few MHz regime and pulse energies above >10 µJ, as were recently demonstrated for a diode-pumped solid-state laser using an Yb:YAG thin-disk gain material [14, 15], therefore do not appear feasible. However, the low saturation fluence is a major advantage for achieving high repetition rates, because Q-switched mode-locking instabilities are strongly reduced (see Section 6.2.2.2). Due to their power-scaling potential, VECSELs are highly attractive to cover the regime of high average power and high repetition rate.

In Figure 6.22, we present an overview of optically pumped mode-locked VECSELs with repetition ranging from MHz to multi-GHz regime (Table 6.1).

In the following sections, we present some specific realizations of passively mode-locked VECSELs, which target high average output power, high repetition rate, short pulse operation, and electrical pumping.

6.3.2
Mode-Locked VECSELs with High Average Output Power

6.3.2.1 Power Scaling of Mode-Locked VECSELs

High-power and efficient ultrafast semiconductor disk lasers with good beam quality are highly desirable for a wide variety of applications. The main requirement for high-power mode-locked VECSELs is to achieve a high-power transverse electromagnetic mode (TEM_{00}) operation because the presence of higher order transverse modes

Table 6.1 Overview of optically pumped VECSELs passively mode locked.

Gain material	Laser wavelength (λ_0)	Pulse duration (τ_p)	Average output power (P_{av})	Repetition rate (f_{rep})	References
InGaAs based					
13 InGaAs/AlGaAsP + intracavity LBO	489 nm	3.9 ps	6 mW	1.88 GHz	[83]
5 InGaAs/GaAs QWs	950 nm	15 ps	950 mW	6 GHz	[67]
		3.9 ps	530 mW		[67]
9 InGaAs/GaAs QWs	950 nm	3.2 ps	213 mW	2 GHz	[84]
13 InGaAs/AlGaAsP	975 nm	3.8 ps	83 mW	1.88 GHz	[85]
7 InGaAs/GaAsP QWs	960 nm	4.7 ps	25 mW	30 GHz	[86]
	957 nm	4.7 ps	2.1 W	4 GHz	[26]
	960 nm	6.1 ps	1.4 W	10 GHz	[87]
	960 nm	5.2 ps	177 mW	30 GHz	[88]
	960 nm	3 ps	100 mW	50 GHz	[27]
	980 nm	9.7 ps	55 mW	21 GHz	[86]
6 InGaAs/GaAsP QWs	1030 nm	13.2 ps	16 mW	328 MHz	[76]
	1035 nm	260 fs	25 mW	1 GHz	[23]
	1036 nm	290 fs	10 mW	3 GHz	[24]
	1040 nm	477 fs	100 mW	1.21 GHz	[89]
12 InGaAs/GaAs QWs	1030 nm	22 ps	20 mW	4.4 GHz	[66]
6 InGaAs/GaAs QWs	1034 nm	486 fs	30 mW	10 GHz	[90]
7 InGaAs/GaAs QWs	1040 nm	15 ps	<100 mW	2.1 GHz HM	[71]
InGaNAs based					
10 GaInNAs/GaAs/GaAsN QWs	1220 nm	5 ps	275 mW	840 MHz	[91]
5 GaInNAs/GaAs QWs	1.308 µm	18.7 ps	57 mW	6.1 GHz	[92]
InGaAsP based					
7 InGaAsP QWs	1.5 µm	6.5 ps	14 mW	1.34 GHz	[93]
20 InGaAsP/InGaAsP QWs	1.554 µm	3.2 ps	120 mW	2.97 GHz	[94]
GaInSb based					
15 GaInSb QWs	2.010 µm	240 ps	80 mW	12.5 GHz	[95]

λ_0: center lasing wavelength, τ_p: measured pulse duration, P_{av}: average output power, f_{rep}: pulse repetition rate, HM: harmonic mode locking.

destabilizes the mode locking. The VECSEL has a one-dimensional heat flow that allows scaling the output power by scaling the mode area on the active region. The power scaling of VECSELs operating in the cw regime is discussed in chapter (see Ch. 2, Section 2.3). As described, a one-dimensional heat flow requires efficient heat removal out of the gain section, which can be achieved by wafer-removal techniques or by bonding an intracavity diamond heat spreader directly on the semiconductor gain chip. For passive mode locking, the first approach is more suitable because it does not suffer from the optical properties of the diamond heat spreader etalon, which typically introduces a modulated spectrum corresponding to the etalon modes and which can destabilize the mode locking.

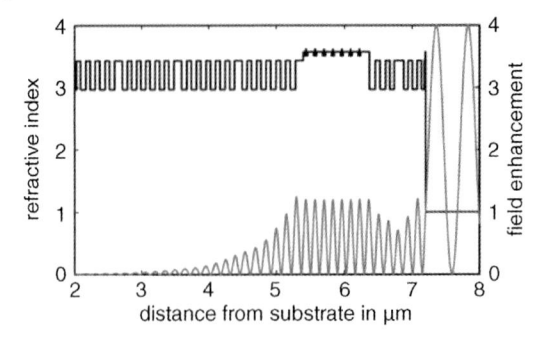

Figure 6.23 VECSEL gain structure with vertical refractive index profile and normalized intensity pattern. The structure consists of a double periodic bottom mirror, a gain section, and an antireflection section.

6.3.2.2 Experimental Results

The highest average output power from an ultrafast VECSEL is currently 2.1 W, which was reported by Aschwanden et al. [26]. The gain structure was grown by MOVPE on a GaAs substrate and consists of three parts: the bottom mirror, the active region, and an antireflective (AR) structure.

Figure 6.23 shows the refractive index profile of the semiconductor gain structure and the on-axis standing wave intensity pattern of the incoming laser beam normalized to the incoming peak intensity. The design is optimized for a laser wavelength around 960 nm (0° angle of incidence) and a pump wavelength of 808 nm (45° angle of incidence). An $Al_{0.2}Ga_{0.8}As/AlAs$ bottom mirror is a 36-pair Bragg reflector optimized not only for high reflectivity at the laser wavelength (99.95%) but also for high reflectivity at the pump wavelength (97%). This results in a double pass through the active region, which leads to an 85% absorption of the pump light. The active region contains seven $In_{0.13}Ga_{0.87}As$ quantum wells placed in the maxima of the standing wave pattern of the laser field. They are separated by spacer layers made of pump-absorbing GaAs and $GaAs_{0.94}P_{0.06}$ layers. The tensile-strained $GaAs_{0.94}P_{0.06}$ layers serve as strain-compensating layers and are positioned on both sides of the QWs. An $Al_{0.2}Ga_{0.8}As/AlAs$ AR section is placed on top toward the external cavity and is also optimized for both the laser and the pump wavelengths: it reflects less than 1% of the laser power and less than 3% of the pump power. The structure was grown in reverse order, meaning that first the etch stop layers and the AR section were grown, followed by the gain section and the bottom mirror. Smaller pieces were then cleaved from the wafer, metallized with Ti-Pt-Au, and, finally, soldered to a copper heat sink. Afterward, the GaAs substrate was removed in a wet chemical etching procedure. The reduced thickness of the semiconductor material ($\approx 7\,\mu m$) leads to low thermal impedance and to a nearly one-dimensional heat flow into the heat sink, which makes the device power scalable; the output power can be doubled by applying twice the pump power to twice the mode area without raising the temperature in the gain structure.

A standard V-shaped laser cavity is used (Figure 6.24), in which the gain structure serves as a folding mirror. It is pumped on a spot with $\approx 175\,\mu m$ radius using a

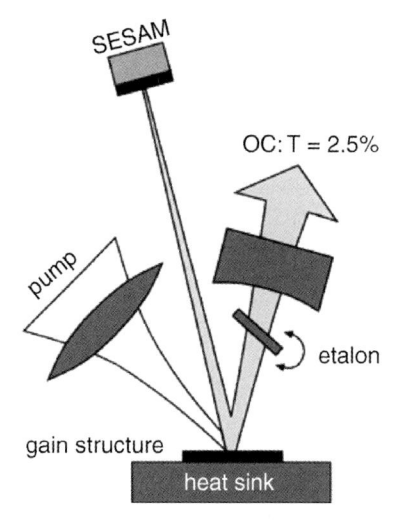

Figure 6.24 Cavity setup for the passively mode-locked high-power VECSEL.

fiber-coupled laser-diode module. The cavity end mirrors are the SESAM and the output coupler with a 38 mm curvature radius and 2.5% transmission. The SESAM contains an 8.5 nm thick $In_{0.15}Ga_{0.85}As$ QW grown with MBE at low temperature. It shows an exciton around 959 nm and has a modulation depth of about 1%. An approximately 20 μm thick uncoated fused silica etalon is inserted near the SESAM.

The output power and pulse duration of the laser were optimized by varying the etalon angle and heat sink temperature of the gain medium. Best performance with 2.1 W average output power in a 4 GHz pulse train has been achieved with 18.9 W pump power and a heat sink temperature of −4 °C. The etalon angle was close to normal incidence.

Figure 6.25a shows the autocorrelation trace of the 4.7 ps pulses. The pulses deviate by a factor of 2 from the transform limit, which is most likely caused by intracavity dispersion (see Section 6.2.2.3). The RF spectrum of the signal from a fast photodiode, shown in Figure 6.3 at the bottom, demonstrates stable mode locking at 4 GHz.

6.3.2.3 Outlook

Achieving multiwatt average output power levels, the ultrafast semiconductor disk lasers have become an attractive alternative to ultrafast ion-doped solid-state fiber lasers in many application areas. The main challenge for further power increase is to achieve sufficiently high power levels in TEM_{00} from a VECSEL gain chip suitable for passive mode locking, which is, for example, achieved by wafer removal (see Ch. 2, Section 2.3). Recently, a VECSEL structure similar to the one described above generated a continuous-wave output power up to 20.2 W in a fundamental transverse mode beam ($M^2 < 1.1$). A diamond heat sink instead of a copper mount was employed for efficient heat removal. It achieved a slope efficiency as high as 49.1% and a maximum optical-to-optical conversion efficiency of 43.2%. We,

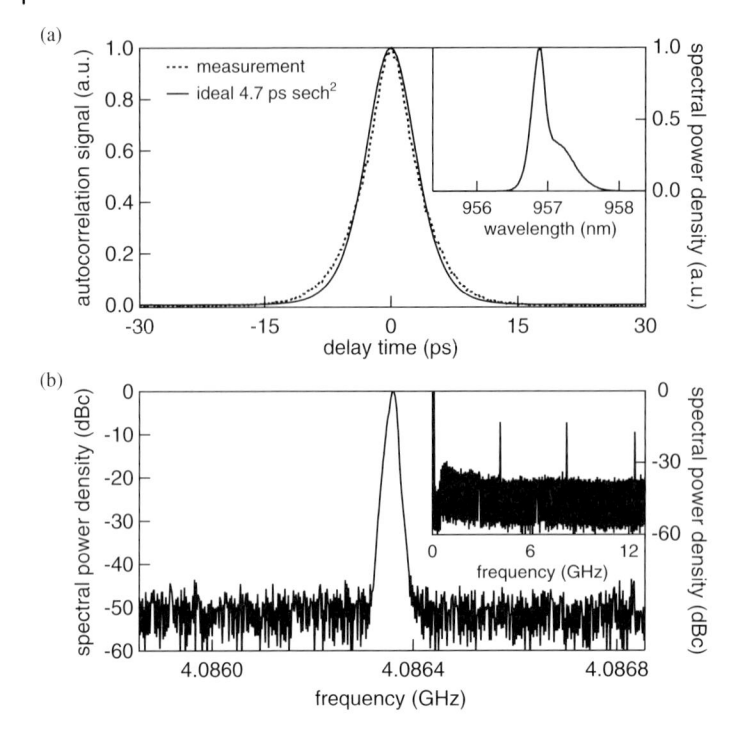

Figure 6.25 (a) Autocorrelation trace of the pulses with 4.7 ps duration obtained from a VECSEL mode locked with a QW-SESAM. The inset shows the optical spectrum centered at 957 nm. (b) RF spectrum of the pulse train at 4 GHz with a 1 MHz span and 10 kHz resolution bandwidth. The inset shows a measurement over a 12 GHz span.

therefore, expect that mode-locked semiconductor disk lasers will exceed 10 W of average power in the near future.

6.3.3
VECSEL Mode Locking at High Repetition Rates

Increasing the repetition rates of a mode-locked laser into the multi-GHz regime raises two main challenges:

1) The intracavity pulse energy becomes low, and achieving stable pulse formation is challenging. This is particularly true for ion-doped bulk or fiber lasers that can suffer from QML instabilities (see Section 6.2.2.2).
2) The cavity design becomes challenging: the repetition rate of an ultrafast laser that employs fundamental mode locking is determined directly by its cavity length. The repetition rate f_{rep} is given by $f_{rep} = c_0/2L$, with L being the optical cavity length and c_0 the speed of light. A repetition rate of 100 MHz requires an optical length of 1.5 m, while a high repetition rate laser operating at 50 GHz needs only an optical cavity length of 3 mm. The high repetition rate lasers have thus very

compact setups; however, this also strongly limits the design flexibility for the laser cavity design.

6.3.3.1 Mode Locking with Similar Area on Gain and Absorber (1:1 Mode Locking)

As shown in Section 6.2.2.2, the condition for stable mode locking is given by

$$\frac{E_{\text{sat,abs}}}{E_{\text{sat,gain}}} = \frac{A_{\text{abs}} F_{\text{sat,abs}}}{A_{\text{gain}} F_{\text{sat,gain}}} < 0.1,$$

with A_{abs} and A_{gain} being the mode areas on absorber and gain, respectively. There are two possibilities to satisfy this inequality. The initial approach based on QW-SESAMs was to design a smaller spot size on the SESAM than on the gain structure. For increasing the repetition rate above 10 GHz, this approach is not suitable anymore because the required longitudinal cavity dimensions for enlarging the beam radius are too long. The solution to this issue is to operate the laser with similar spot sizes on gain and absorber so that no beam enlargement is required. However, this requires a SESAM with very low F_{sat}. As discussed in Section 6.2.3.2, this performance can be achieved with QD saturable absorbers.

6.3.3.2 Mode-Locked VECSELs with up to 50 GHz

At present, the highest repetition rate of a mode-locked VECSEL is 50 GHz, which was presented by Lorenser et al. [27]. The cavity geometry is shown in Figure 6.26.

A major challenge was to design a cavity configuration, which allowed extracting sufficiently high average output power in a single-transverse mode at the small mode radii in the order of 50 μm at cavity lengths around 3 mm. The 3.7 W pump light is incident on the gain structure at a 45° angle in the vertical plane, which allows a more compact cavity in the horizontal plane. The flat output coupler has a transmission of 1.6%. An average output power of 102 mW was obtained at a 50 GHz repetition rate, the pulse characterization is shown in Figure 6.27. The longitudinal modes in the optical spectrum are clearly visible. The modes are spaced exactly by the repetition rate. Taking into account the internal filter of the optical spectrum analyzer, an algorithm computes the position and magnitude of the single modes, indicated in the graph by circles. These circles are fitted with sech^2 function to compute the full-width half-maximum (FWHM) of the optical spectrum. The time–bandwidth product

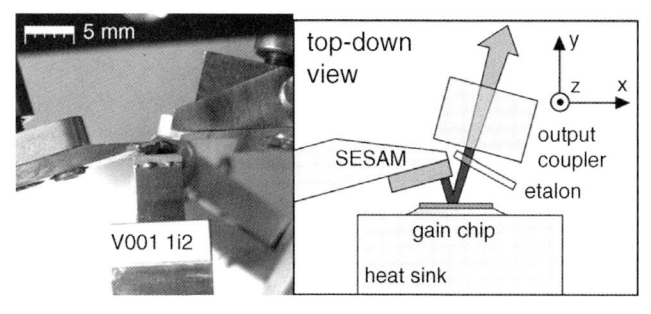

Figure 6.26 Cavity setup of the 50 GHz VECSEL.

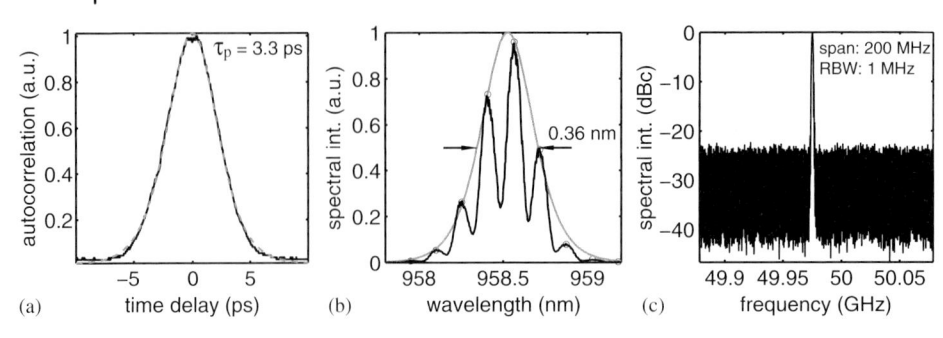

Figure 6.27 Pulse characterization of a 50 GHz VECSEL. The 3.3 ps pulses have an optical spectrum centered at 958.5 nm and a FWHM of 0.36 nm.

(TBP) is 0.38, which is 1.2 times the transform limit. The mode radii on the gain structure and on the SESAM were approximately 62 μm, from which one can calculate the intracavity pulse fluence to be around 1.1 μJ cm^{-2}. The M^2 was 1.25 and 1.13 in the horizontal and vertical axes, indicating that the beam quality was reasonably close to the diffraction limit.

6.3.3.3 Outlook

The ultrafast VECSELs are ideally suited for the operation at high average output power levels and multi-GHz operation because in contrast to DPSSLs, Q-switching instabilities are not a critical issue. With the standard VECSEL-SESAM mode locking approach, up to 100 mW of average output power in 3.3 ps pulses was obtained at a repetition rate of 50 GHz. A key requirement for such performance is a low-saturation SESAM, which is realized by using QD-absorber layers (see Section 6.2.3.2). However, further repetition rate scaling in a folded cavity geometry appears challenging because of the required cavity dimensions. A solution to these issues is the integration of the absorber into the VECSEL gain element, which is described in Section 6.4.

6.3.4
Femtosecond Mode-Locked VECSELs

6.3.4.1 Introduction

Semiconductor gain materials usually exhibit an optical amplification bandwidth >10 nm, which supports femtosecond pulse generation. However, so far, most mode-locked VECSELs operate in the few ps regime with an optical bandwidth below 1 nm. They typically use QW-SESAMs with a recovery time of tens of picoseconds and operate in the in slow saturable absorber regime (see Section 6.2.2), for which the absorber recovery time is long compared to the pulse duration. Normally, the dispersion is optimized for the quasi-soliton mode locking regime (see Section 6.2.2.3), for which shorter and nearly transform-limited pulses are obtained. During the past few years, several research groups achieved the generation of femtosecond pulses by using faster saturable absorbers.

6.3.4.2 Mode Locking Results

The first mode-locked femtosecond VECSEL was already demonstrated in 2002 by Garnache et al. [89]. The short pulses were achieved by careful dispersion control in the gain structure and a SESAM design whose operation utilizes the AC-Stark effect. It contained a single InGaAs/GaAs quantum well with 8 nm thickness, which had a room-temperature excitonic absorption peak at 1025 nm. The high oscillating electric field during a pulse instantaneously shifts the exciton to a shorter wavelength. Therefore, when the laser is operated on the long-wavelength side of the exciton, the absorption of the SESAM is reduced due to this wavelength shift. In the SESAM, the quantum well is placed about 2 nm below the device surface to achieve fast recombination for carriers due to tunneling into surface states. The SESAM was grown by MOVPE at a growth temperature of 735 °C. The laser generated 477 fs pulses with 100 mW average output power at a repetition rate of 1.21 GHz. The output spectrum was centered around 1045 nm.

In 2008, Klopp et al. [24] achieved 290 fs transform-limited pulses with 10 mW average output power from a VECSEL with graded gap barrier design in the gain section. The SESAM used an InGaAs/GaAs quantum well with 10 nm thickness. The top GaAs layer had a thickness of 2 nm and was covered by a SiN coating for surface protection, which also acts as an antireflection coating.

The shortest pulses were reported by Wilcox et al. [23], achieving 260 fs pulses from a VECSEL passively mode locked with a Stark-SESAM. They used a step-index gain structure and an improved SESAM design for this result and confirmed the validity of the Stark mode locking mechanism by numerical simulations. The laser generated 25 mW average output power at a center wavelength of 1035 nm. The optical bandwidth was 4.4 nm, and the output pulses were nearly transform limited (time–bandwidth product of 1.02).

Another interesting option for realizing faster SESAM recovery dynamics is the use of QD absorbers. Using a QD-SESAM with three groups of three layers of InAs/GaAs QDs, Wilcox et al. [96] recently achieved 870 fs pulses with 45 mW average output power at 1027 nm.

6.3.5
Electrically Pumped Mode-Locked VECSELs

For realizing cost-effective, compact ultrafast lasers, implementing optical pumping is challenging and direct electrical pumping appears highly attractive. It makes the bulky pump laser and pump optics redundant and simplifies alignment and packaging. For many applications, for example, optical clocking of microprocessors, this appears to be an essential requirement for mass production and high reliability. However, it is challenging to maintain the excellent properties in terms of high output power and good transverse beam quality. For achieving mode-locked operation, it is essential to achieve single-transverse mode operation because the presence of multiple transverse modes would destabilize the mode locking mechanism. Optical pumping is ideally suited for this task because the carriers are generated exactly where they are needed, that is, in the active region, and with an easily controllable transverse distribution (usually by using a Gaussian pump profile). For electrical pumping, the highest carrier densities are found close to the contacts, and it is important to select an optimized design to combine high output power and good beam quality. Unfortunately, the requirements in terms of electrical and optical properties are opposed, for example, a high doping level for achieving good electrical conductivity can introduce strong optical losses. These challenges are discussed in detail in section (see Ch. 7).

However, EP-VECSEL structures optimized for continuous-wave operation are usually not ideally suited for mode locking. These structures often rely on a strong field enhancement in the gain section, which introduces a narrow gain bandwidth and strong dispersion, leading to relatively long pulse durations. In 2003, Jasim et al. [97] reported the first passive mode locking of an electrically pumped VECSEL. They used the NECSEL structure (see Ch. 7, Section 7.2) in a Z-shaped cavity with a SESAM as one end mirror and achieved 40 mW average output power in 57 ps pulses at 1.1 GHz. By using a SESAM that had a design similar to the active region, but was used reverse biased, Jasim et al. achieved up to 15 ps pulse duration [98].

Realizing shorter pulse durations from electrically pumped VECSELs will most likely require a design that is optimized for broad amplification bandwidth and smoother GDD. Kreuter et al. [99] presented a detailed numerical analysis of EP-VECSEL structures, which takes into account optical losses, current confinement, and device resistance. They discuss a design that balances the conflicting optical and electrical requirements and present guidelines for an optimized design compatible with passive mode locking.

6.4
Mode-Locked Integrated External-Cavity Surface-Emitting Laser (MIXSEL)

6.4.1
Introduction

In an ultrafast semiconductor disk laser, the pulse formation is initiated and maintained with a SESAM (Section 6.2.1). The SESAM is typically used as an end mirror in a Z- or V-shaped cavity geometry, leading to a more complex setup than for a continuous-wave semiconductor disk laser. But since the SESAM is usually manufactured in the same semiconductor material system as the gain element, the integration of the saturable absorber in the gain structure is feasible. The resulting simple, linear cavity setup is easier to manufacture than a folded cavity and bears the potential for higher repetition rates and monolithic integration. This concept was for the first time experimentally demonstrated in 2007 and is referred to as the mode-locked integrated external cavity surface-emitting laser (MIXSEL) [38]. The general concept of the first MIXSEL design is shown in Figure 6.28. The pump light is absorbed in the active region. The intermediate distributed Bragg reflector (DBR) prevents pump light from reaching the saturable absorber, which would otherwise be

6.4 Mode-Locked Integrated External-Cavity Surface-Emitting Laser (MIXSEL)

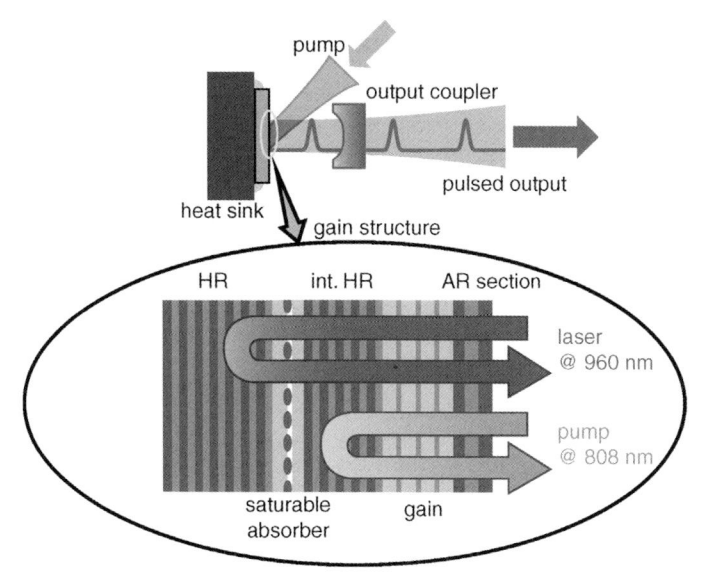

Figure 6.28 MIXSEL setup and semiconductor structure.

saturated from the pump light. The laser light reaches the saturable absorber that starts and stabilizes the mode locking.

6.4.2
Integration Challenges

The key integration challenge of the MIXSEL is achieving suitable saturable absorber parameters for stable passive mode locking. For this, the absorber needs to saturate at lower pulse energies than the gain (see Section 6.2.2). In standard VECSEL-SESAM mode locking, this is usually achieved by strong focusing onto the SESAM (Figure 6.29). The semiconductor structure of a MIXSEL is less than $\approx 10\,\mu m$ thick,

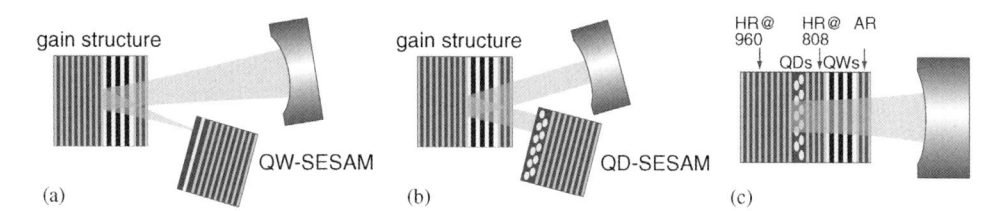

Figure 6.29 (a–c) Evolution from the VECSEL-SESAM mode locking with a large mode area ratios, (b) mode locking with identical mode areas on gain structure and SESAM to the (c) absorber–gain integration in a single device. The MIXSEL contains two high reflectors (HR), a quantum dot (QD) saturable absorber, quantum well (QW) gain, and an antireflection (AR) coating. The intermediate HR is to prevent the pump light from bleaching the saturable absorber.

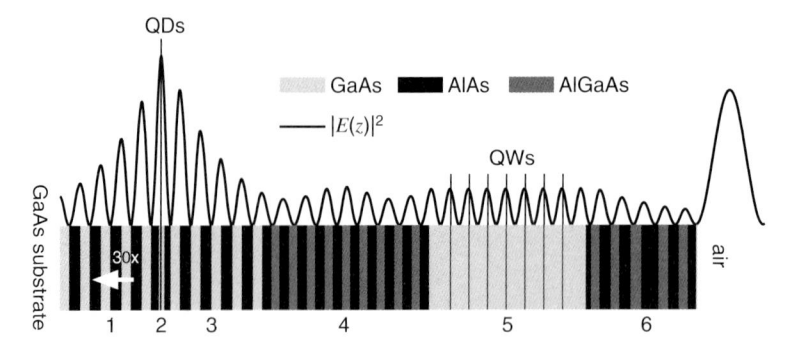

Figure 6.30 The MIXSEL design. The black line is the field enhancement. The structure is designed to have a high field enhancement in the saturable absorber.

which is smaller than the typical pump diameter of several tens to hundreds of micrometers of semiconductor disk lasers. Thus, the beam diameters in the gain and absorber layers are the same. Mode locking with the same spot size on gain and absorber requires the saturation fluence of the saturable absorber to be substantially lower than for the gain. The crucial step toward the MIXSEL realization was the development of low-saturation fluence QD absorbers, which enabled the first passively mode-locked VECSEL with similar mode area on SESAM and gain structure (see Section 6.2.3.2).

In the first MIXSEL, this was achieved by using a single QD absorber layer in a resonant design. The design of the first MIXSEL is shown in Figure 6.30. The structure was designed for a laser wavelength of 955 nm and a pump wavelength of 808 nm. The six building blocks are listed below:

1) Bottom DBR at 955 nm: 30 pair AlAs/GaAs.
2) Saturable absorber: a single QD layer.
3) Intermediate DBR at 955 nm for field enhancement control: 5 pair AlAs/GaAs.
4) Pump DBR (reflective for 808 nm, transmissive for 955 nm): 9 pair AlAs/$Al_{0.2}Ga_{0.8}As$.
5) Active region: $7 \times 7\,nm^2$ $In_{0.13}Ga_{0.87}As$ QWs with GaAs spacer layers.
6) Antireflective section: 11 layer $AlAs/Al_{0.2}Ga_{0.8}As$ with 10 nm GaAs cap layer.

The bottom mirror is designed to reflect the 955 nm laser light. The 30-pair DBR results in a reflectivity of 99.97%. Section 2 contains the QD-based saturable absorber. The QDs are embedded inside a 20 nm thick GaAs layer surrounded by AlAs. The five pairs of AlAs/GaAs Bragg mirror in section 3 enhance the field in the absorber layer to reduce its saturation fluence. Section 4 is a dichroic Bragg mirror that is highly transmissive for the laser wavelength but reflects the pump light. It prevents the saturable absorber to presaturate by the pump light and increases the pump absorption in the active region. The used materials AlAs and $Al_{0.2}Ga_{0.8}As$ do not absorb 808 nm light. The gain section (section 5) consists of seven $In_{0.13}Ga_{0.87}As$ quantum wells with a thickness of 7 nm surrounded by GaAs spacer layers that absorb the pump light. Due to the backreflection of the pump (section 4), 90% of the pump light is absorbed in the active region. The antireflective section (section 6) increases the field enhancement in the gain section; furthermore, it is designed for a larger wavelength tolerance of the field enhancement in the gain.

Combining two different elements, the VECSEL gain structure optimized for high gain, and the QD saturable absorber optimized for low saturation fluence and fast recovery, raises several growth challenges. Different growth temperatures are used for different layers leading to annealing and material quality issues. The low temperature-grown QD layer is located between two high temperature-grown DBRs, and the annealing of the saturable absorber layer changes its properties. This annealing modifies the dot composition and size [100] resulting in a blue shift of the absorption and emission wavelengths, which has to be taken into account.

Furthermore, a high semiconductor growth accuracy has to be achieved for the initial MIXSEL design. For a strong field enhancement inside the structure, even small growth deviations can change the field enhancement significantly and introduce a large amount of GDD. The relative field strength in the absorber section is controlled by the number of pairs in the intermediate distributed Bragg reflector (section 3 in Figure 6.30). The initial MIXSEL design with five intermediate DBR pairs is relatively sensitive to growth inaccuracies. This is illustrated by Figure 6.31, which shows the simulated field enhancement and GDD as function of wavelengths for small deviations of the growth rate (assuming random errors below 1% to the growth rates of the materials GaAs, AlAs, and $Al_{0.2}Ga_{0.8}As$).

6.4.3
Results

The first MIXSEL was grown on a 600 μm thick GaAs substrate using a VEECO GEN III molecular beam epitaxy machine. The QDs were grown at 430 °C, the QWs at 520 °C, and the rest of the structure was grown at 600 °C. This high growth temperature does not affect the QWs, however, it anneals the QDs. The effect of annealing was compensated by initially growing the QDs at a photoluminescence (PL) peak of 1100 nm. The annealing during the growth afterward shifts the absorption

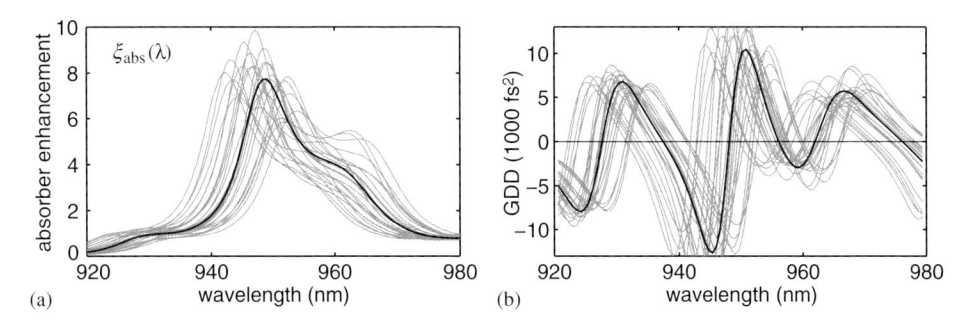

Figure 6.31 Absorber enhancement and GDD as function of the wavelength for a design with five mirror pairs. The black curve is the designed structure, the gray curves correspond to structures with a random growth error of <1%.

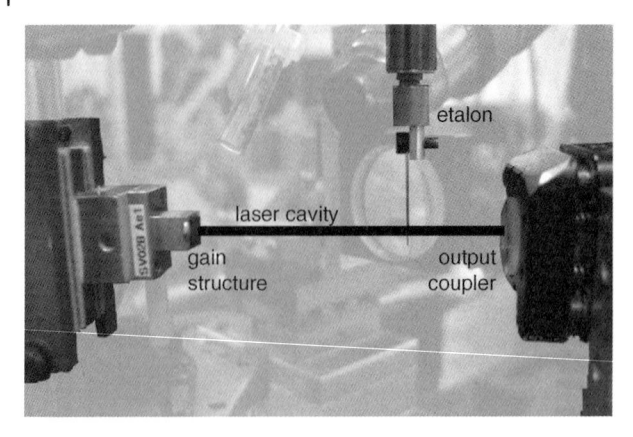

Figure 6.32 MIXSEL cavity. The laser resonator consists of the semiconductor structure and the external output coupler. The etalon selects the lasing wavelength for stable mode locking operation.

peak to our lasing wavelength around 950 nm. The MIXSEL wafer was cleaved in 5×5 mm^2 pieces and soldered to a copper heat sink, which is cooled by two Peltier elements. The continuous-wave pump source delivered 1.5 W at 808 nm and is focused to a circular pump spot with a radius of 80 μm. An output coupler with a radius of curvature (ROC) of 60 mm was used. A 25 μm fused silica etalon was used for wavelength tuning (Figure 6.32).

In the first experimental realization, stable mode locking was obtained with a 5.4 cm long straight cavity and a 0.35% transmission output coupler. The laser delivered 40 mW average power in 35 ps pulses at a 2.8 GHz repetition rate. The optical spectrum had a FWHM of 0.11 nm. The limited output power and low optical-to-optical efficiency is caused by the high temperature increase due to the poor thermal conductivity of the 600 μm thick substrate. In the second realization [101], the setup was optimized by implementing better cooling of the mount, a larger pump spot of 110 μm, and an output coupler with 0.7% transmission. Stable mode-locked operation was achieved with 185 mW average output power and 31.6 ps long pulses at 957 nm (Figure 6.33). The pump power was 4 W. The cavity was 52.4 mm long, corresponding to a pulse repetition rate of 2.86 GHz. The long pulse duration is likely due to the growth deviation that causes larger group delay dispersion and a smaller gain bandwidth. The transverse beam quality was nearly transform limited ($M^2 < 1.1$ in both directions).

6.4.4
Outlook

The performance of the first MIXSEL was limited in average output power by the low thermal conductivity of its 600 μm thick GaAs substrate. We expect that better heat management (for example, by substrate removal) will result in multiwatt average output powers, as previously achieved from the mode-locked VECSELs. Furthermore, the current growth inaccuracies caused a large group delay dispersion and

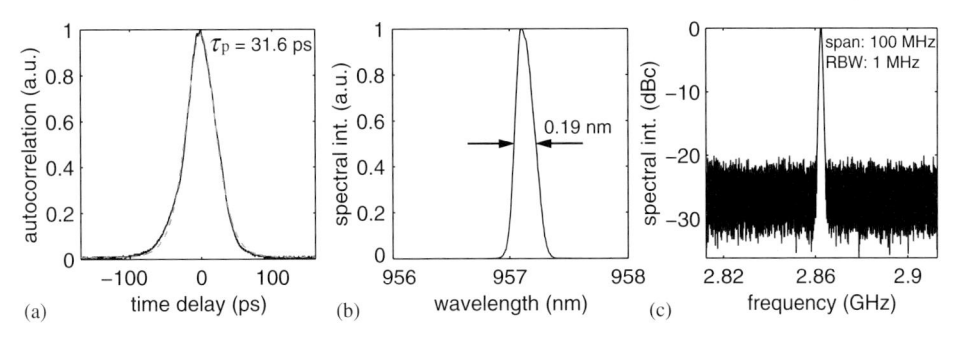

Figure 6.33 Pulse characterization of an optically pumped MIXSEL with 185 mW average output power. (a) Autocorrelation with sech²-fit for a 31.6 ps pulse. (b) Optical spectrum. (c) Microwave spectrum on a 100 MHz span with 1 MHz resolution bandwidth.

a small gain bandwidth, which are likely responsible for the relatively long pulse duration. Designs with less sensitivity toward growth errors are promising to increase the gain bandwidth and reduce the GDD, which will result in shorter pulses (Figure 6.34). If the field enhancement in the absorber section can be reduced to a lower value by using only one intermediate mirror pair (Figure 6.35), the MIXSEL structure becomes substantially less sensitive to growth deviations. Such structures appear feasible using recently developed low-saturation fluence QD-absorber layers [81].

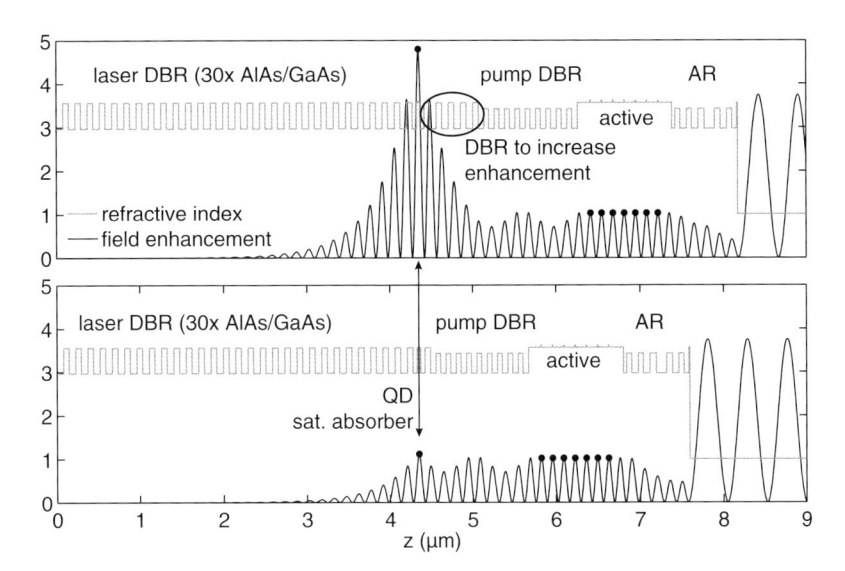

Figure 6.34 Influence of the intermediate mirror on the field enhancement in the absorber section. The gray curve is the refractive index and the black curve is the field enhancement in the structure. The upper graph shows a MIXSEL design with five pairs (black circle) and the bottom graph shows a design with only one pair, which has a reduced absorber enhancement.

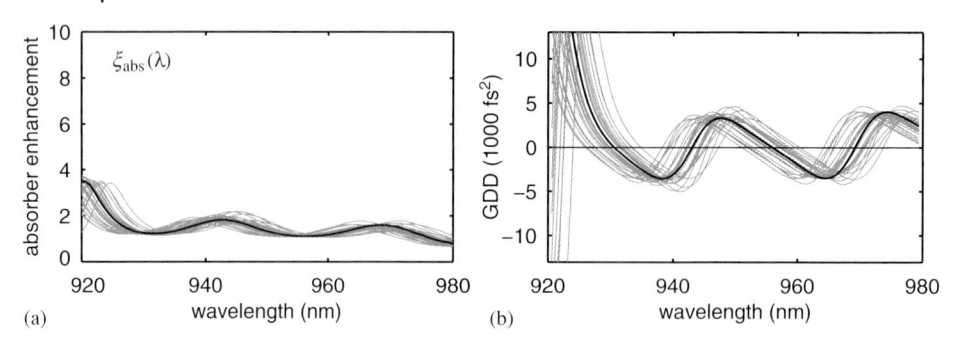

Figure 6.35 Absorber enhancement and GDD as function of the wavelength for a design with one mirror pair. The black curve is the designed structure and the gray curves show the aberration that a <1% error would have.

Furthermore, the simple straight cavity layout bears the potential for a monolithic setup, enabling higher robustness and cost-effective high-volume manufacturing. A possible design concept is shown in Figure 6.36. A transparent wafer into which the curved output couplers are etched is contacted directly onto the MIXSEL semiconductor structure. The pulse mostly propagates in the transparent wafer as the thickness of the semiconductor structure is only ≈10 μm, which is smaller than the total optical cavity length (about 3 mm in the case of a 50 GHz MIXSEL). The finished wafer structure (Figure 6.36a) can then be cut into individual MIXSELs. The MIXSEL can also be optically pumped from the back after the GaAs substrate has been removed.

6.5
Summary and Outlook

Less than one decade after the first demonstration of a passively mode-locked semiconductor disk laser, these sources have become a viable alternative to other ultrafast laser systems due to their advantages in terms of wavelength flexibility, power scaling, short pulse generation, and cost-effective manufacturing. Numerous research groups contributed to the large technology progress. Many different semiconductor gain materials were exploited, leading to a large spectral coverage of ultrafast VECSELs from the visible region up to 2 μm wavelength. Ultrafast VECSELs are ideally suited for the operation at high repetition rates with high average output power, which is important for commercially attractive applications such as optical clocking and interconnects in computers. Repetition rates up to 50 GHz were already realized with 100 mW average output power. Power scaling into the multiwatt regime made ultrafast semiconductor disk lasers attractive for numerous applications currently relying on more complex and expensive laser sources. The key pulse formation mechanisms were investigated and understood; examples are the quasi-soliton theory or the Stark-SESAM mode locking mechanism. The latter technique enabled femtosecond pulse generation with less than 300 fs duration,

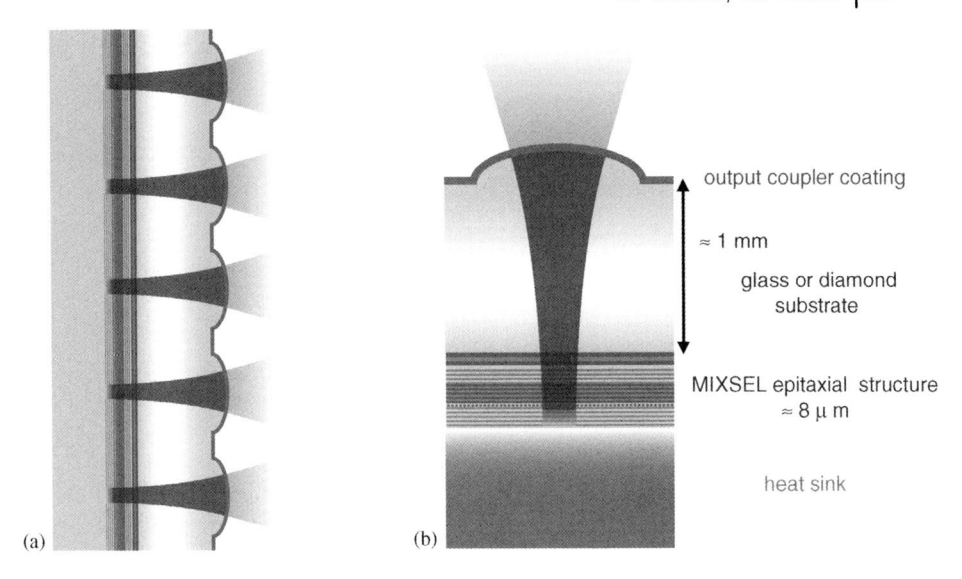

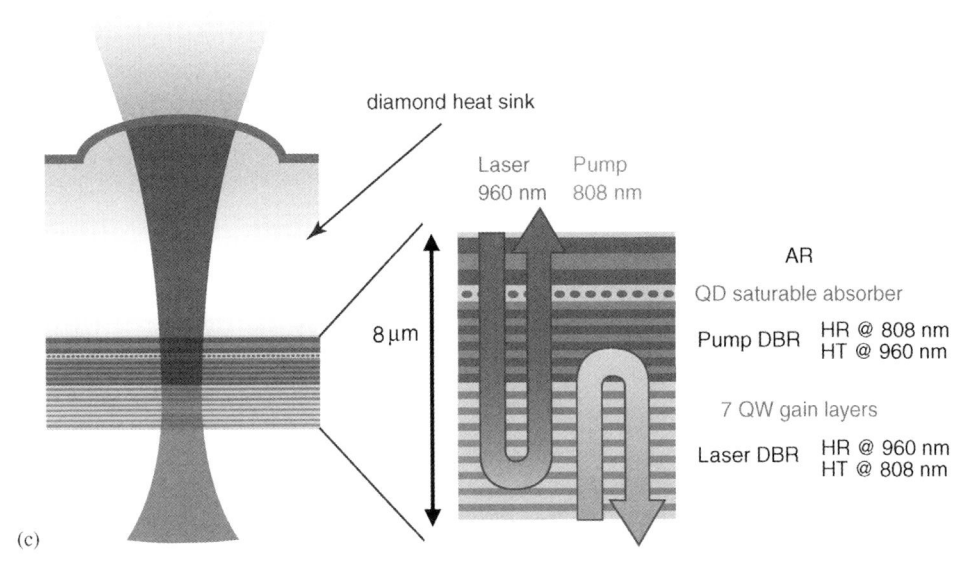

Figure 6.36 Wafer-scale integration concept. (a) Semiconductor MIXSEL structure wafer glued to a transparent wafer into which the curved output coupler is etched. (b) Wafer-based MIXSEL mounted onto a heat sink. Ideally suited for electrical pumping or integrated optical pumping. (c) Wafer-based MIXSEL for backside optical pumping when transparent external cavity wafer is also used as a heat sink (for example, with diamond).

which is shorter than for any other semiconductor laser without external pulse compression. Even passive mode locking of electrically pumped VECSELs was achieved. Finally, in the MIXSEL, the saturable absorber for pulse formation was integrated into the semiconductor gain structure, leading to a particularly simple setup that has a large potential for cost-effective mass production.

Given the tremendous progress achieved during the past years, we expect that ultrafast semiconductor lasers will continue to increase their performance and penetrate into numerous new application areas. An important research task is to extend the femtosecond operation toward higher power levels than the currently demonstrated 100 mW, which will make these lasers more attractive for applications relying on high peak power levels. Another task is to further reduce the pulse duration. Initial results already showed the suitability of semiconductor disk lasers for the generation of sub-100 fs pulses [25]. The ultrafast disk lasers have a large potential for wavelength tunability, which has to be explored. Wavelength-flexible sources may replace Ti:sapphire lasers in many cases such as multiphoton imaging in biology and medicine. The MIXSEL concept has to demonstrate its suitability to support higher repetition rates and similar average output power levels than SESAM-mode-locked semiconductor disk lasers. The electrically pumped ultrafast VECSEL with an optimized design for the generation of shorter pulses than 15 ps have to be realized. The final step toward even lower cost and more compact ultrafast semiconductor surface-emitting lasers will be electrically pumped MIXSELs. This would result in devices ideally suited for many applications such as telecommunications, optical clocking, frequency metrology, microscopy, laser display – anywhere where the current ultrafast laser technology is considered to be too bulky or expensive.

References

1 Keller, U. (2003) Recent developments in compact ultrafast lasers. *Nature*, **424**, 831–838.
2 Zewail, A.H. (1988) Laser femtochemistry. *Science*, **242**, 1645–1653.
3 Telle, H.R., Steinmeyer, G., Dunlop, A.E., Stenger, J., Sutter, D.H., and Keller, U. (1999) Carrier-envelope offset phase control: a novel concept for absolute optical frequency measurement and ultrashort pulse generation. *Appl. Phys. B*, **69**, 327–332.
4 Jones, D.J., Diddams, S.A., Ranka, J.K., Stentz, A., Windeler, R.S., Hall, J.L., and Cundiff, S.T. (2000) Carrier-envelope phase control of femtosecond mode-locked lasers and direct optical frequency synthesis. *Science*, **288**, 635–639.
5 Apolonski, A., Poppe, A., Tempea, G., Spielmann, C., Udem, T., Holzwarth, R., Hänsch, T.W., and Krausz, F. (2000) Controlling the phase evolution of few-cycle light pulses. *Phys. Rev. Lett.*, **85**, 740–743.
6 Ell, R., Morgner, U., Kärtner, F.X., Fujimoto, J.G., Ippen, E.P., Scheuer, V., Angelow, G., Tschudi, T., Lederer, M.J., Boiko, A., and Luther-Davies, B. (2001) Generation of 5-fs pulses and octave-spanning spectra directly from a Ti:sapphire laser. *Opt. Lett.*, **26**, 373–375.
7 Gallmann, L., Sutter, D.H., Matuschek, N., Steinmeyer, G., Keller, U., Iaconis, C., and Walmsley, I.A. (1999) Characterization of sub-6-fs optical pulses with spectral phase interferometry for direct electric-field reconstruction. *Opt. Lett.*, **24**, 1314–1316.
8 Sutter, D.H., Steinmeyer, G., Gallmann, L., Matuschek, N., Morier-Genoud, F., Keller, U., Scheuer, V., Angelow, G., and

Tschudi, T. (1999) Semiconductor saturable-absorber mirror-assisted Kerr-lens mode-locked Ti:sapphire laser producing pulses in the two-cycle regime. *Opt. Lett.*, **24**, 631–633.

9 Keller, U. (2007) Ultrafast solid-state lasers, in *Landolt-Börnstein. Laser Physics and Applications. Subvolume B: Laser Systems. Part I* (eds G. Herziger, H. Weber, and R. Proprawe), Springer Verlag, Heidelberg, pp. 33–167.

10 Keller, U., Weingarten, K.J., Kärtner, F.X., Kopf, D., Braun, B., Jung, I.D., Fluck, R., Hönninger, C., Matuschek, N., and Aus der Au, J. (1996) Semiconductor saturable absorber mirrors (SESAMs) for femtosecond to nanosecond pulse generation in solid-state lasers. *IEEE J. Sel. Top. Quantum Electron.*, **2**, 435–453.

11 Keller, U., Miller, D.A.B., Boyd, G.D., Chiu, T.H., Ferguson, J.F., and Asom, M.T. (1992) Solid-state low-loss intracavity saturable absorber for Nd:YLF lasers: an antiresonant semiconductor Fabry–Perot saturable absorber. *Opt. Lett.*, **17**, 505–507.

12 Innerhofer, E., Südmeyer, T., Brunner, F., Häring, R., Aschwanden, A., Paschotta, R., Keller, U., Hönninger, C., and Kumkar, M. (2003) 60 W average power in 810-fs pulses from a thin-disk Yb:YAG laser. *Opt. Lett.*, **28**, 367–369.

13 Brunner, F., Innerhofer, E., Marchese, S.V., Südmeyer, T., Paschotta, R., Usami, T., Ito, H., Kurimura, S., Kitamura, K., Arisholm, G., and Keller, U. (2004) Powerful red–green–blue laser source pumped with a mode-locked thin disk laser. *Opt. Lett.*, **29**, 1921–1923.

14 Marchese, S.V., Baer, C.R.E., Engqvist, A.G., Hashimoto, S., Maas, D.J.H.C., Golling, M., Südmeyer, T., and Keller, U. (2008) Femtosecond thin disk laser oscillator with pulse energy beyond the 10-microjoule level. *Opt. Express*, **16**, 6397–6407.

15 Neuhaus, J., Bauer, D., Zhang, J., Killi, A., Kleinbauer, J., Kumkar, M., Weiler, S., Guina, M., Sutter, D.H., and Dekorsy, T. (2008) Subpicosecond thin-disk laser oscillator with pulse energies of up to 25.9 microjoules by use of an active multipass geometry. *Opt. Express*, **16**, 20530–20539.

16 Südmeyer, T., Marchese, S.V., Hashimoto, S., Baer, C.R.E., Gingras, G., Witzel, B., and Keller, U. (2008) Femtosecond laser oscillators for high-field science. *Nat. Photon.*, **2**, 599–604.

17 Limpert, J., Röser, F., Schreiber, T., and Tünnermann, A. (2006) High-power ultrafast fiber laser systems. *IEEE J. Sel. Top. Quantum Electron.*, **12**, 233–244.

18 Krainer, L., Paschotta, R., Lecomte, S., Moser, M., Weingarten, K.J., and Keller, U. (2002) Compact Nd:YVO$_4$ lasers with pulse repetition rates up to 160 GHz. *IEEE J. Quantum Electron.*, **38**, 1331–1338.

19 Oehler, A.E.H., Südmeyer, T., Weingarten, K.J., and Keller, U. (2008) 100 GHz passively mode-locked Er:Yb:glass laser at 1.5 µm with 1.6-ps pulses. *Opt. Express*, **16**, 21930–21935.

20 Hartl, I., McKay, H.A., Thapa, R., Thomas, B.K., Ruehl, A., Dong, L., and Fermann, M.E. (2009) Fully stabilized GHz Yb-fiber laser frequency comb. Advanced Solid-State Photonics (ASSP), Denver, CO.

21 Plant, J.J., Gopinath, J.T., Chann, B., Ripin, D.J., Huang, R.K., and Juodawlkis, P.W. (2006) 250 mW, 1.5 µm monolithic passively modelocked slab-coupled optical waveguide laser. *Opt. Lett.*, **31**, 223–225.

22 Hoogland, S., Dhanjal, S., Tropper, A.C., Roberts, J.S., Häring, R., Paschotta, R., Morier-Genoud, F., and Keller, U. (2000) Passively mode-locked diode-pumped surface-emitting semiconductor laser. *IEEE Photon. Technol. Lett.*, **12**, 1135–1137.

23 Wilcox, K.G., Mihoubi, Z., Daniell, G.J., Elsmere, S., Quarterman, A., Farrer, I., Ritchie, D.A., and Tropper, A. (2008) Ultrafast optical Stark mode-locked semiconductor laser. *Opt. Lett.*, **33**, 2797–2799.

24 Klopp, P., Saas, F., Zorn, M., Weyers, M., and Griebner, U. (2008) 290-fs pulses from a semiconductor disk laser. *Opt. Express*, **16**, 5770–5775.

25 Wilcox, K.G., Mihoubi, Z., Elsmere, S., Quarterman, A., Farrer, I., Ritchie, D.A., and Tropper, A. (2009) 70-fs transform-limited pulses emitted by InGaAs/GaAs

quantum well laser. Advanced Solid-State Photonics (ASSP), Denver, CO, p. ME4.

26 Aschwanden, A., Lorenser, D., Unold, H.J., Paschotta, R., Gini, E., and Keller, U. (2005) 2.1-W picosecond passively mode-locked external-cavity semiconductor laser. *Opt. Lett.*, **30**, 272–274.

27 Lorenser, D., Maas, D.J.H.C., Unold, H.J., Bellancourt, A.-R., Rudin, B., Gini, E., Ebling, D., and Keller, U. (2006) 50-GHz passively mode-locked surface-emitting semiconductor laser with 100 mW average output power. *IEEE J. Quantum Electron.*, **42**, 838–847.

28 Keller, U. and Tropper, A.C. (2006) Passively modelocked surface-emitting semiconductor lasers. *Phys. Rep.*, **429**, 67–120.

29 Maas, D. (2009) Modelocking of semiconductor vertical emitters: from VECSEL to MIXSEL. Dissertation, ETH Zurich, Nr. 18206, Hartung-Gorre Verlag, Konstanz.

30 Moore, G.E. (1998) Cramming more components onto integrated circuits. *Proc. IEEE*, **86**, 82–85 (Reprinted from *Electronics*, April 19, 1965, 114–117).

31 Fukatsu, K., Shiba, K., Suzuki, Y., Suzuki, N., Anan, T., Hatakeyama, H., Yashiki, K., and Tsuji, M. (2008) 30 Gb/s over 100-m MMFs using 1.1-μm range VCSELs and photodiodes. *IEEE Photon. Technol. Lett.*, **20**, 909–911.

32 Keeler, G.A., Nelson, B.E., Agarwal, D., Debaes, C., Helman, N.C., Bhatnagar, A., and Miller, D.A.B. (2003) The benefits of ultrashort optical pulses in optically interconnected systems. *IEEE J. Sel. Top. Quantum Electron.*, **9**, 477–485.

33 Cassan, E., Marris, D., Rouviere, M., Vivien, L., and Laval, S. (2005) Comparison between electrical and optical global clock distributions for CMOS integrated circuits. *Opt. Eng.*, **44**, 105402–105410.

34 Miller, D.A.B. (2000) Rationale and challenges for optical interconnects to electronic chips. *Proc. IEEE*, **88**, 728–749.

35 Huang, D., Swanson, E.A., Lin, C.P., Schuman, J.S., Stinson, W.G., Chang, W., Hee, M.R., Flotte, T., Gregory, K., Puliafito, C.A., and Fujimoto, J.G. (1991) Optical coherence tomography. *Science*, **254**, 1178–1181.

36 Fujimoto, J.G. (2003) Optical coherence tomography for ultrahigh resolution in vivo imaging. *Nat. Biotechnol.*, **21**, 1361–1367.

37 Swanson, E.A., Izatt, J.A., Hee, M.R., Huang, D., Lin, C.P., Schuman, J.S., Puliafito, C.A., and Fujimoto, J.G. (1993) In-vivo retinal imaging by optical coherence tomography. *Opt. Lett.*, **18**, 1864.

38 Maas, D.J.H.C., Bellancourt, A.-R., Rudin, B., Golling, M., Unold, H.J., Südmeyer, T., and Keller, U. (2007) Vertical integration of ultrafast semiconductor lasers. *Appl. Phys. B*, **88**, 493–497.

39 Set, S.Y., Yaguchi, H., Tanaka, Y., and Jablonski, M. (2004) Ultrafast fiber pulsed lasers incorporating carbon nanotubes. *IEEE J. Sel. Top. Quantum Electron.*, **10**, 137–146.

40 Schibli, T., Minoshima, K., Kataura, H., Itoga, E., Minami, N., Kazaoui, S., Miyashita, K., Tokumoto, M., and Sakakibara, Y. (2005) Ultrashort pulse-generation by saturable absorber mirrors based on polymer-embedded carbon nanotubes. *Opt. Express*, **13**, 8025–8031.

41 Fong, K.H., Kikuchi, K., Goh, C.S., Set, S.Y., Grange, R., Haiml, M., Schlatter, A., and Keller, U. (2007) Solid-state Er:Yb:glass laser mode-locked by using single-wall carbon nanotube thin film. *Opt. Lett.*, **32**, 38–40.

42 Stankov, K.A. (1988) A mirror with an intensity-dependent reflection coefficient. *Appl. Phys. B*, **45**, 191–195.

43 Fermann, M.E. (1994) Ultrashort-pulse sources based on single-mode rare-earth-doped fibers. *Appl. Phys. B*, **58**, 197–209.

44 Spence, D.E., Kean, P.N., and Sibbett, W. (1991) 60-fsec pulse generation from a self-mode-locked Ti:sapphire laser. *Opt. Lett.*, **16**, 42–44.

45 Keller, U. (2004) Ultrafast solid-state lasers. *Prog. Opt.*, **46**, 1–115.

46 Brovelli, L.R., Keller, U., and Chiu, T.H. (1995) Design and operation of antiresonant Fabry–Perot saturable semiconductor absorbers for

mode-locked solid-state lasers. *J. Opt. Soc. Am. B*, **12**, 311–322.

47 Hönninger, C., Paschotta, R., Morier-Genoud, F., Moser, M., and Keller, U. (1999) Q-switching stability limits of continuous-wave passive mode locking. *J. Opt. Soc. Am. B*, **16**, 46–56.

48 Schibli, T.R., Thoen, E.R., Kärtner, F.X., and Ippen, E.P. (2000) Suppression of Q-switched mode locking and break-up into multiple pulses by inverse saturable absorption. *Appl. Phys. B*, **70**, S41–S49.

49 Grange, R., Haiml, M., Paschotta, R., Spuhler, G.J., Krainer, L., Golling, M., Ostinelli, O., and Keller, U. (2005) New regime of inverse saturable absorption for self-stabilizing passively mode-locked lasers. *Appl. Phys. B*, **80**, 151–158.

50 Chow, W.W. and Koch, S.W. (1999) *Semiconductor-Laser Fundamentals: Physics of the Gain Materials*, Springer.

51 Siegman, A.E. (1986) *Lasers*, University Science Books, Mill Valley, California.

52 Haiml, M. (2001) Optimized non-stoichiometric semiconductor materials for applications in ultrafast nonlinear optics. Dissertation, ETH Zurich, Nr. 14411.

53 Thoen, E.R., Koontz, E.M., Joschko, M., Langlois, P., Schibli, T.R., Kärtner, F.X., Ippen, E.P., and Kolodziejski, L.A. (1999) Two-photon absorption in semiconductor saturable absorber mirrors. *Appl. Phys. Lett.*, **74**, 3927–3929.

54 Haiml, M., Grange, R., and Keller, U. (2004) Optical characterization of semiconductor saturable absorbers. *Appl. Phys. B*, **79**, 331–339.

55 Maas, D.J.H.C., Rudin, B., Bellancourt, A.-R., Iwaniuk, D., Marchese, S.V., Südmeyer, T., and Keller, U. (2008) High precision optical characterization of semiconductor saturable absorber mirrors. *Opt. Express*, **16**, 7571–7579.

56 Marchese, S. (2007) Towards high field physics with high power thin disk laser oscillators. Dissertation, ETH Zurich, Nr. 17583, Hartung-Gorre Verlag, Konstanz.

57 Spence, D.E., Kean, P.N., and Sibbett, W. (1991) 60-fsec pulse generation from a self-mode-locked Ti:sapphire laser. *Opt. Lett.*, **16**, 42–44.

58 Keller, U., 'tHooft, G.W., Knox, W.H., and Cunningham, J.E. (1991) Femtosecond pulses from a continuously self-starting passively mode-locked Ti:Sapphire laser. *Opt. Lett.*, **16**, 1022–1024.

59 Brabec, T., Spielmann, C., Curley, P.F., and Krausz, F. (1992) Kerr lens mode locking. *Opt. Lett.*, **17**, 1292–1294.

60 Kärtner, F.X. and Keller, U. (1995) Stabilization of soliton-like pulses with a slow saturable absorber. *Opt. Lett.*, **20**, 16–18.

61 Jung, I.D., Kärtner, F.X., Brovelli, L.R., Kamp, M., and Keller, U. (1995) Experimental verification of soliton modelocking using only a slow saturable absorber. *Opt. Lett.*, **20**, 1892–1894.

62 Kärtner, F.X., Jung, I.D., and Keller, U. (1996) Soliton mode-locking with saturable absorbers. *IEEE J. Sel. Top. Quantum Electron.*, **2**, 540–556.

63 Paschotta, R. and Keller, U. (2001) Passive mode locking with slow saturable absorbers. *Appl. Phys. B*, **73**, 653–662.

64 New, G.H.C. (1974) Pulse evolution in mode-locked quasi-continuous lasers. *IEEE J. Quantum Electron.*, **10**, 115–124.

65 Arthurs, E.G., Bradley, D.J., and Roddie, A.G. (1973) Buildup of picosecond pulse generation in passively mode-locked rhodamine dye lasers. *Appl. Phys. Lett.*, **23**, 88–90.

66 Hoogland, S., Dhanjal, S., Tropper, A.C., Roberts, S.J., Häring, R., Paschotta, R., and Keller, U. (2000) Passively mode-locked diode-pumped surface-emitting semiconductor laser. *IEEE Photon. Technol. Lett.*, **12**, 1135–1138.

67 Häring, R., Paschotta, R., Aschwanden, A., Gini, E., Morier-Genoud, F., and Keller, U. (2002) High-power passively mode–locked semiconductor lasers. *IEEE J. Quantum Electron.*, **38**, 1268–1275.

68 Paschotta, R., Häring, R., Keller, U., Garnache, A., Hoogland, S., and Tropper, A.C. (2002) Soliton-like pulse-shaping mechanism in passively mode-locked surface-emitting semiconductor lasers. *Appl. Phys. B*, **75**, 445–451.

69 Zeller, S.C., Südmeyer, T., Weingarten, K.J., and Keller, U. (2007) Passively

modelocked 77 GHz Er:Yb:glass laser. *Electron. Lett.*, **43**, 32–33.

70 Spühler, G.J., Weingarten, K.J., Grange, R., Krainer, L., Haiml, M., Liverini, V., Golling, M., Schon, S., and Keller, U. (2005) Semiconductor saturable absorber mirror structures with low saturation fluence. *Appl. Phys. B*, **81**, 27–32.

71 Saarinen, E.J., Harkonen, A., Herda, R., Suomalainen, S., Orsila, L., Hakulinen, T., Guina, M., and Okhotnikov, O.G. (2007) Harmonically mode-locked VECSELs for multi-GHz pulse train generation. *Opt. Express*, **15**, 955–964.

72 Hoffmann, M., Maas, D.J.H.C., Sieber, O., Wittwer, V.J., Bellancourt, A.-R., Rudin, B., Barbarin, Y., Golling, M., Südmeyer, T., and Keller, U. (2009) First experimental verification of soliton-like pulse-shaping mechanisms in passively mode-locked VECSELs. Conference on Lasers and Electro-Optics (CLEO), Munich, Germany, CB14.12, FRI.

73 Kopf, D., Zhang, G., Fluck, R., Moser, M., and Keller, U. (1996) All-in-one dispersion-compensating saturable absorber mirror for compact femtosecond laser sources. *Opt. Lett.*, **21**, 486–488.

74 Chilwell, J. and Hodgkinson, I. (1984) Thin-films field-transfer matrix theory of planar multilayer waveguides and reflection from prism-loaded waveguides. *J. Opt. Soc. Am. A*, **1**, 742–753.

75 Qasaimeh, O., Zhou, W.D., Phillips, J., Krishna, S., Bhattacharya, P., and Dutta, M. (1999) Bistability and self-pulsation in quantum-dot lasers with intracavity quantum-dot saturable absorbers. *Appl. Phys. Lett.*, **74**, 1654–1656.

76 Garnache, A., Hoogland, S., Tropper, A.C., Gerard, J.M., Thierry-Mieg, V., and Roberts, J.S. (2001) Pico-second passively mode locked surface-emitting laser with self-assembled semiconductor quantum dot absorber. CLEO/Europe-EQEC, postdeadline paper.

77 Rafailov, E.U., White, S.J., Lagatsky, A.A., Miller, A., Sibbett, W., Livshits, D.A., Zhukov, A.E., and Ustinov, V.M. (2004) Fast quantum-dot saturable absorber for passive mode-locking of solid-state lasers. *IEEE Photon. Technol. Lett.*, **16**, 2439–2441.

78 Lagatsky, A.A., Bain, F.M., Brown, C.T.A., Sibbett, W., Livshits, D.A., Erbert, G., and Rafailov, E.U. (2007) Low-loss quantum-dot-based saturable absorber for efficient femtosecond pulse generation. *Appl. Phys. Lett.*, **91**, 231111.

79 Maas, D.J.H.C., Bellancourt, A.R., Hoffmann, M., Rudin, B., Barbarin, Y., Golling, M., Südmeyer, T., and Keller, U. (2008) Growth parameter optimization for fast quantum dot SESAMs. *Opt. Express*, **16**, 18646–18656.

80 Solomon, G.S., Trezza, J.A., and Harris, J.J.S. (1995) Effects of monolayer coverage, flux ratio, and growth rate on the island density of InAs islands on GaAs. *Appl. Phys. Lett.*, **66**, 3161–3163.

81 Bellancourt, A.-R., Barbarin, Y., Maas, D.J.H.C., Shafiei, M., Hoffmann, M., Golling, M., Südmeyer, T., and Keller, U. (2009) Low saturation fluence antiresonant quantum dot SESAMs for MIXSEL integration. *Opt. Express*, **17**, 9704–9711.

82 Jiang, W., Derickson, D.J., and Bowers, J.E. (1993) Analysis of laser pulse chirping in mode-locked vertical-cavity surface-emitting lasers. *IEEE J. Quantum Electron.*, **QE-29**, 1309–1318.

83 Casel, O., Woll, D., Tremont, M.A., Fuchs, H., Wallenstein, R., Gerster, E., Unger, P., Zorn, M., and Weyers, M. (2005) Blue 489-nm picosecond pulses generated by intracavity frequency doubling in a passively mode-locked optically pumped semiconductor disk laser. *Appl. Phys. B*, **81**, 443–446.

84 Häring, R., Paschotta, R., Gini, E., Morier-Genoud, F., Melchior, H., Martin, D., and Keller, U. (2001) Picosecond surface-emitting semiconductor laser with >200 mW average power. *Electron. Lett.*, **37**, 766–767.

85 Casel, O., Woll, D., Tremont, M.A., Fuchs, H., Wallenstein, R., Gerster, E., Unger, P., Zorn, M., and Weyers, M. (2005) Blue 489-nm picosecond pulses generated by intracavity frequency doubling in a passively mode-locked

optically pumped semiconductor disk laser. *Appl. Phys. B*, **81**, 443–446.

86 Lorenser, D., Unold, H.J., Maas, D.J.H.C., Aschwanden, A., Grange, R., Paschotta, R., Ebling, D., Gini, E., and Keller, U. (2004) Towards wafer-scale integration of high repetition rate passively mode-locked surface-emitting semiconductor lasers. *Appl. Phys. B*, **79**, 927–932.

87 Aschwanden, A., Lorenser, D., Unold, H.J., Paschotta, R., Gini, E., and Keller, U. (2005) 10-GHz passively mode-locked surface emitting semiconductor laser with 1.4-W average output power. *Appl. Phys. Lett.*, **86**, 131102.

88 Rudin, B., Maas, D.J.H.C., Lorenser, D., Bellancourt, A.-R., Unold, H.J., and Keller, U. (2006) High-performance mode-locking with up to 50 GHz repetition rate from integrable VECSELs. Conference on Lasers and Electro-Optics (CLEO), Long Beach, CA.

89 Garnache, A., Hoogland, S., Tropper, A.C., Sagnes, I., Saint-Girons, G., and Roberts, J.S. (2002) <500-fs soliton pulse in a passively mode-locked broadband surface-emitting laser with 100-mW average power. *Appl. Phys. Lett.*, **80**, 3892–3894.

90 Hoogland, S., Garnache, A., Sagnes, I., Roberts, J.S., and Tropper, A.C. (2005) 10-GHz train of sub-500-fs optical soliton-like pulses from a surface-emitting semiconductor laser. *IEEE Photon. Technol. Lett.*, **17**, 267–269.

91 Rautiainen, J., Korpijärvi, V.-M., Puustinen, J., Guina, M., and Okhotnikov, O.G. (2008) Passively mode-locked GaInNAs disk laser operating at 1220 nm. *Opt. Express*, **16**, 15964–15969.

92 Rutz, A., Liverini, V., Maas, D.J.H.C., Rudin, B., Bellancourt, A.-R., Schön, S., and Keller, U. (2006) Passively modelocked GaInNAs VECSEL at centre wavelength around 1.3. *Electron. Lett.*, **42**, 926.

93 Hoogland, S., Paldus, B., Garnache, A., Weingarten, K.J., Grange, R., Haiml, M., Paschotta, R., Keller, U., and Tropper, A.C. (2003) Picosecond pulse generation with a 1.5 μm passively modelocked surface emitting semiconductor laser. *Electron. Lett.*, **39**, 846.

94 Lindberg, H., Sadeghi, M., Westlund, M., Wang, S., Larsson, A., Strassner, M., and Marcinkevicius, S. (2005) Mode locking a 1550 nm semiconductor disk laser by using a GaInNAs saturable absorber. *Opt. Lett.*, **30**, 2793–2795.

95 Härkönen, A., Rautiainen, J., Orsila, L., Guina, M., Rößner, K., Hümmer, M., Lehnhardt, T., Müller, M., Forchel, A., Fischer, M., Koeth, J., and Okhotnikov, O.G. (2008) 2-μm mode-locked semiconductor disk laser synchronously pumped using an amplified diode laser. *IEEE Photon. Technol. Lett.*, **20**, 1332–1334.

96 Wilcox, K.G., Butkus, M., Farrer, I., Ritchie, D.A., Tropper, A., and Rafailov, E.U. (2009) Subpicosecond quantum dot saturable absorber mode-locked semiconductor disk laser. *Appl. Phys. Lett.*, **94**, 3.

97 Jasim, K., Zhang, Q., Nurmikko, A.V., Mooradian, A., Carey, G., Ha, W., and Ippen, E. (2003) Passively modelocked vertical extended cavity surface emitting diode laser. *Electron. Lett.*, **39**, 373–375.

98 Jasim, K., Zhang, Q., Nurmikko, A.V., Ippen, E., Mooradian, A., Carey, G., and Ha, W. (2004) Picosecond pulse generation from passively modelocked vertical cavity diode laser at up to 15 GHz pulse repetition rate. *Electron. Lett.*, **40**, 34–35.

99 Kreuter, P., Witzigmann, B., Maas, D.J.H.C., Barbarin, Y., Südmeyer, T., and Keller, U. (2008) On the design of electrically-pumped vertical-external-cavity surface-emitting lasers. *Appl. Phys. B*, **91**, 257–264.

100 Malik, S., Roberts, C., Murray, R., and Pate, M. (1997) Tuning self-assembled InAs quantum dots by rapid thermal annealing. *Appl. Phys. Lett.*, **71**, 1987–1989.

101 Bellancourt, A.-R., Maas, D.J.H.C., Rudin, B., Golling, M., Südmeyer, T., and Keller, U. (2009) Modelocked integrated external-cavity surface emitting laser (MIXSEL). *IET Optoelectron.*, **3**, 61–72.

7
External-Cavity Surface-Emitting Diode Lasers*

Aram Mooradian, Andrei Shchegrov, Ashish Tandon, and Gideon Yoffe

7.1
Introduction

The early pioneering work of Soda *et al.* [1] on electrically pumped surface-emitting diode lasers has opened up numerous applications for these devices in telecommunications, information storage, and display. These lasers were limited to a few mW of cw power output that was adequate for single-mode and multimode fiber communications as well as CD playback and high-speed backplane communication links. The progress in materials development using MOCVD and MBE growth with low-loss, high-reflectivity semiconductor mirrors together with external-cavity mode control has allowed single spatial mode and single-frequency operation at power levels greater than 500 mW cw from a single emitter. We call these devices E-VECSELs (electrically pumped vertical-external-cavity semiconductor lasers). Note that the trademarked acronym NECSEL™ was used to describe much of this work carried out at Novalux, Inc. In addition, these devices are generally not prone to catastrophic damage as are the edge-emitting diode lasers. The peak power levels in excess of 4 W in a TEM_{00} mode at 500 mW of average power at 980 nm have been demonstrated as described below. An early work by Coldren and Corzine [2] and the Ulm University group [3] on surface-emitting diode lasers with monolithic cavities and emitting diameters in excess of 50 μm demonstrated cw multimode power levels in the range of 50 to a few hundred milliwatts. More recently, monolithic cavity devices have produced about 3 W cw multimode from a 300 μm diameter aperture [4].

Optically pumped, external-cavity surface-emitting semiconductor lasers have produced comparable or greater single-mode power [5]. However, direct electrical pumping offers some significant advantages for many applications. The series resistance in electrically pumped devices limits the output power in the present designs for a single emitter. High peak power from optically pumped devices is limited by catastrophic damage in the edge-emitting pump diodes. This limitation could be overcome by optical pumping using pulsed or Q-switched lasers.

---

*This chapter is dedicated to the memory of Arvydas Umbrasas.

Semiconductor Disk Lasers. Physics and Technology. Edited by Oleg G. Okhotnikov
Copyright © 2010 WILEY-VCH Verlag GmbH & Co. KGaA, Weinheim
ISBN: 978-3-527-40933-4

Refinements in the design of electrically pumped devices will allow output power levels in a single beam of several watts as described below.

High brightness is defined as power in the lowest order spatial mode in a single frequency. As such, this allows efficient coupling into single-mode optical fibers and efficient nonlinear frequency conversion. More than 500 mW cw has been coupled into a single-mode optical fiber at a wavelength of 980 nm. These lasers can be highly manufacturable and lend themselves to array operation. There are numerous other applications for these devices, including laser radar, sensors, medical therapy, and medical diagnostics. Nonlinear frequency conversion of these electrically pumped devices is described in detail subsequently.

7.2
Device Design and Performance

Figure 7.1 is an example of a representative design that highlights the key operating parameters required for efficient, high-brightness operation. Heat is extracted through the p-mirror that is designed for high reflectivity and the laser beam is extracted through the n-substrate, the n-mirror, and the external mode-controlling mirror. While devices without an intermediate n-mirror have been operated, the presence of the n-mirror can significantly reduce the threshold and improve the operating efficiency in these low gain lasers.

The large refractive index differences available in the GaAlAs material system enable high-reflectivity mirrors to be fabricated from relatively thin epitaxially grown quarter-wave stacks. At 980 nm, the refractive indices of GaAs and AlAs are 3.52 and

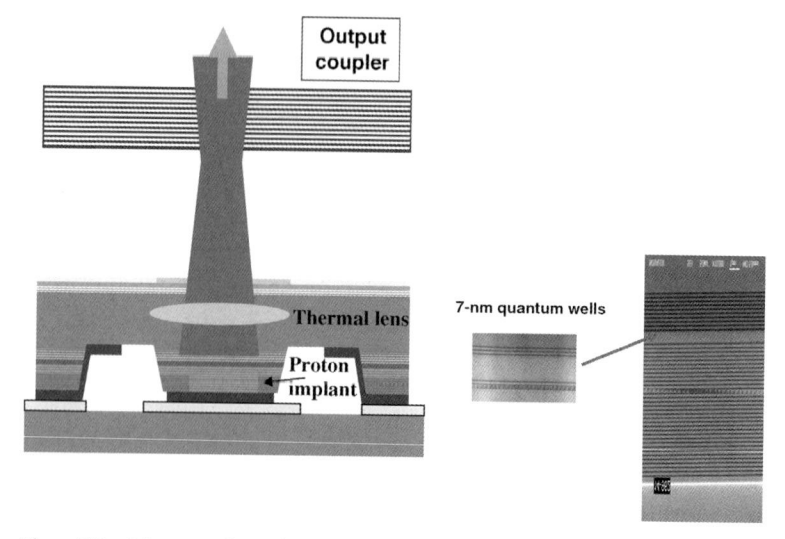

Figure 7.1 Schematic of a single-sided chip design. A proton implant confines the electrical current to help reduce threshold. A top aperture helps to confine the output to a single spatial mode. An electron micrograph of epitaxial layers is shown with 7 nm quantum wells and n- and p-mirrors.

2.97, respectively. The mirrors can be designed using standard thin-film optical filter theory [6]. For example, a 90% mirror can be realized by 10.5 pairs of AlAs and $Al_{0.14}Ga_{0.86}As$ layers, a total of 1.6 μm thick, sandwiched between GaAs layers, and 99% can be realized with 19.5 pairs.

One key difference between optically pumped and electrically pumped surface-emitting diode lasers is the electrical resistance primarily due to the resistance of the mirrors. The layers are made of GaAlAs alloy compositions with different bandgaps and refractive indexes. Schottky barriers at the interface between layers in the mirrors are the largest contributors to the series resistance. The series resistance can be especially high in the p-mirror where the large valence band offset creates a high barrier for holes, which leads to strongly nonohmic behavior [7]. These barriers have been reduced by composition grading and spike or modulation doping at the interface, which reduced the series resistance by two orders of magnitude in early work, without affecting the reflectivity of the mirror [8].

The p-mirror needs to be designed for very high reflectivity while paying attention to electrical and thermal resistance. There is an obvious trade-off concerning the number of layers in the mirror. The metal of the bottom electrode can be used to enhance the reflectivity of the stack if an unalloyed contact is used. The final epitaxial layer comprises heavily doped p-type GaAs to allow a good ohmic contact to be made. If the ohmic metal is to contribute to the reflectivity, the thickness of the GaAs cap layer needs to be designed carefully to ensure phase matching for the metal and epitaxial mirrors. If a typical titanium–gold ohmic contact is used, the reflectivity for a 20-pair stack can be as high as 99.6% if the GaAs cap is 110 nm thick, or as low as 97.7% for a thickness of 45 nm.

The gain region into which the carriers are injected consists of several quantum wells that are about 7 nm thick. The effective gain of the quantum wells can be enhanced by a factor of ∼2 by placing them only at the peaks of the optical standing wave [9]. In a simple vertical-cavity device, there are typically between one and three quantum wells located at the center of a one-wavelength cavity. For external-cavity devices with greater optical absorption loss in the resonator, more quantum wells are used for greater gain. To maximize the overlap between the quantum wells and the optical standing wave, the cavity can be extended to a few wavelengths in length with the quantum wells clustered around the standing wave peaks. For GaInAs wells, strain compensation is required as described below.

A simplified expression for the power efficiency of a diode laser is given by

$$\eta = \eta_{slope}(I-I_{th})/(VI+I^2R), \tag{7.1}$$

where η is the overall power efficiency, η_{slope} is the actual slope power efficiency, I is the operating current, I_{th} is the threshold current, V is the operating voltage, and R is the device resistance. Figure 7.2 shows the calculated power efficiency for a typical 150 μm gain diameter device as a function of drive current. Also shown is the measured slope efficiency for pulsed operation (3 μs at 10% duty cycle). The power efficiency at 4 A peak was 20%. The overall power efficiency for these vertical-cavity devices drops off at high current levels in contrast to edge-emitting diode lasers.

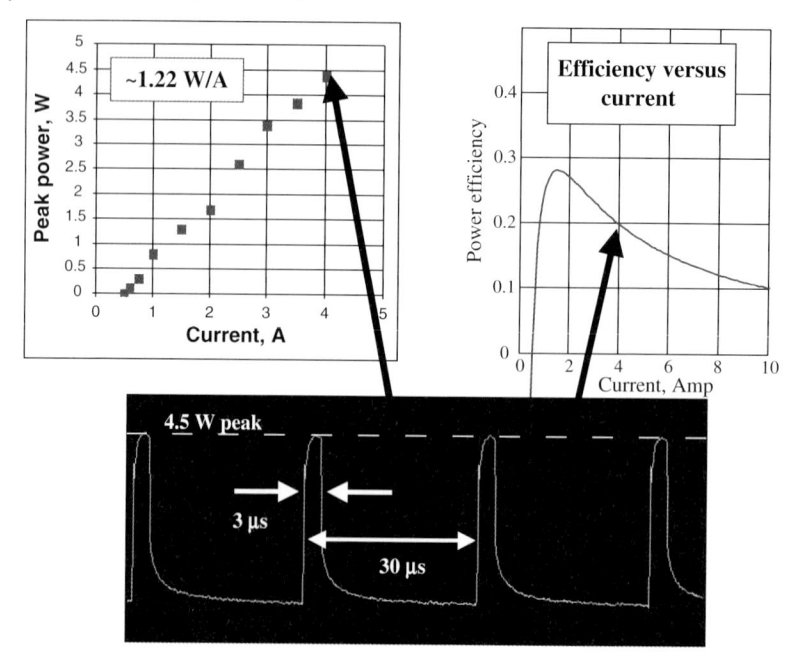

Figure 7.2 Performance of an external-cavity device with a 1.22 W A^{-1} slope efficiency operated at 10% duty cycle. The power efficiency was 20% at 4 A. The output was about 1.7 times diffraction limited at maximum power. Detector time response causes the tail.

While there is no catastrophic damage in these lasers, the peak power is usually limited by their I^2R loss. Despite this, there can be sufficient optical intensity, especially in the laser cavity, to be efficiently frequency doubled as described below.

Optical intracavity losses also limit the efficiency at low current levels. These losses include absorption of the laser field that penetrates the mirrors as well as the bulk absorption in the substrate. The mirror losses are minimized by step doping with the highest doping at the end of the device where the laser field intensity has dropped off significantly. Figure 7.3 shows the calculated laser field intensity in a three-mirror device.

The optical absorption in GaAs substrates as a function of n-type doping is shown in Figure 7.4. The data are at 980 nm and includes published data [10] and measurements taken from commercial substrates from a number of suppliers. To avoid defects propagating into the growth region from the substrate, wafers were grown with a thick (100 µm) layer prior to the epitaxial growth for high yield of devices.

The original substrate was subsequently removed. The absorption at 980 nm below 10^{17} cm^{-3} is attributed to band edge impurity absorption [10], while from 10^{17} cm^{-3} to 10^{18} cm^{-3}, an intraband Γ to X transition dominates, and above 10^{18} cm^{-3} the true free carrier absorption starts to dominate. Note that the curve is not linear as the Fermi level increases through the bandgap and the GaAs becomes degenerate at room temperature above $n = 10^{17}$ cm^{-3}.

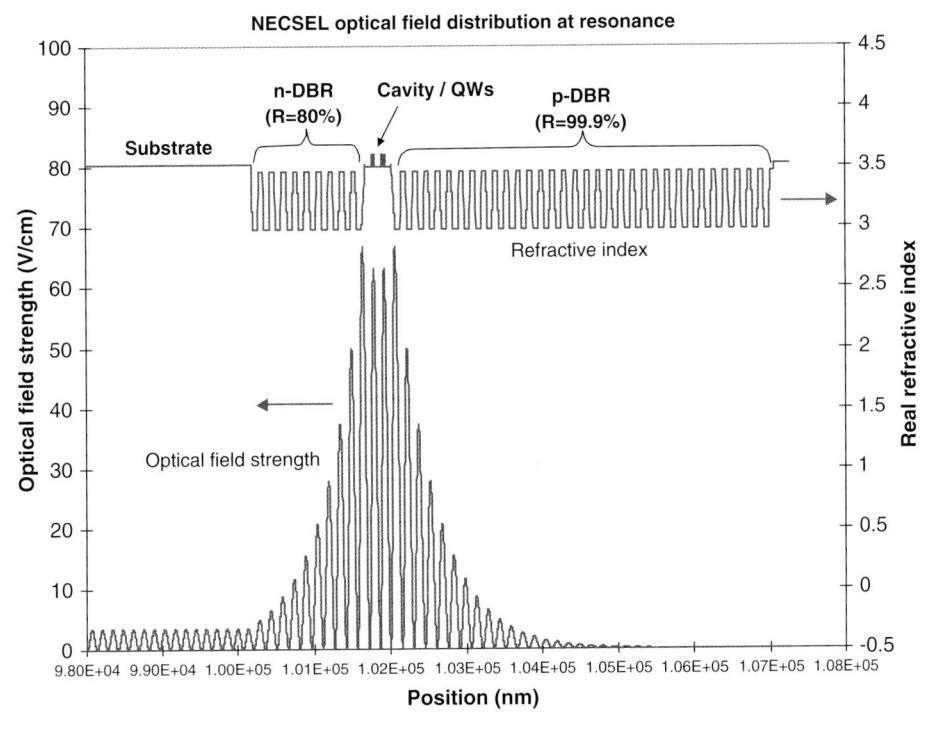

Figure 7.3 Calculated laser field intensity.

The best efficiency for external-cavity devices was obtained with "substrate" n-doping levels of 0.5 to 1×10^{17} cm^{-3} and was verified by experimental results. These devices had "substrate" thicknesses of around 50 μm. Wafers of 4 in. could be processed at these thickness levels. The series resistance for 150 μm gain diameter

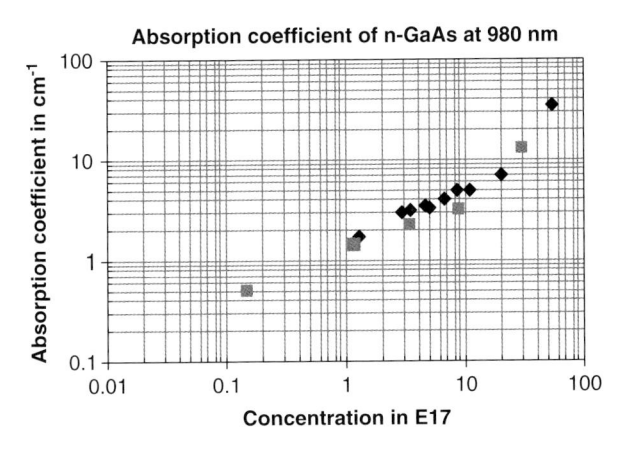

Figure 7.4 Absorption coefficient of n-type GaAs at 980 nm. Diamond data are from Spitzer and Whelan [10] and squares are for commercial wafers.

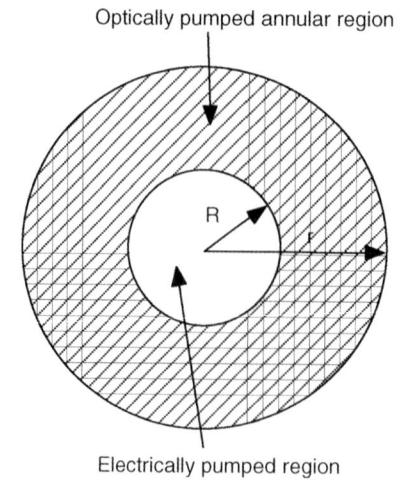

Figure 7.5 Electrically and optically pumped regions.

devices was measured to be close to 1 Ω and was consistent with the calculated device resistance using a finite element analysis. The finite element analysis calculates not only the series resistance but also the carrier distribution.

Large-area, vertical-cavity semiconductor lasers can be limited in their power efficiency by the increasing amount of in-plane-amplified spontaneous emission light that can direct energy out of the primary pump region (region R in Figure 7.5). This has been reported specifically by the group at Ulm University [3] in the literature but has not been analyzed in detail. They discovered that the efficiency of surface-emitting diode lasers decreased as the current injection diameter of the device increased. This effect occurs despite the fact that the laser continues to operate in the vertical direction. The Ulm group attributed the reduction in efficiency to be due, in part, to lateral in-plane amplified spontaneous emission that directed energy out of the primary gain region. Since these diode lasers are quasi-two-level systems, the lateral gain will amplify the in-plane spontaneous emission and optically pump an annual region adjacent to the primary gain region [11]. Detailed analysis of this effect is beyond the scope of this article; however, the observation of this effect is described below.

Much of the spontaneous emission is expected to be index guided by the design of the structure. This magnitude of gain would rapidly saturate the gain and the ASE power would pump an annular region around the electrically excited region. The annular region would be optically pumped to inversion (bleached) out to a radius r, where the light intensity would become insufficient to pump above inversion. This radial distance is proportional to the amount of injected current in the first region. This radial pumping does not have a sharp cutoff but decays exponentially with distance (Figure 7.6). In the case where the lateral gain is quite large, much of the injected energy would be directed in the radial direction with the exception of some spontaneous emission that escapes out of the plane of the device.

This effect will occur in devices with larger initial diameters, typically 50 μm or greater. In addition, larger diameter devices will provide more lateral power to invert a larger annular volume. The external mirror in a VECSEL would be adjusted to match the new inverted mode waist diameter, and much of the lateral in-plane flow of energy would be extracted vertically by the external resonator in a single TEM_{00} mode. The above effect is not significant for quantum well lasers having only a few wells with injected carrier diameters of less than about 20 μm operating in the cw regime. Note that smaller diameter devices have lower power available to bleach an annular ring.

The power dependence on the in-plane pump power becomes quadratic for pulsed devices when thermal limitations are not significant. In this case, the effective width of the annular ring can become much greater than the diameter of the pumped region. The limitation for pulsed excitation bleaching occurs when the bleaching time of the annular ring becomes greater than the pumping time. In this case, there will be a more complicated transient response of the device. The laser can be driven at the cavity relaxation frequency to provide gain-switched pulses with high peak power at high repetition rate. Conversely, an intracavity loss modulator could also be used to avoid the difficulty of using a high current pulser at high repetition rate. An optically pumped VECSEL could also be driven by a high-energy pulsed or Q-switched laser. A 1 mm diameter device could produce peak power levels of about 50 kW in a single TEM_{00} beam. Pulsed excitation also offers the possibility of having structures with an even larger number of quantum wells to produce gains that would allow use of variable reflector output couplers and unstable resonators.

The effect of this lateral pumping is to eliminate most of the losses associated with the quasi-two-level system of the semiconductor laser, which in turn allows the

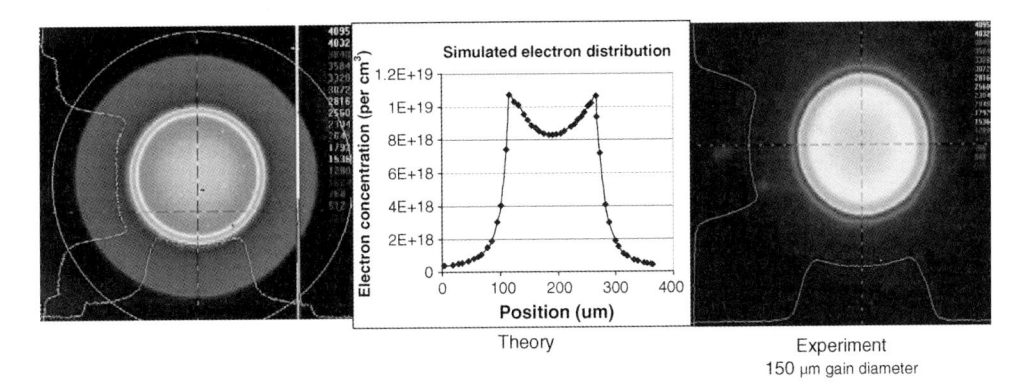

Figure 7.6 Left shows the experimental intensity profile of the spontaneous emission from a 150 μm injection diameter device at 1 A of injection current. Note that the emission profile is cut off by the top electrical aperture of the device. The top current aperture is about 1.95 times the current injection aperture and allows the TEM_{00} mode to expand to extract most of the excitation energy above threshold. The measured emission at low current on the right side (<50 mA) for the same 150 μm device shows no lateral spreading of the excitation and is in close agreement with the finite element-calculated distribution shown without the presence of lateral in-plane pumping.

fundamental spatial mode to expand and more efficiently extract power in the TEM_{00} mode. This is demonstrated experimentally in Figure 7.2 in which the slope quantum efficiency (the fraction of injected carriers that are extracted in the laser mode) is approaching 90%, a record for such devices. In this example, the output beam is near TEM_{00} with an M^2 of 1.7. These results are in contrast to the work reported by Hadley et al. [12] in which their design had more than a factor of 40 less peak power and an order of magnitude less efficiency.

7.3
Mode Control, Cavity Design, and Thermal Lensing

Normal operation of these devices gives rise to a parabolic thermal profile underneath the active region due to resistive heating in the device. Since the refractive index of these materials increases a function of temperature, the thermal profile leads to a gradient in the refractive index as well. This variation in the refractive index is a thermal lens that focuses the beam as it makes a round-trip in the external cavity. This lens helps stabilize the resonator and improve the confinement factor for the gain. The focal length of the lens depends upon the heating of the substrate and therefore depends on the operating bias point of the device. By making measurements on a cavity that involve varying the cavity length and reflectivity of the output coupler, one can determine the focal length of the thermal lens. Figure 7.7 illustrates

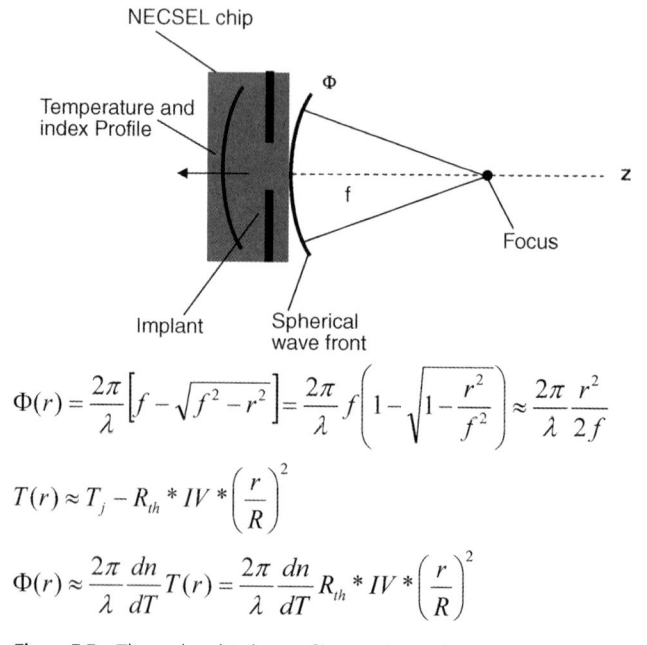

$$\Phi(r) = \frac{2\pi}{\lambda}\left[f - \sqrt{f^2 - r^2}\right] = \frac{2\pi}{\lambda} f\left(1 - \sqrt{1 - \frac{r^2}{f^2}}\right) \approx \frac{2\pi}{\lambda} \frac{r^2}{2f}$$

$$T(r) \approx T_j - R_{th} * IV * \left(\frac{r}{R}\right)^2$$

$$\Phi(r) \approx \frac{2\pi}{\lambda} \frac{dn}{dT} T(r) = \frac{2\pi}{\lambda} \frac{dn}{dT} R_{th} * IV * \left(\frac{r}{R}\right)^2$$

Figure 7.7 Thermal and index profile in a device that gives rise to a thermal lens.

the principal and a quick calculation of the focal length of the thermal lens [13, 38]. This simple calculation reveals a focal length of ~10 mm for the thermal lens, which is to first order a good estimate (as confirmed by cavity experiments).

The mode radii w_1, w_2, and w_0 (mode waist radius in cavity) at the two ends of the cavity for a device with a thermal lens having an equivalent radius of curvature a and an output mirror radius of b as a function of cavity length x are given by

$$w_1 = \left(\frac{\lambda a}{\pi}\right)^{1/2} \left(\frac{b-x}{a-z}\frac{x}{a+b-x}\right)^{1/4}, \tag{7.2}$$

$$w_2 = \left(\frac{\lambda b}{\pi}\right)^{1/2} \left(\frac{a-x}{b-x}\frac{x}{a+b-x}\right)^{1/4}, \tag{7.3}$$

$$w_0 = \left(\frac{\lambda}{\pi}\right)^{1/2} \left(\frac{x(a-x)(b-x)(a+b-x)}{(a+b-2x)^2}\right)^{1/4}. \tag{7.4}$$

Figure 7.8 shows the calculated mode radii as a function of cavity length for a specific set of cavity parameters. Use of a top circular aperture on the chip can serve both as a spatial mode filter and as an electrical contact. Devices with 150 μm current apertures operating at maximum power levels had a TEM$_{00}$ mode quality, M^2, of 1.05 in the best case, while on average fundamental mode, quality was typically 1.1–1.15. Figure 7.9 shows the output beam profile for such a device at 500 mW cw. More than 600 mW cw was demonstrated into a single-mode fiber.

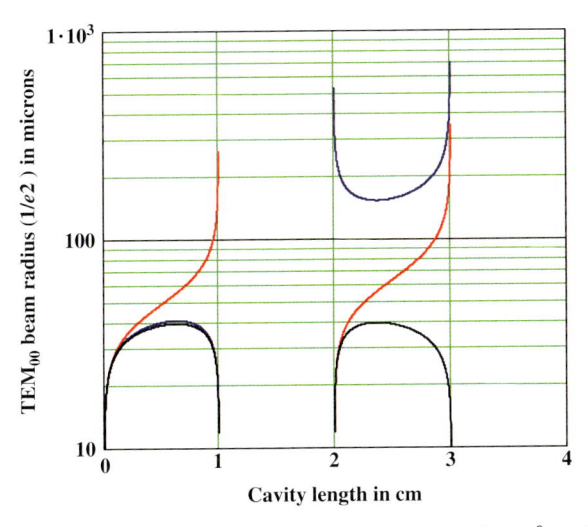

Figure 7.8 Calculation of the mode radii on a log scale (1/e^2) at the chip (red), intracavity waist (black dashed), and output mirror (blue) as a function of cavity length for $a = 1$ cm and $b = 2$ cm at a wavelength of 980 nm [14].

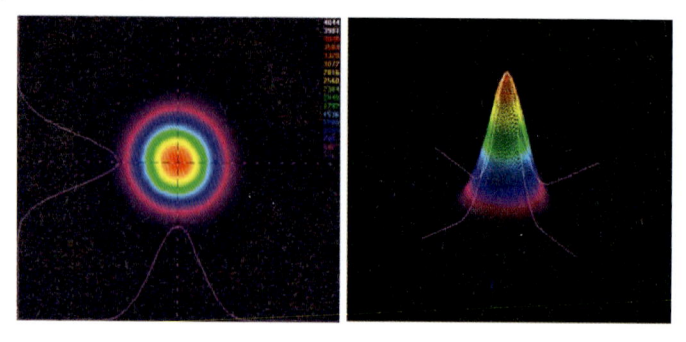

Figure 7.9 Mode profile for a device operating at 500 mW cw.

In its simplest version, a linear cavity is formed by an E-VECSEL chip and a spherical end mirror. While this is a three-mirror cavity due to the presence of p-mirror (highly reflective) and n-mirror (partially reflective) in the E-VECSEL chip, regions of cavity stability and mode radii can be estimated from the well-known two-mirror "g_1g_2-model" [14]. A more comprehensive three-mirror approach was developed by Shchegrov [15], and the reader is referred to this reference or the follow-up paper by Yang et al. [16].

The results from a two-mirror model are illustrated in Figure 7.8 for the cavity formed by a chip that has a thermal lens, modeled as an ideal thin lens, and a spherical end mirror. The plotted curves show the eigenmode radii ($1/e^2$) w_0, w_1, and w_2 for the mode waist, the mode on the chip, and the mode on the end mirror, respectively. Figure 7.8 shows that the cavity supports two families of solutions for eigenmodes, often referred to as the "short-cavity solution" and the "long-cavity solution."

The short-cavity solution offers a compact laser architecture with typical cavity length values measured in several millimeters. For example, an E-VECSEL chip with the thermal lens that has the focal length $f = 10$ mm and the eigenmode radius on the chip $w_1 = 50$ μm (this should closely match the implant area radius) will operate in the single spatial mode at the wavelength $\lambda = 1$ μm with the cavity length $L \approx 4$ mm, when a flat-end mirror is chosen. When optical elements such as nonlinear crystals are present in the cavity, the required physical cavity length is longer since the physical lengths occupied by these elements have to be divided by their indices of refraction so that the result will add up to the required 4 mm value. A spherical mirror will allow slightly longer cavity length, depending on the mirror radius of curvature - in the example just considered, the end mirror with the radius of curvature $R = 20$ mm would extend the air cavity length to 5 mm. The flat-mirror cavity architecture is perhaps the most useful case for the short-cavity laser eigenmode branch. As can be seen from the original demonstration (Figure 7.29), this compact design has only flat optical surfaces. This is a significant advantage for low-cost, high-volume applications because laser alignment can be done passively. In addition, this design is scalable to many emitters and, therefore, can be used for scaling up power levels by increasing the number of emitters in a single-array chip and using a larger aperture nonlinear crystal. In principle, this design is compatible in a wafer-scale architecture – but this goal may not be the most practical because of the practical limits in nonlinear crystal aperture size and the possible need to accommodate other elements that may not be easily scalable.

The long-cavity solution always requires some curvature for the end-mirror surface. Using the example of the E-VECSEL chip with the thermal lens that has the focal length $f = 10$ mm and the eigenmode radius on the chip $w_1 = 50$ µm and the spherical end mirror with the radius of curvature $R = 20$ mm, we can estimate the required cavity length to be $L \approx 22.5$ mm. This computed value (as well as those shown for the short-cavity solution) was in fact observed experimentally – therefore, the simple two-mirror model seems to be adequate in predicting cavity behavior, at least its design geometry.

The short-cavity solution offers scalable, quasi-monolithic architecture suitable for low-cost, high-volume packaging designs developed in semiconductor industry. In contrast, the long-cavity solution provides more space and design flexibility (e.g., elements such as polarizers, etalons, and so on can be easily accommodated) and is more reminiscent of solid-state laser engineering designs. This illustrates the crossover nature of E-VECSEL that combines features of semiconductor and solid-state lasers.

The spectral output could be in a single frequency when operated in a fundamental transverse mode. Polarization in many devices was oriented along a crystal axis but usually was random with one of the two degenerate polarization modes being selected by a crystal strain or an intracavity polarizer with weak polarization. The single-frequency output could be tuned by temperature with axial mode jumps at the long cavity frequencies. Figure 7.10 shows such a tuning with mode jumps at about 15 GHz.

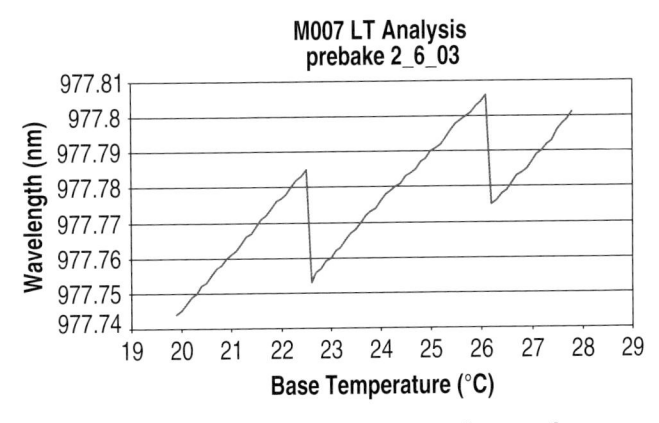

Figure 7.10 Thermal tuning and mode jumps as a function of temperature of a single-frequency external-cavity device.

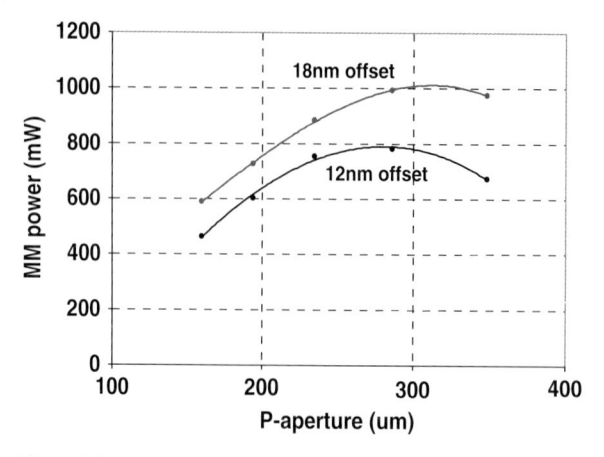

Figure 7.11 Maximum cw power as a function of device aperture.

Figure 7.11 shows the multimode roll-off power for external-cavity devices as a function of the injection current diameter at a heat sink temperature of 25 °C. The offset is the difference between the peak of the threshold gain and the internal Fabry–Perot peak. The roll-off power is defined as the maximum output power as a function injection current. These devices were fabricated from the same wafer for a given offset. The maximum multimode power achieved from a 150 μm aperture external-cavity device was 1200 mW cw. This device was 15 times diffraction limited in a single dimension.

An early work by Martinsen [17] described a scheme to control or eliminate the thermal lens at all power levels. This involved forcing unidirectional heat flow by adjusting the diameter of the soldered heat sink to be about 1.25 times the heat-generating gain region. Experimental measurement of the radial temperature profile is shown in Figure 7.12 for various attach diameters demonstrating the afocal characteristic at 1.25 diameter of the attach compared to the gain region.

7.4
High-Power Arrays and Multielement Devices

These surface-emitting lasers can be scaled to increase power. Two dimensional arrays of VCSELs (no external cavities) [18] as well as more than one surface-emitting gain element in a single cavity have been demonstrated. Figure 7.13 shows the near-field intensity pattern of a 15 × 15 element array operating at a cw laser output at 980 nm with more than 80 W cw. In their work, Seurin et al. [19] demonstrated 230 W cw from a 5 × 5 mm² array operating at 15 °C and 980 nm. While these were not external-cavity devices, the uniformity of the laser output is a demonstration of the quality of the material across a large area. Figure 7.14 shows the cw output as a function of total current to the array.

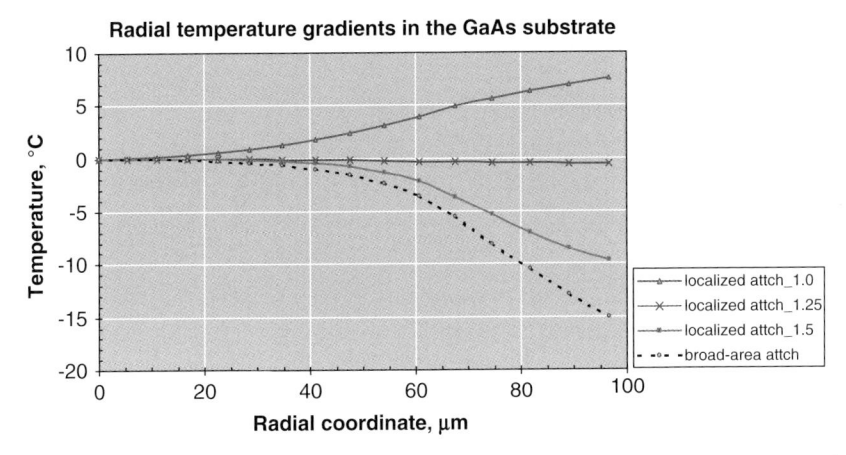

Radial temperature gradients in the GaAs substrate

Figure 7.12 Measured temperature profile for a 150 μm diameter device operating at a typical high power level for different diameters of the heat sink die attach.

The uniformity of the output over a 5 × 5 mm² array makes possible the combining of the output from adjacent elements in series [20] to increase the power in a single beam. Two similar gain elements have been combined in a single cavity, demonstrating a doubling of the laser output.

The demonstrated high output power and wavelength uniformity from 225 emitters in a 5 × 5 mm² array (VCSELs) make them ideal candidates for coherent coupling [21]. Prior work by Nabors [22] on coherent locking of a linear array of

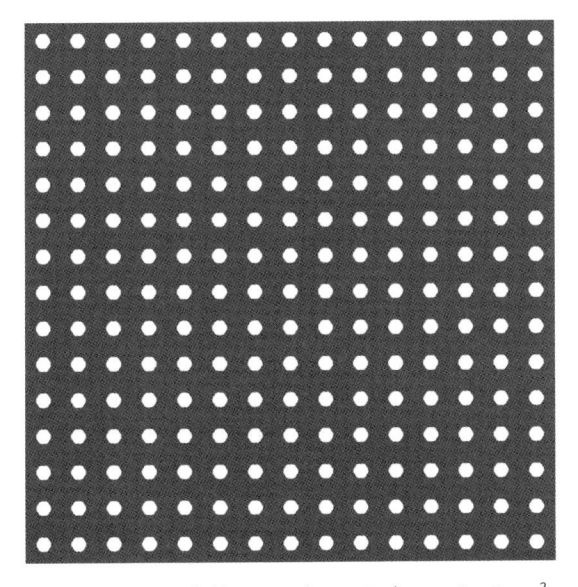

Figure 7.13 Near-field image of 15 × 15 element 5 × 5 mm² array operating at 80 W cw.

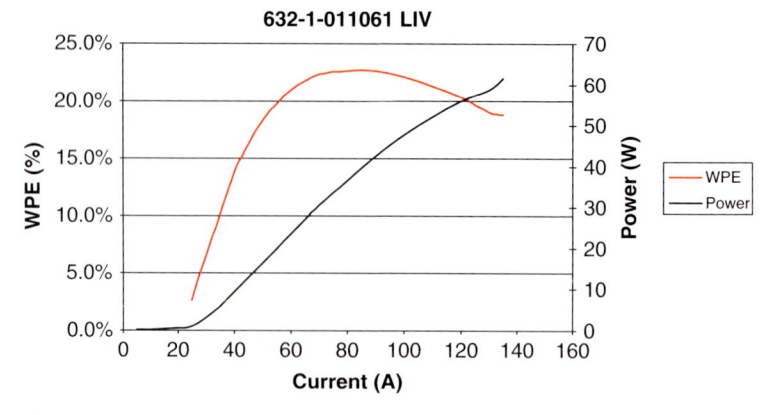

Figure 7.14 CW output power and total wall-plug efficiency at 980 nm for 15 × 15 element array as a function of total current.

diode-pumped solid-state lasers demonstrated efficient locking of an array of circularly symmetric beams using an external cavity with a spatial filter.

7.4.1
Design of the Chip

Meeting the gain requirements for the active region is specifically challenging for these devices. For GaAs-based lasers, longer wavelengths are achieved by increasing the indium content in the InGaAs QWs. This leads to a larger strain field in the active region. Relaxation of the strain field takes place via formation of dislocations and defects that increase nonradiative recombination in the active region. In addition, strain in the active region reduces the density of states in the valence bands, thus increasing the penetration of the hole quasi-Fermi level into the bands. While this is of great advantage for high-speed telecom lasers (the differential gain is enhanced) [2, 23–27], a reduced density of states leads to lower peak gain, a disadvantage for high power operation. In addition, both the free-carrier absorption loss and the Auger recombination are more pronounced at longer wavelengths. A combination of higher defect count, lower gain, and higher loss mechanisms limits the power output of longer wavelength devices. The longer wavelength devices tend to have deeper wells and hence the capture versus escape cross section is expected to be favorable for these devices [28, 29]. The T_0 for devices where defects do not limit performance tend to be higher for longer wavelength lasers [23–27]. This is of great benefit for high power operation where there is significant gain suppression due to higher temperatures. However, in most practical long-wavelength devices, the formation of defects limits the high-temperature performance due to increased recombination.

The device structure of an E-VECSEL is similar to that of a VCSEL with lowered reflectivity for the output coupler. The E-field profile for our typical device is shown in Figure 7.15. The higher gain requirements for the device are satisfied by using a multiwavelength cavity with multiple sets of QWs [30]. There are some subtle design

E-field profile for a 2QW-4RPG-NECSEL with external cavity

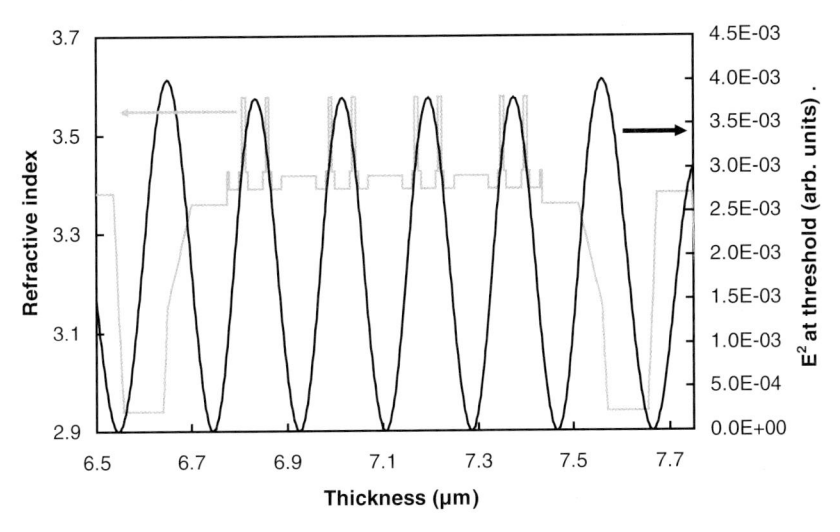

Figure 7.15 Detail of E-field and index profile in active region of laser.

features that enable the efficient operation of the device. The substrate needs to have reasonable conductivity (for electrical access to the active region) coupled with low free-carrier loss (the intracavity field is significant in this region). The strained QWs in the active region need to be separated from each other to incorporate strain dilution for improved performance, however not at the cost of reducing the confinement factor. The barrier regions adjoining the InGaAs QWs are GaAs to prevent group-V sublattice intermixing between the barriers and QWs. The barrier regions not adjacent to the QWs are tensile-strained GaAsP and provide strain balancing for the entire cavity [31].

The main loss mechanism that a designer needs to optimize in standard SEL is doping-dependent free-carrier/impurity absorption in the DBRs. This loss is higher for p-type layers in the 850 nm wavelength range and tends to increase with wavelength. At longer wavelengths of ∼1300 nm, the p-type and n-type carrier losses are both high and comparable. It is therefore crucial to optimize the doping in the DBR mirrors, especially those close to the cavity where the electric field strength is significant. The design calls for a careful balance of free-carrier loss (improved by lower doping) with electrical conductivity and carrier distribution (improved by higher doping). In general, the n-DBR tends to have acceptable conductivity for moderate doping levels of $\sim 2 \times 10^{18}$ cm^{-3} due to the high mobility and low effective mass of electrons. The focus in most devices turns on the p-DBR where optimization proves crucial to device performance. Figure 7.16 illustrates the general concept behind this optimization where the doping at the interfaces between the high- and low-index layers has been alternated. The high-doping regions overlap with the nulls of the E field, while the low-doped regions overlap with the E field peaks. In addition, the general doping is lowest in the DBR pairs adjacent to the active region, then steps

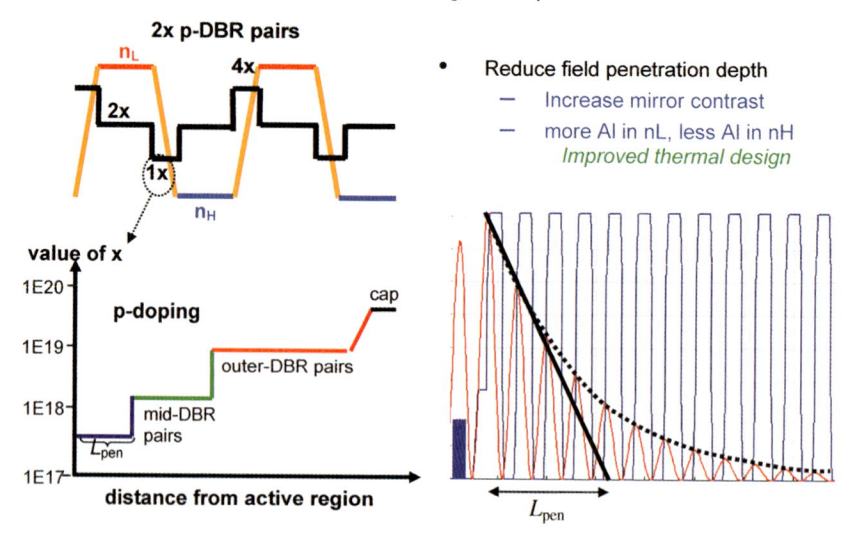

Figure 7.16 General concept for optimizing doping levels in p-DBR.

up to a medium level for the next few pairs, and finally steps up to high levels in DBRs that are removed from the active region (where the E-field strength is low).

The results of DBR calculations comparing optical loss and series resistance are shown in Table 7.1. We analyzed both simple DBRs (one DBR pair repeated several times) as well as composite DBR structures (using the doping scheme illustrated

Table 7.1 Experimental matrix to optimize doping levels in p-DBR.

DBR description	High Al doping	Low Al doping	Pairs (L, M, H)	Resistance (100 μm device)	Relative optical loss
20X p-DBR linear grades – high doping (H)	2.E+19	6.E+18		0.24	5.80
20X p-DBR linear grades – mid-doping (M)	6.E+18	2.E+18		0.37	2.30
20X p-DBR linear grades – low doping (L)	6.E+18	1.E+18		0.69	2.20
20X p-DBR uniparabolic grades – mid-doping (M)	4.E+18	2.E+18		0.46	1.80
20X p-DBR uniparabolic grades – low doping (L)	1.E+18	4.E+17		1.00	1.00
32X Composite p-DBR with spikes at null (L+M+H)			4, 10, 18	0.37	1.40
32X Composite p-DBR with no spikes (L+M+H)			4, 10, 18	0.35	1.30

7.4 High-Power Arrays and Multielement Devices

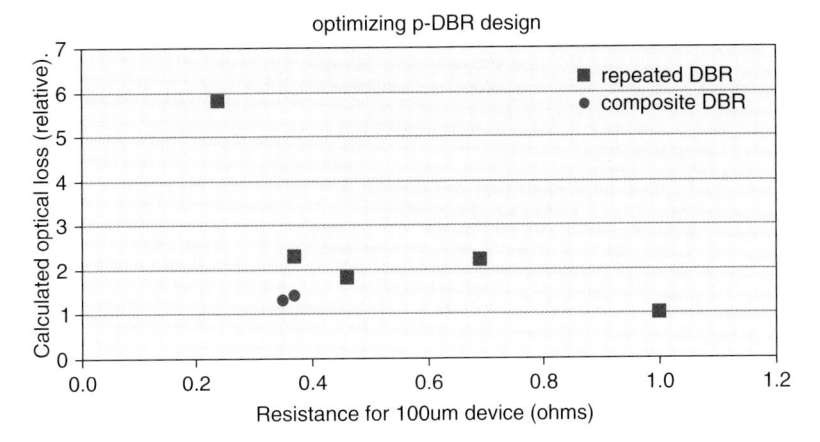

Figure 7.17 Correlation of optical loss with doping levels in p-DBR.

in Figure 7.16). The series resistance was measured on actual device structures and the results matched well with calculations done using a thermionic emission model in conjunction with a Poisson solver. The grading in the DBRs as well as the spike doping at alternate interfaces helps smooth out the band discontinuities, thereby assisting in carrier transport. The optical loss was calculated using a 1D transmission matrix field calculator. Figure 7.17 shows the general correlation between series resistance and optical loss in devices. The data argue in favor of the composite doping scheme, which is what we implemented in our devices.

We performed a 2D simulation of current spreading in both the n- and p-DBRs assuming asymmetric conductivities (in-plane versus across interfaces) for the DBRs. The final doping profile used in our devices was derived from these calculations and is shown overlapped with the field profile in Figure 7.18. Ideally, if the DBR were not used for current injection, the alternating high- and low-index layers would have no grades between them; they would be abrupt. This would lead to lower thermal

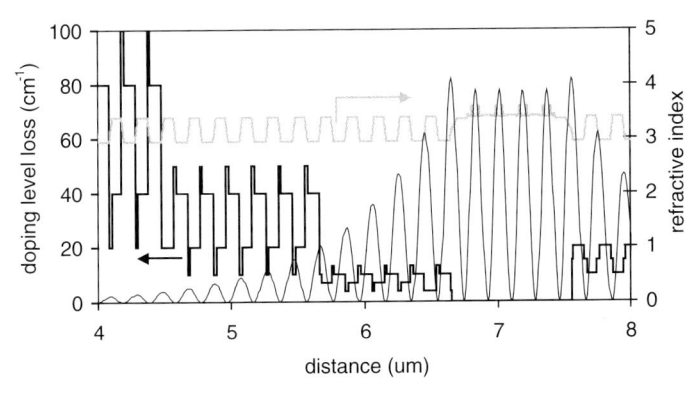

Figure 7.18 E-field, index, and doping profile in active region of NECSEL.

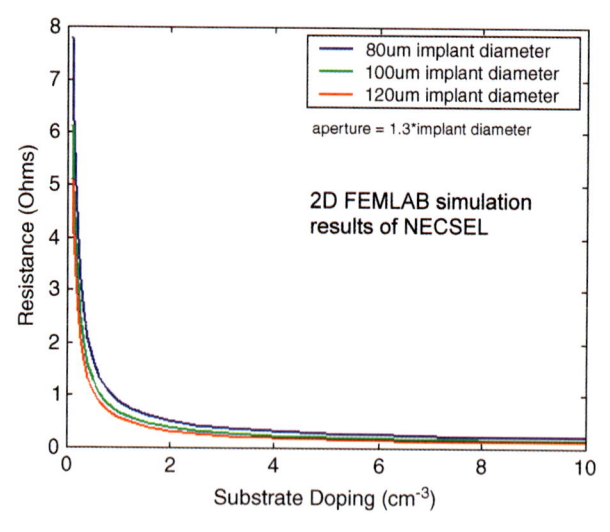

Figure 7.19 Dependence of substrate doping on series resistance of NECSEL.

resistance and higher reflectivity (which would reduce the number of DBR pairs required). However, such a scheme leads to extremely high series resistance due to large band discontinuities at each interface. We used 200 A grades between the high- and low-index layers to reduce the series resistance in our DBRs, while maintaining good reflection and thermal properties.

The doping in the substrate needs to be large enough to lower resistive losses, but low enough to keep optical losses low. Figure 7.19 shows the results of a 2D simulation of resistive losses in the substrate as a function of the doping level. It is clear that given the size of our devices (~100 μm aperture) and the thickness of our substrates (~100 μm), a doping level of 1×10^{17} cm^{-3} or greater is good enough.

A low-doped, high-bandgap substrate would be ideally suited to reduce the two major sources of absorption (subband and free carrier). This substrate would have to be lattice matched to the epitaxial layers that constitute the device and have low defect density. Commercially available GaAs and AlGaAs substrates do not satisfy all of the conditions above. The lower doped substrates when available, tend to have a high etch pit density, while the low etch pit density substrates have high doping levels. We therefore developed a novel substrate design that helped us significantly improve the performance of our devices.

The substrate preparation method is outlined in Figure 7.20. We start with a highly doped GaAs substrate that has a low etch pit density. We then epitaxially deposit (MOCVD) a thin etch-stop layer on top of the GaAs substrate. This etch-stop layer has a composition significantly different from that of the following epitaxial layer. The next step is to deposit a thick (~100 μm) AlGaAs layer with low Al content and mild (~2×10^{17} cm^{-3}) doping. Given the large area devices and the thickness of the substrate, such a doping level is sufficient to prevent current crowding without compromising the transmission properties. Figure 7.21 shows the results of 2D simulations of substrate spreading resistance as well as the metal contact/implant

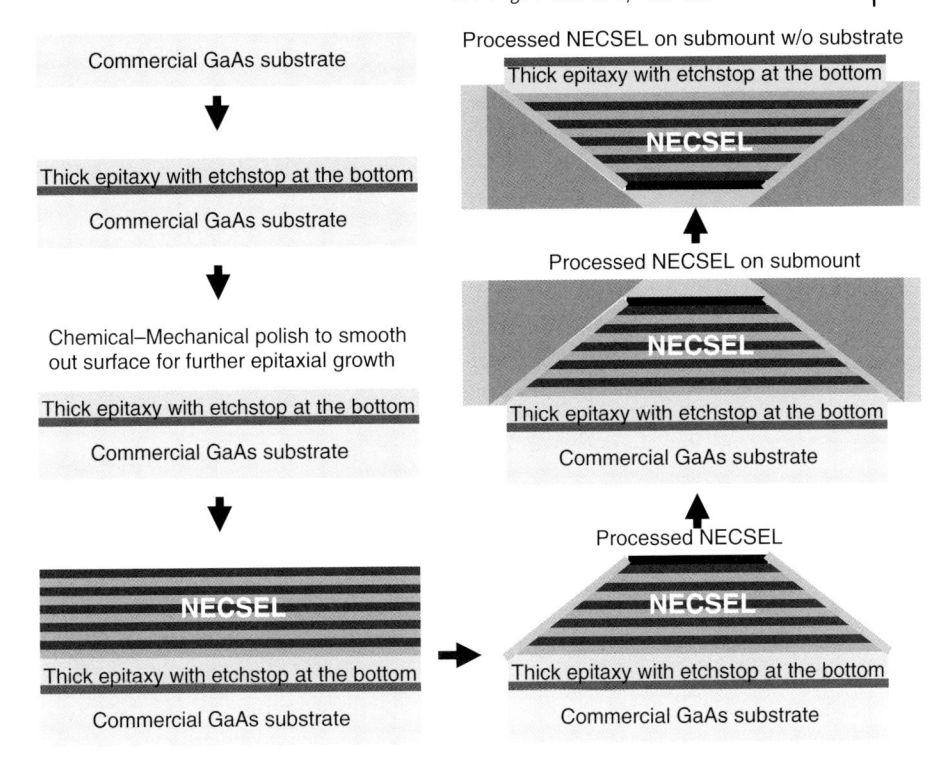

Figure 7.20 Reducing substrate-dependent loss.

ratio on a 100 μm aperture device with a 60 μm thick substrate. After the growth of the thick epitaxial layer (at high growth rates in excess of 15 μm h^{-1}), the surface of the film is fairly nonspecular. A mechanical–chemical combination polish is then employed to render the surface smooth for the growth of the device. Once the device growth and process is completed, the device is flip-chip bonded to a carrier that aids mechanical stability and provides heat sinking. The original substrate that has high absorption losses is then removed to enable improved operation of the device.

One of the major challenges facing the device designer is the strain in the active region for GaAs-based devices. As we move from blue (~415 nm) to green (~532 nm) to red (~620 nm) emission, the IR (nonfrequency doubled) emission needs to be tuned from ~830 to ~1064 to ~1240 nm. The indium mole fraction (and hence the strain) in the InGaAs QWs must be continually increased for longer wavelength operation [32]. Figure 7.22 shows the crystal strain that is induced due to the difference in the native lattice constants of the InGaAs and GaAs layers. The well-known Matthew Blakeslee thermodynamic critical layer thickness is plotted for different In mole fractions in the InGaAs layer. It is possible to obtain high-quality crystalline films that exceed the strain limits imposed by the thermodynamic limit. In reality, one can grow coherent layers beyond the "critical thickness" limit by using kinetics to prevent the system from reaching equilibrium. This is accomplished by optimizing the growth conditions during MOCVD or MBE of these layers [33].

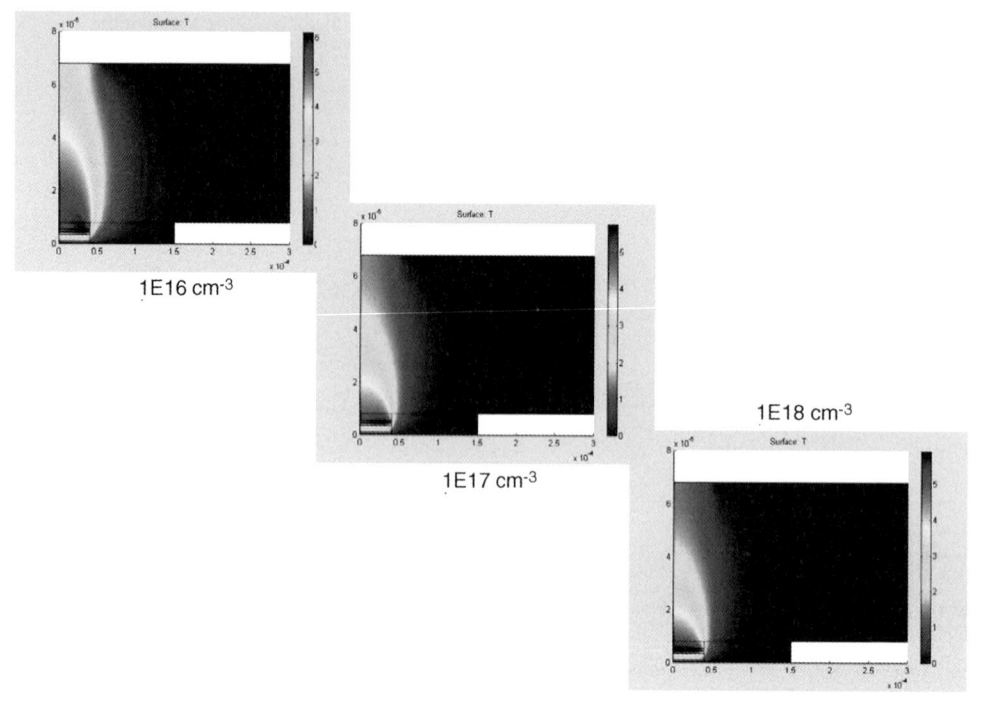

282 | *7 External-Cavity Surface-Emitting Diode Lasers*

1E16 cm⁻³

1E17 cm⁻³

1E18 cm⁻³

Figure 7.21 Temperature profiles for different substrate doping (current ~140 mA).

InGaAs

GaAs

Compressive strain = $\dfrac{a_1 - a_0}{a_0}$

h_c

InGaAs/GaAs QWs
Matthew Blakeslee (thermodynamic)

h_c (Å)

In (fraction)

InGaAs/GaAs QWs in NECSELs

- Mathews Blakeslee model assumes system to be in thermodynamic equilibrium.
- In reality, one can grow coherent layers beyond the "critical thickness" limit by using kinetics that prevent system from reaching equilibrium.

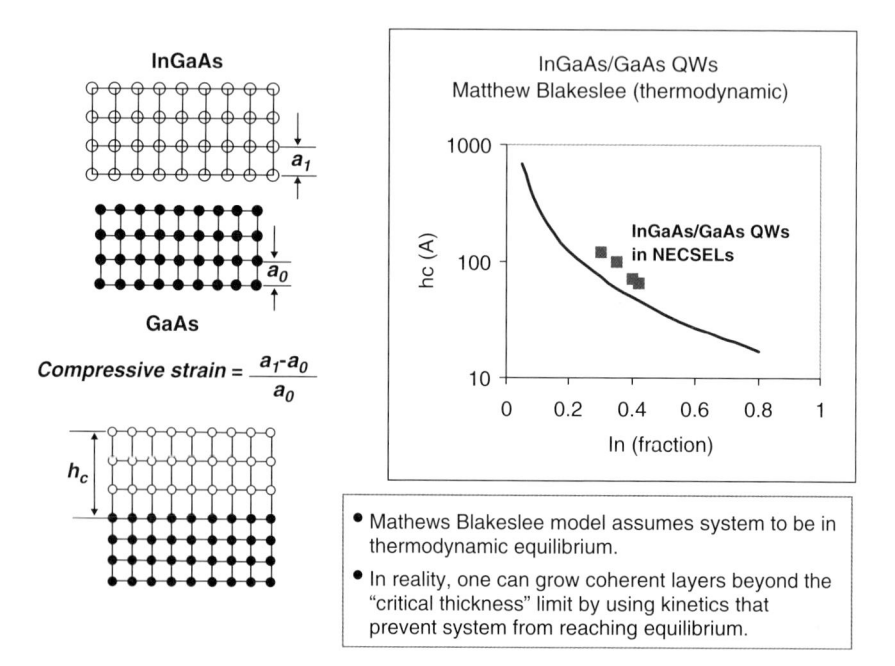

Figure 7.22 Built-in strain in lattice-mismatched layers for GaAs-based lasers.

7.4 High-Power Arrays and Multielement Devices

In the simplest case, system relieves strain via island formation:
Let us say we have a **substrate** and **adsorbate** with identical crystal structure,

lattice constants $\begin{cases} \sigma_A & \text{adsorbate} \\ \sigma_S & \text{substrate} \end{cases}$ mismatch $\varepsilon = \dfrac{\sigma_A - \sigma_S}{\sigma_S}$

initial coherent growth → species segregation/accumalation
enforced in first layers   far from the substrate

strain relaxation:

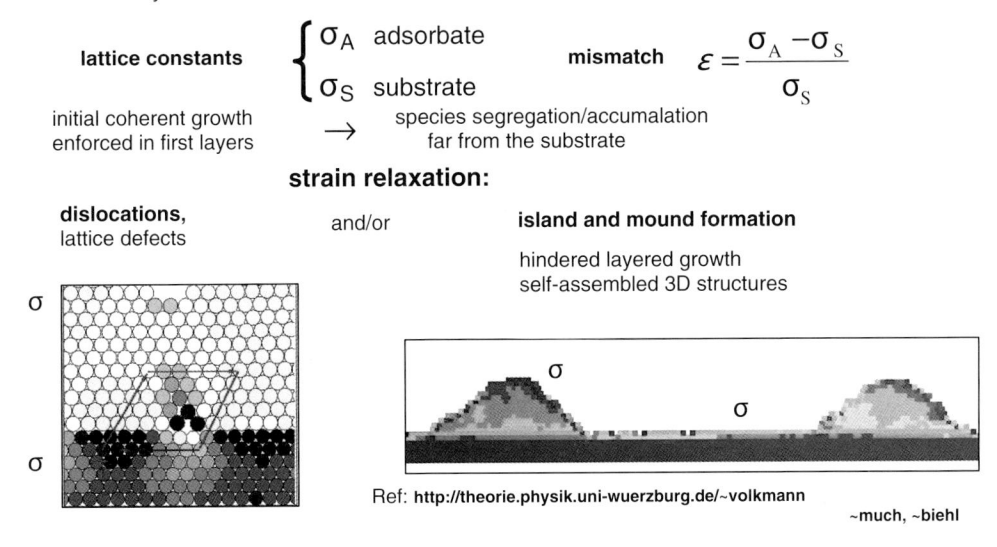

dislocations,   and/or   **island and mound formation**
lattice defects   hindered layered growth
                self-assembled 3D structures

Ref: http://theorie.physik.uni-wuerzburg.de/~volkmann

~much, ~biehl

Figure 7.23 Island and dislocation formation as strain release mechanisms.

The strained active region may relieve strain via a series of mechanisms such as dislocation, defect, or island formation. These mechanisms are active both during and after (under temperature cycling) the growth of the active region. Island formation is a common challenge in the growth of strained layers facing designers and crystal growers [33]. These islands cause bad morphology (lead to scattering), act as nonradiative recombination centers (increase transparency), and reduce gain (lower strength of matrix elements). Figure 7.23 [34] illustrates the formation of 3D islands in strained layer structures. As the strain builds up in the epitaxial layers, close to the critical layer thickness, the chemical potential of the surface layers switches signs favoring the formation of islands. In dislocated islands, strain relief arises by forming interfacial misfit dislocations. In these cases, the reduction in strain energy by the formation of a dislocation more than compensates the increase in surface energy associated with creating the islands.

Figure 7.24 shows the net strain profile of a typical 1230 nm (for 615 nm second harmonic generation (SHG) emission) active region. The net strain in the active region is typically compressive. The net compressive strain depends on the P (phosphorus) content in the intermediate GaAsP barrier layers. These tensile-strained GaAsP barriers provide strain relief by compensating the compressive strain in the InGaAs quantum wells. While this seems to be a viable solution for pushing the InGaAs toward longer wavelength operation, there are many challenges. While the net strain can be well compensated by the tensile strained barriers, the local strain fields tend to be very high and give rise to dislocations and defects. In addition, the material quality of the GaAsP degrades fairly rapidly as the P content is increased. From a thermodynamic consideration, the compressively strained QWs prefer lower

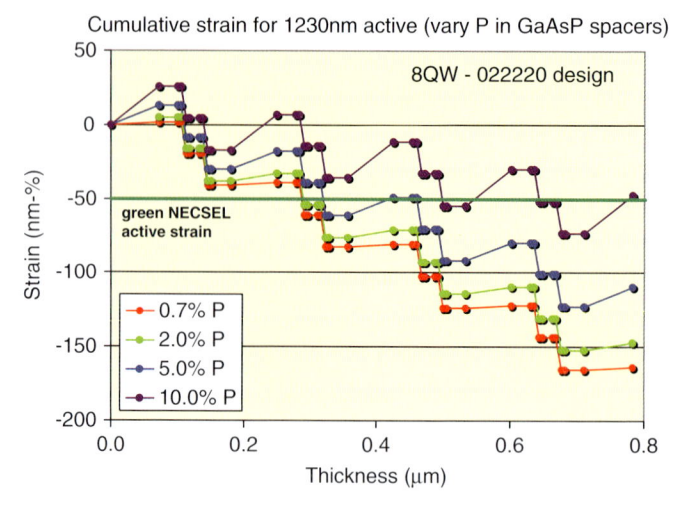

Figure 7.24 Net strain profile as a function of thickness for InGaAs/GaAsP active regions.

growth temperatures, whereas the GaAsP barriers exhibit better crystal quality when grown at higher temperatures. The temperature cycling during growth of the cavity region when coupled with the local strain fields poses challenge and thus careful optimization is needed for high-performance devices.

It is clear that longer wavelength emission poses challenges due to strain constraints in GaAs-based devices. One approach is to switch the material system to InP and make use of InGaAsP and AlInGaAs active regions [35, 36] that can be lattice matched all the way from 1100 to 1700 nm. The obvious challenge here is the availability of good DBR technology. More recently, groups have employed intracavity contacts for the operation of long wavelength InP surface-emitting lasers. Figure 7.25 illustrates the layer design of an optimized long-wavelength InP-based structure that can be used for high power emission for frequency doubling to red emission [37].

The entire device structure can be monolithically grown where the bottom DBRs are epitaxial AlInGaAs/InP and the top DBRs are dielectrically deposited AlN/a-Si stack that has high reflectivity and thermal conductivity. The current injection into the active region occurs via intracavity contacts on its either sides. To reduce free-carrier loss and increase electrical conductivity, the p-side intracavity contact consists of an n-type current-spreading layer and an $n+/p+$ tunnel junction that "converts" electrons to holes. The aperture to the device is formed by an undercut etch to the tunnel junction. The air-gap formed due to etching of the tunnel junction causes the InP:Zn and InP:Si layer to contact each other. However, if there is sufficiently thick p-cladding layer, the n–p–n contact layers form back-to-back diodes and block leakage currents. This is the reason why the tunnel junction is placed at the fourth field null from the active region, as shown in Figure 7.26, even though to first order one would like to place the TJ as close to the active region as possible and minimize the p-cladding thickness and maximize the n-cladding thickness. As can be seen in the field profile, the TJ needs to be placed at the null of the standing wave to reduce free-carrier absorption loss.

7.4 High-Power Arrays and Multielement Devices

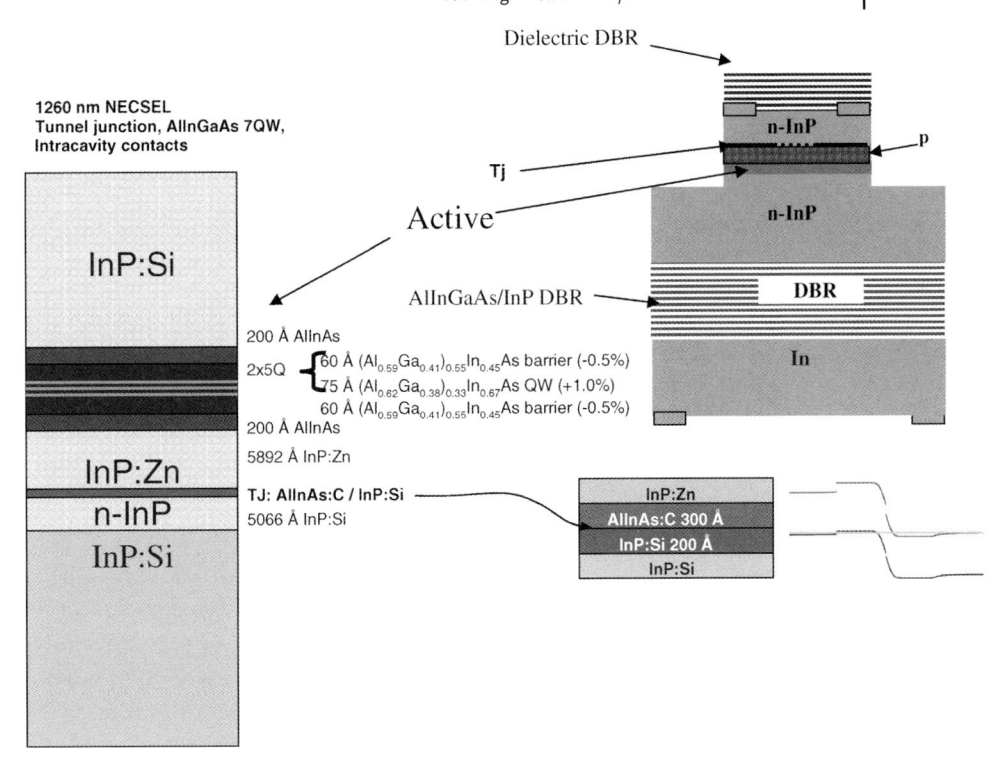

Figure 7.25 Device structure for an optimized InP-based long-wavelength NECSEL.

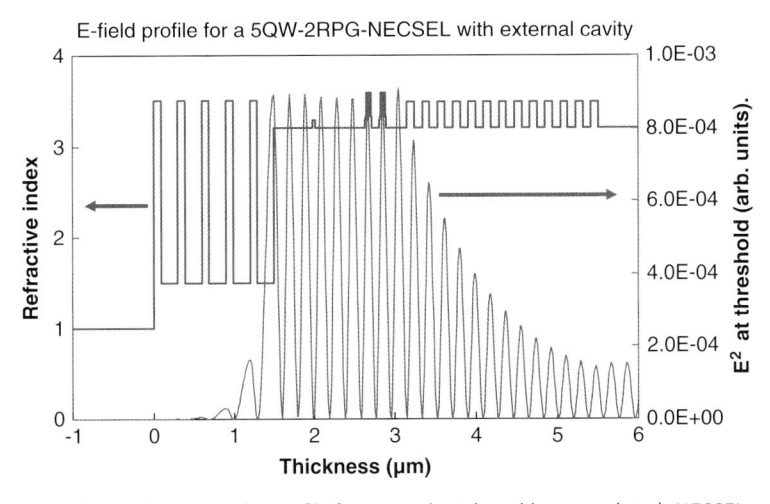

Figure 7.26 Field and index profile for a typical InP-based long-wavelength NECSEL.

7.5
Carrier Dynamics

Devices with three or more quantum wells were studied with six wells being determined to be the optimum for cw or pulsed operation. Injected electrons have 20 times the mobility of holes and the holes must occupy most of the wells to take advantage of the allowed gain. In six-well, 150 μm devices, the wells can be rapidly filled for injection currents of less than 1 A. Population of the wells with holes can occur by the recombination radiation that circulates in the cavity. Even with full pumping of the wells, ballistic transport of carriers can occur at current levels above a few hundred mA in 150 μm diameter devices. Figure 7.27 shows the measured total spontaneous light emission from a 150 μm device without an external mirror. The observed departure from linearity can be explained, in part, due to saturation of the wells. Increased Auger recombination may also contribute. With a near top-hat gain distribution, a fundamental spatial mode will burn a hole in the gain distribution, preventing a fraction of the injected carriers from participating in the stimulated emission. This effect can be partly overcome by the presence of the lateral pumping as described above.

Another phenomenon that affects the spectral gain is shown in Figure 7.28, which shows the spectral distribution of emission for a six-well device. The fringes are due to the thickness of the chip, about 70 μm. The absorption edge of GaAs limits the short-wavelength emission. The absorption edge shift together with the fringe shift is consistent with a 70 °C rise in chip temperature at 1 A, a 16-fold increase in power. The long-wavelength shift would require a temperature rise of more than 300 °C supporting the interpretation of bandgap lowering with increasing current density.

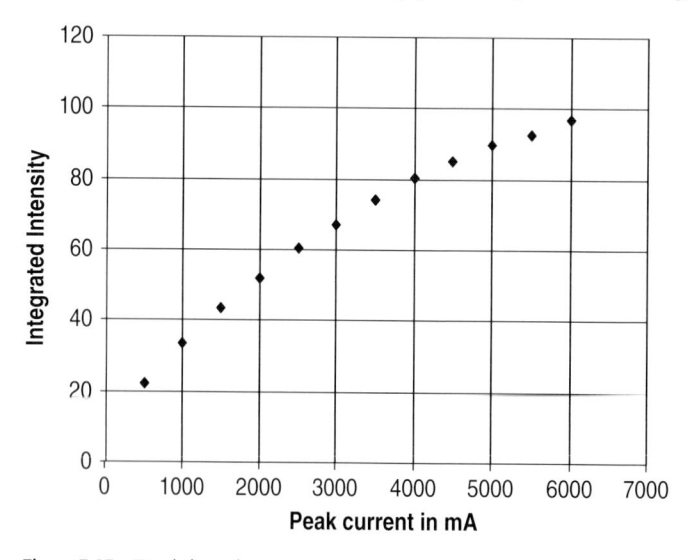

Figure 7.27 Total electroluminescent emission intensity as a function of peak drive current at a low duty cycle. This saturation of emission is related to the saturation of the quantum wells in addition to ballistic transport of injected carriers.

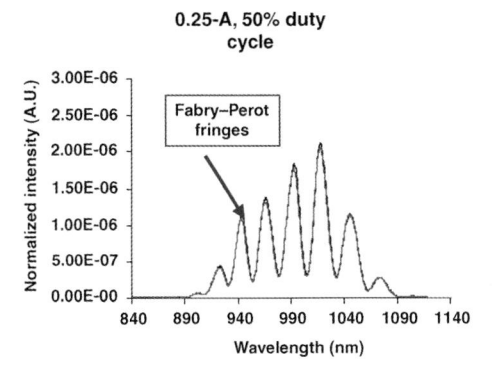

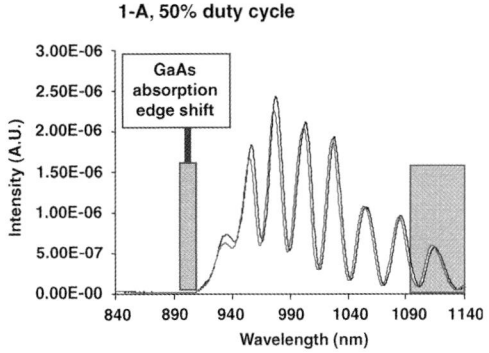

Figure 7.28 The electroluminescent spectra are shown in a nonlasing device without an internal n-Bragg mirror at two average power drive levels. Long wavelength shift is consistent with bandgap reduction (long wavelength shaded section). Data are shown nearly overlapped for two similar devices.

7.5.1
Mode Locking

Numerous studies have been made on optically pumped, mode-locked surface-emitting semiconductor lasers using semiconductor saturable absorbing mirrors and are described elsewhere in this book. Using a reverse-biased gain element described here as a saturable absorber [41], Zhang et al. [39, 40] have demonstrated mode locking with pulse repetition rates of up to 20 GHz. Two identical gain elements were used in a single cavity: one biased in the forward direction provided the gain, while the other element was reverse biased and operated as a saturable absorber. The loss in this element was adjusted by tuning the voltage and moving the bandgap to higher energy. The fast carrier sweep out in this element allowed the relatively high mode locking rates. The forward bias of this element added gain and increased the output power. Output coupling was through an intracavity beam splitter. This element could also serve as an amplitude modulator for the mode-locked output. The increased spectral width for these short pulses together with the high peak power levels would be useful for frequency doubling with reduced speckle output.

7.6
Nonlinear Optical Conversion with Surface-Emitting Diode Lasers: Design and Performance

7.6.1
Visible Laser Sources: Applications and Requirements

The preceding sections describe design and operation characteristics of electrically pumped, vertical-extended-cavity surface-emitting lasers (E-VECSELs). As was

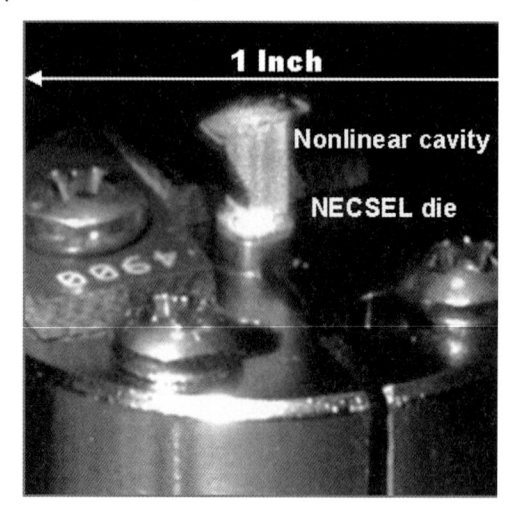

Figure 7.29 Experiment on an E-VECSEL chip + nonlinear crystal (KNbO$_3$) laser cavity architecture, generating 488 nm output. The cavity for 976 nm is stabilized by the thermal lens in the E-VECSEL chip. The end mirror is flat and is provided by the coating on KNbO$_3$ (from Ref. [42]).

shown, it is possible to scale up the power and control the laser emission spectrum and high beam quality by optimizing the design of the E-VECSEL semiconductor chip and the extended cavity. This was done in a wide range of near-infrared wavelengths. These properties of electrically pumped E-VECSELs made them very attractive for nonlinear frequency conversion to visible wavelengths.

Early work on frequency-doubled E-VECSELs [42, 43] provided concept demonstrations of intracavity second harmonic generation, with the main focus on the second harmonic wavelength of 488 nm that corresponded to 976 nm fundamental laser, developed and optimized for telecom EDFA applications [44]. Architectures demonstrated in these research efforts are illustrated in Figure 7.29 [42] and Figure 7.30 [43], and they provided a starting point to intensive application-focused development efforts led by Novalux team over the next several years.

Visible laser sources are of interest in a variety of applications such as bioanalytical instrumentation, semiconductor inspection, confocal microscopy, digital imaging and reprographics, and projection displays. In terms of technical requirements to commercial products, visible laser sources can be split into several broad groups. The first group of interest to E-VECSELs is instrumentation-grade lasers. These usually require milliwatts or tens of milliwatts of power, high stability, single spatial mode with the beam quality factor $M^2 < 1.2$, and reliable operation inside other systems and instruments. In most applications such as flow cytometry and DNA sequencing, such lasers need to operate in continuous-wave mode with very low noise. Particular wavelengths of interest are 488, 532, and 460 nm, with 488 nm being the most popular due to the multidecade legacy of argon ion lasers.

The second group includes lasers that are supposed to serve as improved illumination sources, where improvements come from spectral purity, efficiency,

7.6 Nonlinear Optical Conversion with Surface-Emitting Diode Lasers: Design and Performance

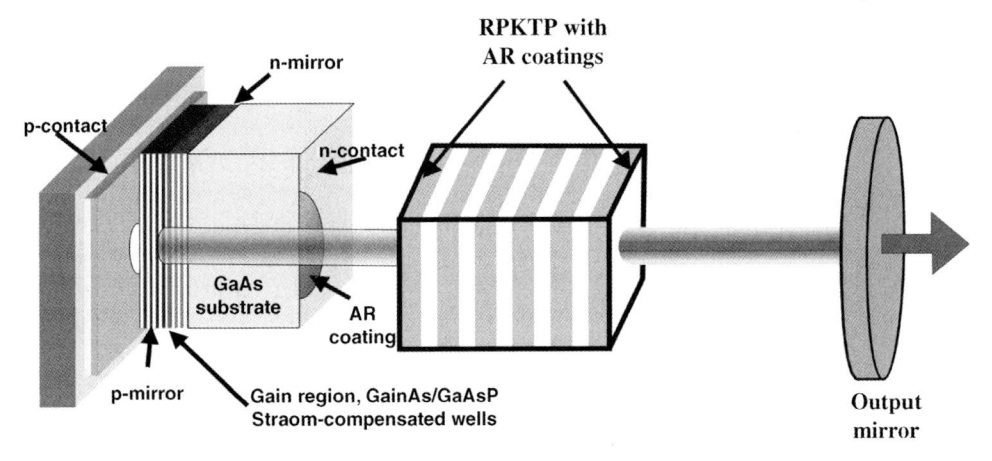

Figure 7.30 Laser cavity architecture from Ref. [43]. The 976 nm E-VECSEL chip is shown on the left, the spherical end mirror is shown on the right. The nonlinear crystal used for 488 nm generation was periodically poled KTP (PPKTP).

lifetime, polarization, and other properties provided by lasers. The primary application for laser illumination sources is projection displays. This market, in turn, includes handheld devices, business front projectors, head-up displays, rear-projection television, cinema, and many other segments. The wavelength of interest is defined by the need to access maximum color space available to human eye by mixing red (R), green (G), and blue (B) laser colors. While some variations are permitted, the range of wavelength for a blue, green, and red laser source would be 465 ± 10, 532 ± 10, and 632 ± 10 nm, respectively. Target power levels depend on the desired luminous flux of the projector, for example, a miniprojector, embedded in a handheld device, would require tens of milliwatts per color, while a rear-projection television would require power level of several watts per color.

7.6.2
Cavity Design Optimization and Trade-Offs for Second Harmonic Generation with Surface-Emitting Diode Lasers

The basic approach of obtaining efficient second harmonic generation into the visible wavelength range is well known. The general requirements for obtaining efficient SHG are high power density in the fundamental frequency beam, high beam quality, efficient and properly chosen nonlinear converter (crystal), and angular and frequency composition of the fundamental beam that is within the acceptance bandwidth of the nonlinear crystal. These requirements can be achieved via proper design optimization of E-VECSEL chip, nonlinear crystal, and external cavity.

For the continuous-wave operation and Gaussian spatial mode, the power in the generated second harmonic wave can be expressed as

$$P_{2\omega} = \frac{16\pi^2 d_{\text{eff}}^2}{\varepsilon_0 c \lambda_\omega^3 n_\omega n_{2\omega}} P_\omega^2 e^{-\alpha L} L h(\sigma, \beta, \varkappa, \xi, \mu). \tag{7.5}$$

Here $P_{2\omega}$ is the power in the generated second harmonic beam (frequency 2ω), P_ω is the power in the fundamental frequency beam, d_{eff} is the nonlinear coefficient for the frequency doubling crystal, L is its length, n_ω and $n_{2\omega}$ are indices of refraction of the fundamental and SHG beams, respectively, λ_ω is the wavelength of the fundamental beam in vacuum, and α is the loss factor accounting for loss at both wavelengths. The function h is expressed in terms of integrals that have to be evaluated numerically and the dimensionless parameters in the argument of this function are $\varkappa = \alpha b/2$, $\sigma = b\Delta k/2$, $b = \varrho/\theta_0$, $\xi = L/b$, $\mu = (L - 2f)/L$, $b = 2\pi n_\omega w_0^2/\lambda_\omega$ is the confocal parameter, w_0 is the beam waist radius, $\theta_0 = \lambda_\omega/2\pi n_\omega w_0$ is the far-field beam divergence, and f describes the position of the focus. For negligible loss and focus positioned in the center of the crystal, we have $\varkappa = \mu = 0$.

Equation (7.5) applies to the type I SHG process, in which the polarization states of the two fundamental frequency beams participating in the SHG process are the same, and the integral form of the function h of Eq. (7.5) was first derived by Boyd and Kleinmann [45]. A similar expression was derived by Zondy [48], which describes the type II SHG process in which states of the two fundamental frequency beams participating in the SHG process are orthogonal to each other. In many practical cases, the fundamental beam diameter is not changing significantly over the length of the nonlinear crystal. In this case of loose focusing ($\xi \ll 1$), second harmonic power is given by a simple analytical expression, applicable to both type I and type II phase-matching SHG processes:

$$P_{2\omega} = \frac{8\pi d_{\text{eff}}^2}{\varepsilon_0 c \lambda_\omega^2 n_\omega^2 n_{2\omega} w_0^2} P_\omega^2 L^2. \tag{7.6}$$

Since average power levels demonstrated with E-VECSELs have been limited to 1 W, useful SHG output levels (milliwatts and higher) can only be obtained in the intracavity architecture. In this case, one takes advantage of the large power levels of the fundamental frequency beam inside the laser cavity. For continuous-wave operation, a single-transverse-mode E-VECSEL that delivers ~0.5 W in output power with the end mirror reflectivity ~90% will have 5 W of circulating power inside the laser cavity. In intracavity SHG, it is customary to "close" the cavity by choosing high-reflectivity (~100%) coating for the end mirror. In this case, intracavity power of the fundamental frequency beam will be significantly higher – in many practical cases, approximately twice the power levels for the optimally outcoupled cavity, that is, 10 W for the example mentioned above. These ballpark numbers are illustrative for practical E-VECSEL design configurations with gain aperture having diameters in the range of 100–150 μm and n-mirror reflectivity in the range of 65–85%.

In general, Eqs (7.5) and (7.6) can apply to the instantaneous peak power levels of the fundamental and SHG beams. Therefore, a further increase in SHG efficiency can be obtained by operating the E-VECSEL chip in the pulsed regime via either direct current modulation [46] or mode locking (see, for example, the work by Zhang et al. [39, 40], only the fundamental frequency design was demonstrated there) as long as the frequency and angular composition of the fundamental beam remain within the acceptance bandwidth of the nonlinear crystal. The design flexibility with cw or pulsed mode of operation in a compact and efficient architecture is one of the key advantages provided by the E-VECSEL architecture and was successfully used in a number of research and development efforts.

In the review of work on frequency-doubled E-VECSELs, the laser cavity design for stable spatial mode operation that can accommodate efficient intracavity SHG is one of the key topics. Like solid-state laser crystals, E-VECSEL chips allow different and complex cavity designs (see, for example, Ref. [40]), including L-, V-, and Z-shaped cavities, cavities involving multiple E-VECSEL chips, and so on. However, most work was done on the simplest, compact, and practically useful linear cavity. Therefore, we are going to review it in some detail here.

An important parameter for E-VECSEL design with intracavity SHG is the reflectivity value for the n-mirror. As discussed in the previous sections describing the design of the fundamental-frequency, infrared E-VECSEL, the n-mirror is introduced mainly to limit the losses due to GaAs substrate on which the epitaxial E-VECSEL structure is grown and, possibly, the other losses too in the external cavity. While it is not obvious that the range of useful n-mirror reflectivity values should be the same for the fundamental frequency and intracavity second harmonic E-VECSEL design, they turn out to be close to each other and lie in the range of $R_n \sim 70–85\%$. This has been confirmed both theoretically and experimentally during the course of development efforts by Novalux team during 2001–2006. At the first glance, it may seem that the intracavity SHG design may benefit from reducing the reflectivity of the n-mirror, for example, to $\sim 50\%$ (this is done by reducing the number of high-index and low-index pairs in the n-mirror stack), because this would lead to the increase in the intracavity power for the fundamental frequency beam, and, therefore, to a better SHG conversion efficiency. However, this approach is incorrect since the maximum SHG power that can be extracted from a nonlinear crystal is limited by the maximum power that can be outcoupled from the fundamental frequency laser [47] Therefore, an E-VECSEL chip with its n-mirror properly optimized for maximum infrared power extraction in the TEM_{00} beam is usually also well optimized for intracavity SHG – the rest has to be done via optimizing SHG process itself.

7.6.3
Nonlinear Crystals Used in Intracavity Frequency Conversion

The choice and optimization of a frequency-doubling crystal has been one of the central tasks in the development of E-VECSEL for visible wavelength applications. According to Eqs (7.5) and (7.6), the nonlinear coefficient d_{eff} of the frequency doubler is one of the key parameters that defines SHG efficiency. The early work on frequency-doubled E-VECSELs was done for the fundamental wavelengths around 976 nm [42, 43]. The primary reason for this was the availability of mature and efficient 976 nm semiconductor chip developed for EDFA telecom market during 1998–2002, and the lucky coincidence is that the corresponding SHG wavelength of 488 nm matches one of the most widely used lines of the argon ion laser and is in great demand for instrumentation applications. In early 2000s, the nonlinear crystals available for frequency doubling of 976 nm semiconductor chip were limited to LBO (lithium triborate) and KNbO$_3$ (potassium niobate) crystals.

The limited choice of nonlinear crystals is due to constraints of "phase matching" that has to be achieved in the process of SHG between the fundamental and the second harmonic beams. This can be done only for certain crystallographic directions and not all nonlinear crystals happen to have such a direction for the target wavelengths. Even if the desired crystallographic cut exists, the efficiency of the second harmonic process is often limited by the low-effective nonlinearity for this crystallographic orientation and/or spatial walk-off of the participating fundamental and second harmonic beams.

After 2003, the rapid development of E-VECSEL material systems and their expansion into the wavelength range 900–1260 nm was accompanied by the rapid progress in periodically poled nonlinear crystals: periodically poled KTP (PP-KTP), periodically poled lithium niobate (PPLN), and periodically poled lithium tantalate (PPLT). The process of efficient second harmonic generation is achieved via the "quasi-phase matching" process, in which the phase synchronism between the fundamental and second harmonic beams is achieved by the periodic inversion of ferroelectric domain (i.e., the periodic "flipping" of the sign of nonlinear coefficient). This allows accessing the largest crystallographic component of nonlinearity and designing a walk-off-free SHG process, which is normally aligned with the principal crystallographic axes (for both direction of propagation and polarization orientations). With respect to PPLN and PPLT, the initial work was done on congruent lithium niobate and lithium tantalite materials that are easier for obtaining a high-quality domain structure. However, congruent materials suffer from reliability problems such as photorefractive damage (PRD) or green-induced infrared absorption (GRIIRA) or blue-induced infrared absorption effects (BLIIRA). The effects of GRIIRA and BLIIRA have the same physics but have different names to point out the color of the second harmonic beam that causes the absorption in the infrared. The easiest way to overcome this problem is to elevate the temperature of nonlinear crystal to $>200\,°C$, and this was what exactly done in early experiments. However, this approach becomes impractical in low-cost designs for commercial lasers. The better solution to reliability limitations was provided by MgO-doped or stoichiometric lithium niobate or lithium tantalate crystals. Their periodically poled versions are often abbreviated as PP-MgO:LN and PP-SLT, and the problem of creating high-quality periodic domain structure of 4–8 μm for the blue–green SHG range was recently resolved. Regarding the periodically poled KTP, the limiting reliability mechanism is called "gray tracking" and it manifests itself as the formation of absorption centers in the optical beam area. As is the case with the bulk KTP, gray tracking phenomena can be greatly reduced by improving the material purity.

A detailed discussion about the physics of nonlinear processes and material properties can be found in the book by Risk et al. [49]. For the purposes of this discussion, we will focus only on the characteristics most relevant for E-VECSEL designs with intracavity second harmonic generation. Table 7.2 summarizes several SHG configurations for wavelengths of practical interest and provides only the essential information to enable a researcher or engineer in the field to make the

Table 7.2 Summary of practically important second harmonic configurations for E-VECSELs.

Crystal	SHG process	Walk-off angle (mrad)	Poling period (μm)	Nonlinearity (d_{eff}, pm V^{-1})	SHG power for 10 W, ⌀100 μm IR beam for 3 mm crystal (mW)
LBO	976(o) + 976(o) → 488(e)	10	n/a	0.6	0.3
KTP	1064(e) + 1064(o) → 532(e)	4	n/a	3.6	6.5
KNbO$_3$	976(e) + 976(e) → 488(o)	8	n/a	10	33
KNbO$_3$	920(e) + 920(e) → 460(o)	18	n/a	11	43
PP-KTP	976 + 976 → 488	0	6.8	10	52
PP-SLT	976 + 976 → 488	0	6.1	11	41
PP-MgO:LN	920 + 920 → 460	0	4.3	16	97
PP-MgO:LN	976 + 976 → 488	0	5.3	16	86
PP-MgO:LN	1064 + 1064 → 532	0	7.0	16	72
PP-MgO:LN	1260 + 1260 → 630	0	11.5	16	52

first-cut evaluation of efficiency and suitability of the candidate nonlinear crystal for the target wavelength. The data are collected from the database and software SNLO [50] and other relevant information such as the details of crystal cuts, phase-matching bandwidths, and so on can be found in this source developed and maintained by Arlee Smith. The parameters are given for the temperature of 35 °C. The nonlinearity coefficients for the periodically poled materials are maximum theoretical values and in reality are always smaller, especially for short poling periods.

The letters "o" and "e" in the parentheses describing the bireferingent phase matching correspond to the ordinary and extraordinary optical waves, respectively. The last column of the table is a quick estimate of how much SHG power can be obtained in a 3 mm long nonlinear crystal with 10 W power for the intracavity fundamental-frequency, continuous-wave beam with 100 μm diameter. This estimate is obtained with the aid of Eq. (7.6), and it ignores some important effects such as spatial walk-off and the finite spectral bandwidth of the fundamental beam. While walk-off effects are important as can be seen from the values provided in Table 7.2, the SHG power values in the same table provide an easy-to-understand absolute and relative metrics of SHG output. The parameters for the power (10 W) and diameter (100 μm) of the intracavity beam represent "ballpark" numbers that illustrate performance of E-VECSEL devices; therefore, the estimated SHG levels are representative of what can be and has been achieved with real E-VECSELs with nonlinear crystals listed. It is clear that LBO crystals, widely used with higher power laser systems (solid-state, fiber, or optically pumped semiconductor lasers), are not very useful for E-VECSELs. The most advantageous nonlinear crystals for E-VECSELs are clearly periodically poled devices, which can be engineered for the desired wavelengths and can offer very high coefficients of nonlinear conversion, especially PP-MgO:LN crystals. Not surprisingly, the best results, reviewed in the next two sections, for instrumentation and display lasers were obtained with nonlinear crystals that have the highest nonlinear conversion: $KNbO_3$, PP-KTP, and PP-MgO:LN. Not much work has been done with PP-SLT because of the higher cost and limited availability of these crystals, and the fact is that their only advantage over PP-MgO:LN is their higher damage threshold, which becomes important for power levels approaching ~100 W for the beam diameters of interest. In summary, the wavelength range and power levels offered by E-VECSEL technology find an ideal match in the recently developed periodically poled SHG crystals, especially PP-MgO:LN, which at the time of writing this review was commercially available from at least two companies: Spectralus (United States) and HC Photonics (Taiwan).

We will next discuss the technical challenges and solutions for designing visible E-VECSEL sources for instrumentation applications and projection display applications.

7.6.4
Low-Noise, High Mode Quality, Continuous-Wave Visible Laser Sources for Instrumentation Applications

Having reviewed the general aspects of intracavity SHG with E-VECSELs, we now turn to technical details driven by specific applications in instrumentation markets such as flow cytometry, confocal microscopy, and semiconductor inspection. In most cases, these applications require continuous-wave blue–green lasers with power levels of 5–50 mW, high beam quality, very high beam pointing stability, and very low noise. For several decades, the only usable blue–green laser platform was the argon ion laser that had set standards in stability and efficiency but could not overcome intrinsic limitations in low efficiency, bulk size, and lifetime of about 5000 h. The Novalux team had demonstrated the good capabilities of the E-VECSEL platform for instrumentation markets and developed 488 [42], 460 [51], and 532 nm [52] visible E-VECSEL lasers based on the common laser cavity design with only significant adjustments in the design of semiconductor E-VECSEL chip and nonlinear crystal parameters to make them optimized for different wavelengths.

The "long cavity" solution discussed earlier proves to be advantageous for achieving high beam quality (with the beam quality factor $M^2 < 1.2$) and in providing space for frequency-selective elements. The key requirement for very low noise levels (typically <0.1%) can be achieved in E-VECSEL architecture only by forcing the laser to operate in the single longitudinal mode of the external laser cavity. Figure 7.31 [42] illustrates qualitatively that the modes of the external cavity "feel" the frequency selection due to the Fabry–Perot peak of the active E-VECSEL cavity formed by the p- and n-doped mirror stacks. While this frequency selection may often be adequate for the single longitudinal mode operation with the optimally outcoupled end mirror, it is usually not sufficient selector for the case of the highly reflective end mirror that is usually used in the intracavity SHG design. Furthermore, the SHG process itself provides additional loss at the wavelength where the single longitudinal mode operation is to be achieved. Therefore, reliable single longitudinal mode design requires the introduction of additional frequency-selecting elements into the external cavity.

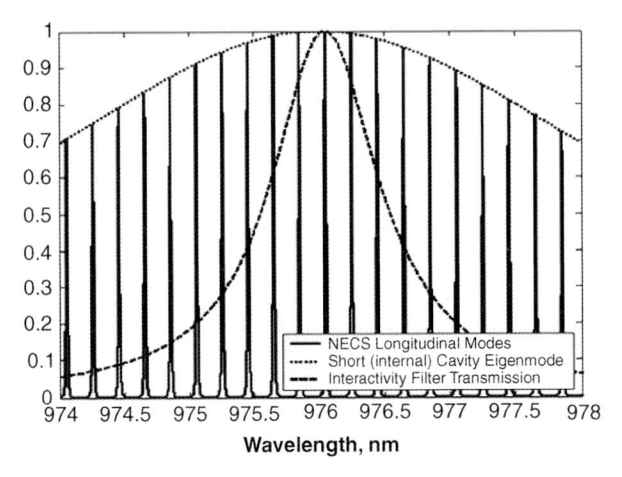

Figure 7.31 Illustration of the longitudinal mode spectrum in the composite E-VECSEL cavity and the wavelength stabilization mechanism (from Ref. [42]).

The results for the 488 nm E-VECSEL source [42] were based on the use of a traditional tilted etalon that provides spectral filtering in the external cavity (see Figure 7.31) that leaves the laser running in only one longitudinal mode that matches with the filter transmission peak. In some cases, however, it turned out that the free spectral range provided by standard fused silica etalon is not large enough (e.g., a 100 μm fused silica etalon has a free spectral range of 3.3 nm at 976 nm wavelength) to suppress lasing in adjacent transmission peaks of the etalon. This situation is not acceptable because it leads to a high noise in laser operation due to mode hopping and to the loss of SHG power because the wavelength separation of two adjacent etalon peaks is usually comparable to or larger than the spectral bandwidth of the SHG process. To solve this problem, designs based on the use of birefringent filters or thin-film interference filters were successfully explored. One of the measured spectral characteristics of the interference filters is shown in Figure 7.32. The details of the design for this filter can be found in the patent by Shchegrov [53] and such filters were shown to provide low enough loss and high enough selectivity to ensure reliable operation of frequency-doubled E-VECSELs in a single longitudinal mode regime.

The spectral behavior of a frequency-doubled 976 nm E-VECSEL, operating in a single longitudinal mode regime, is shown in Figure 7.33 [42]. As can be seen, all unwanted modes are suppressed to the level of ∼40 dB. In this particular case, the only lasing mode was designed to be maintained over the range of currents from 540 to 600 mA (Figure 7.34, from Ref. [42]) at the fixed temperature of the E-VECSEL diode of 100 μm gain aperture and $KNbO_3$ nonlinear crystal, which was antireflection coated for 976 and 488 nm on both ends. The maximum blue (488 nm) power was 8 mW for that configuration and was outcoupled through the end mirror that had high-reflection coating for 976 nm and antireflection coating for 488 nm. In later results obtained with improved E-VECSELs, PP-MgO:LN nonlinear crystals, and optimized spectral filters, power levels of 100 mW were achieved for the same current

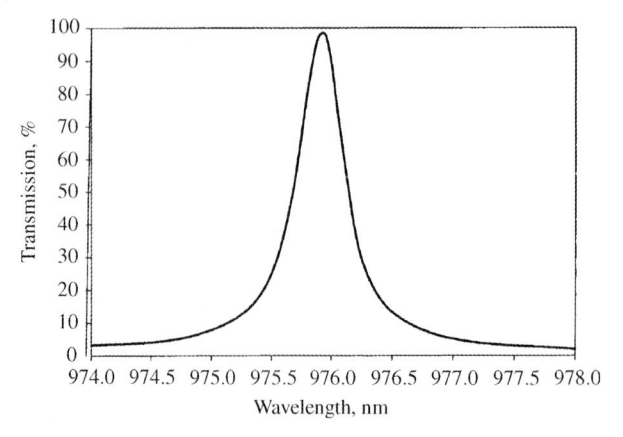

Figure 7.32 Experimentally measured transmission curve of the interference filter from Ref. [52]. This filter shows peak transmission >99.5% and narrow bandwidth ∼0.4 nm that enforces single longitudinal mode operation in the external cavity.

levels for all of the three "laser instrumentation" wavelengths (488, 460, and 532 nm) that were under development at Novalux during 2002–2005.

The low-noise operation and power stability of visible E-VECSEL laser source are illustrated in Figure 7.35. It must be noted that the extremely low noise (<0.002%) operation of the blue laser source is the direct result of the true single longitudinal mode operation. The high power stability required not only addressing all the aspects of the semiconductor chip design, optical cavity design, and nonlinear crystal choice but also providing the stable mechanical frame for the laser cavity and components as well as electronics feedback loop. These details go beyond the scope of this review.

In summary, visible E-VECSEL sources were shown to provide a competitive and flexible platform with the notable advantages of compact size and "designable" wavelength that are often impossible to achieve with gas or solid-state lasers.

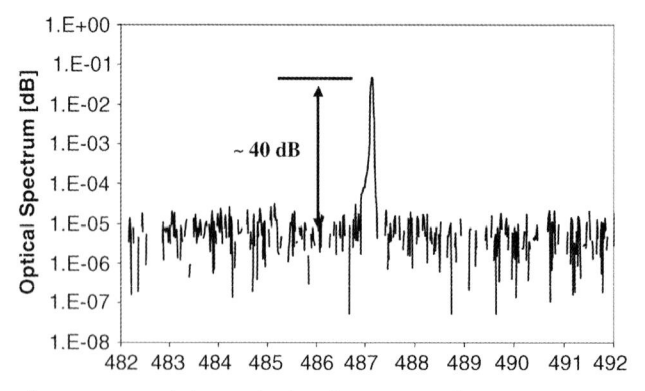

Figure 7.33 Single longitudinal mode spectrum of a 488 nm intracavity frequency-doubled E-VECSEL of Ref. [42]. Better than 40 dB side mode suppression was observed with an optical spectrum analyzer.

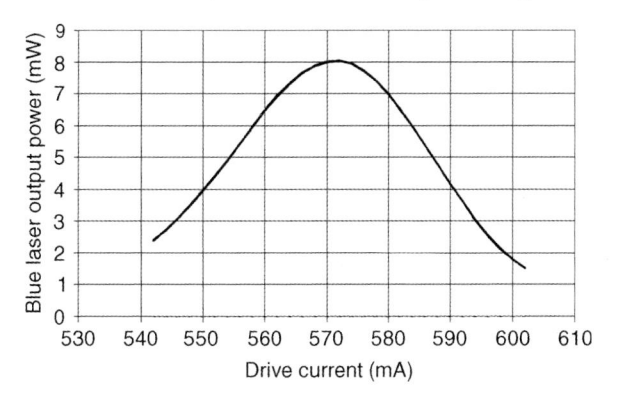

Figure 7.34 Power–current characteristics of a 488 nm, single longitudinal, single-transverse mode, intracavity frequency-doubled E-VECSEL of Ref. [42].

7.6.5
Compact Visible Sources Scalable to Array Architecture

In the past few years, the rapid development of projection display technologies created a demand for efficient and scalable RGB (red, green, and blue) light sources. Because E-VECSEL is one of the few platforms that naturally lend itself to the array architecture, it has been one of the leading contenders in the emerging projection display applications that rely on new-generation light sources: LED or lasers.

The set of key requirements for RGB lasers is quite different from those in the instrumentation market. The most important parameters driving technical design decisions are ultracompact size, low cost, and high efficiency. For these applications, the "short cavity" design for intracavity SHG discussed in the previous sections becomes highly advantageous. The main advantage of the "short cavity" architecture

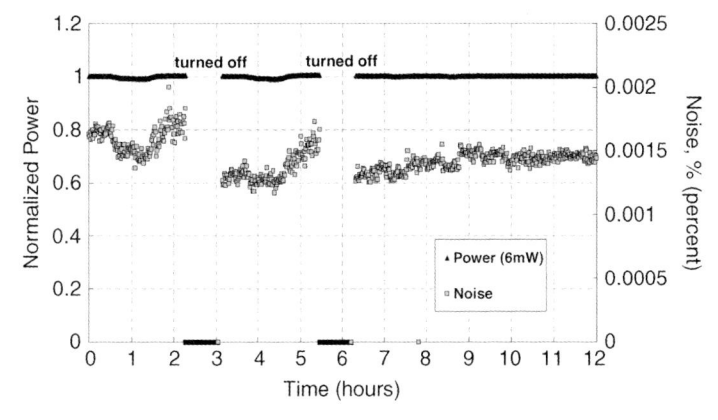

Figure 7.35 Power and noise of the 488 nm laser source of Ref. [42], tested on a hot plate. The plate temperature in the first two windows of operation in cycled 20–45 °C, in the third window of operation, it is not controlled (normal room temperature).

is that the cavity is built using flat (as opposed to spherical or cylindrical) optics components only, enabling low-cost passive alignment and scalability to higher power levels via the use of array architecture.

The single longitudinal mode operation that is highly desirable for some instrument applications is not required for display light sources. Furthermore, it is actually undesirable since it also results in a highly coherent laser source and creates unwanted speckle in the projected picture. Thus, the E-VECSEL design with intracavity SHG concentrated on achieving multilongitudinal mode operation with several longitudinal modes running within the acceptance bandwidth of nonlinear crystal. This was aided by spectral chirping that occurred during pulsed operation. Nearly all the results were obtained with PP-MgO:LN crystals that provide highest conversion efficiency and can be manufactured on a wafer scale for high-volume applications. The spectral acceptance bandwidth for PP-MgO:LN is approximately $\Delta\lambda \sim 2$ nm per 1 mm of crystal length for 1064 $\rightarrow$ 532 nm wavelength conversion, with the bandwidth value decreasing inversely proportionally to the crystal length. Therefore, for example, a 5 mm long PP-MgO:LN crystal would have ~ 0.4 nm SHG bandwidth and would require an intracavity spectral filter allowing laser operation in such a spectral window.

An attractive wavelength control solution was found in the volume Bragg grating (VBG) technology [54] that allows using a frequency-selective volume grating as an end mirror. Figure 7.36 [53] shows the measured reflectivity of the VBG element chosen for E-VECSEL spectral narrowing as a function of the wavelength. One can observe that the VBG spectral filter provides the desired window that can be used to match the SHG acceptance bandwidth for PP-MgO:LN. Another wavelength control design was based on the interference filter, which can also be combined with the end mirror [53].

The 1064 nm cavity layout with VBG developed by the Novalux team is shown in Figure 7.37 [53, 55]. The laser cavity consists of the following elements: (i) E-VECSEL semiconductor chip; (ii) a plane-parallel optical plate coated to lock the desirable

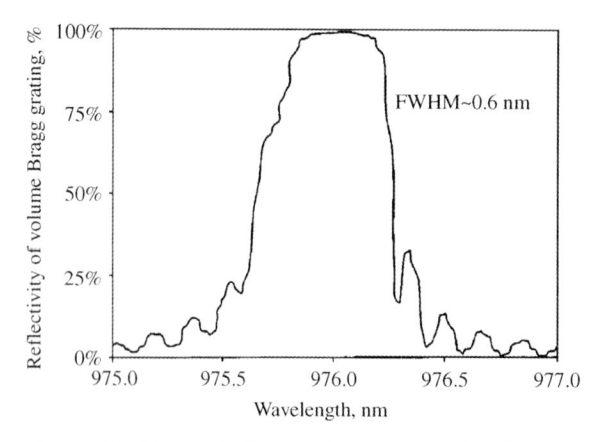

Figure 7.36 Measured reflectivity characteristics of a volume Bragg grating (VBG) element from Ref. [53].

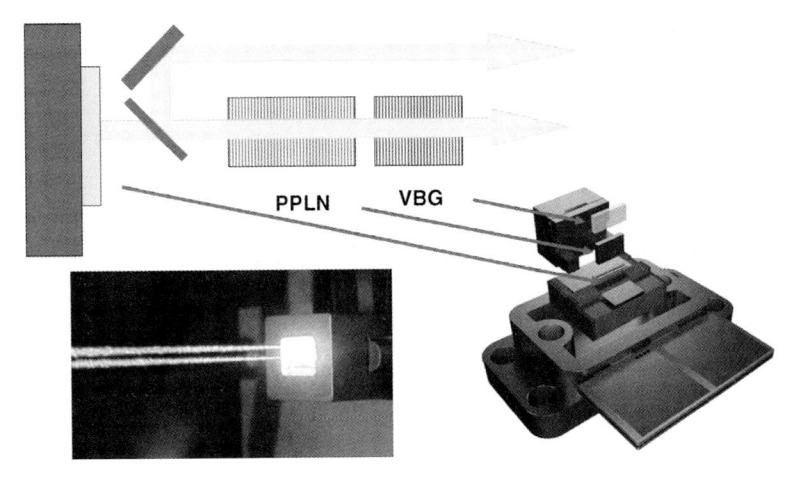

7.6 Nonlinear Optical Conversion with Surface-Emitting Diode Lasers: Design and Performance

Figure 7.37 Layout of the optical cavity is shown containing the PPLN nonlinear crystal, volume Bragg grating, and a dichroic polarizing optical assembly to extract the backward-going second harmonic beam. An exploded assembly is also shown together with both the forward and the backward beams exiting the device (from Ref. [53, 56]).

p-polarization for the intracavity beam and to reflect the backward-generated 532 nm SHG beam (another identical plate shown on the top is used to turn this 532 nm to be collinear with the forward-generated 532 nm beam; (iii) the PP-MgO:LN nonlinear crystal with the poling period $\sim 6.9\,\mu m$, and (iv) the VBG element coated for high transmission at 532 nm. As mentioned, the laser cavity is stabilized by the thermal lens in the E-VECSEL chip and it has only flat optics elements. Therefore, output power can be scaled up by increasing the number of emitters and using array architecture. Figure 7.37 also shows a green laser array during operation in a package (lower left picture), and the package itself (lower right picture) illustrating the assembly of all optical elements. To achieve optimum SHG conversion, the temperature of the PP-MgO:LN crystal was controlled by a resistive heater.

A photograph illustrating the properties of the optical output of laser array with the design illustrated in Figure 7.37 is shown in Figure 7.38. The spatially separated beams seen in Figure 7.38 are obtained with the aid of a lens. If no lens is used, the second harmonic beams overlap in the far field and add up to the Gaussian beam profile displayed in the inset picture of Figure 7.38. The array architecture has the additional advantage of speckle reduction with the speckle contrast decreasing as $1/\sqrt{N}$, where N is the number of emitters in the array.

Since the laser sources for projection display do not have to run in continuous-wave regime, pulsed operation via direct current modulation can be used to increase the SHG efficiency for E-VECSELs. Since the SHG process is sensitive to the peak optical power levels, it is beneficial to drive the laser into the regime where intracavity power levels of several watts in cw mode are converted into tens of watts peak power levels in pulsed operation. The possibility to operate E-VECSELs in the pulsed regime provides design flexibility and allows to keep compact size of the laser cavity without using longer nonlinear crystals or even to use shorter PP-MgO:LN crystals to achieve

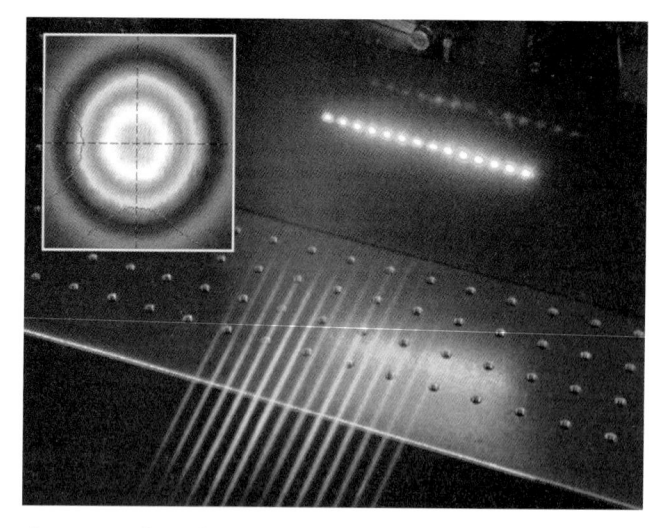

Figure 7.38 Photo of a 532 nm E-VECSEL array. Separate beams, representing near field of the array source, are obtained with a lens. *Inset*: far-field intensity pattern of the overall array obtained at 24″ from the source (without lens) (from Ref. [56]).

better temperature tolerances and lower cost in mass production. A discussion of modeling and experiments on pulsed versus cw operation for frequency-doubled E-VECSELs can be found in Ref. [46].

Based on architectures shown in Figures 7.37 and 7.38, E-VECSELs were shown to achieve visible output power levels of 100 mW for a single emitter. For array E-VECSELs, several watts of visible output was demonstrated in a platform common for red, green, and blue colors [56, 57]. Figure 7.39 shows a power–current

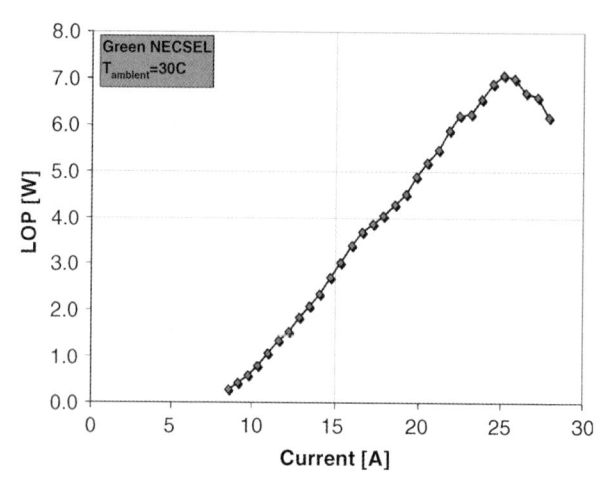

Figure 7.39 Average output power at 532 nm from a frequency-doubled E-VECSEL array operating at 30 °C ambient as a function of peak current into a 24-element linear array. Chip size was $8 \times 1\,mm^2$ (from Ref. [56]).

characteristic for a green laser array of 24 emitters with the maximum 532 nm output reaching 7 W level. An output power of 5 W at 465 nm (blue) and 2 W at 620 nm (red) has also been demonstrated for frequency-doubled E-VECSEL devices in strained GaInAs material. These power levels are suitable for use in rear-projection or front-projection displays. Novalux' team and its partners had demonstrated successful integration of these visible laser arrays in rear-projection television and cinema projectors. As expected, the color gamut and overall picture quality obtained from the laser light sources were superior to those provided by the conventional lamp or LED technology.

Acknowledgments

We acknowledge the following former employees, consultants, coworkers, and advisors to Novalux who contributed to the development of this technology: Glen Carey, Stepan Essaian, Wonill Ha, Sascha Hallstein, Eric Honea, James Harrison, James Keszenheimer, Professor James Harris, Kent Choquette, William Krupke, Professor Hiroshi Takuma, Professor Arto Nurmikko, K. Jasim, Q. Zhang, Professor E. P. Ippen, E. U. Rafailov, Professor Alan Wilner, Mitch Jansen, Neil McKinnon, Arvydas Umbrasas, Yae Okuno, Joachim Krueger, Sui Lim, James Dudley, Frank Hu, Chistopher Kocot, Giorgio Giaretta, Renata Carico, Rene Dato, Eva M. Strzelecka, Professor John G. McInerney, Alan Lewis, Dickey Lee, Jason P. Watson, Keith Kennedy, Hailong Zhou, Brad Cantos, William R. Hitchens, David L Heald, Vincent Doan, Professor Kevin Lear, Professor Anthony Siegman, Rashit Nabiev, Professor Sir Wilson Sibbet, Frank Patterson Todd Brehmer, Tin Nguyen, Natalia Simanovskaia, Robert Martinsen, Gary Oppedahl, John Green, Charles Amsden, Michael Liebman, Serguei Anikitchev and Greg Niven.

A portion of this work was supported by DARPA.

References

1. Soda, H., Iga, K., Kitahara, C., and Suematsu, Y. (1979) GaInAsP/InP surface emitting injection lasers. *Jpn. J. Appl. Phys.*, **18**, 2329–2330.
2. Coldren, L. and Corzine, S. (1995) *Diode Lasers and Photonic Integrated Circuits*, John Wiley & Sons, Inc.
3. Grabherr, M. *et al.* (1998) Bottom-emitting VCSELs for high cw optical output power. *IEEE Photon. Technol. Lett.*, **10** (8), 1061–1063.
4. Princeton Optronics web site. http://www.princetonoptronics.com/. 2009.
5. Kuznetsov, M., Hakimi, F., Sprague, R., and Mooradian, A. (1997) High power (>0.5 W CW) diode pumped vertical-external-cavity surface emitting semiconductor lasers with circular TEM_{00} beams. *Photon. Technol. Lett.*, **9**, 1063–1065.
6. MacLeod, H.A. (2001) *Thin-Film Optical Filters*, 3rd edn, Institute of Physics Publishing.
7. Yoffe, G.W. (1991) Rectification in heavily doped p-type GaAs/AlAs heterojunctions. *J. Appl. Phys*, **70**, 1081.
8. Tai, K. *et al.* (1990) Drastic reduction of series resistance in doped semiconductor distributed Bragg reflectors for surface-emitting lasers. *Appl. Phys. Lett.*, **56** (25), 2496–2498.

9 Corzine, S.W. et al. (1989) Design of Fabry–Perot surface-emitting lasers with a periodic gain structure. *IEEE J. Quantum Electron.*, **25** (6), 1513–1524.

10 Spitzer, W.G. and Whelan, J.M. (1965) Infrared absorption and electron effective mass in n-type gallium arsenide. *Phys. Rev.*, **114** (1), 59–62.

11 Mooradian, A. (2001) High power laser devices. US Patent 6,243,407; Heald, D.L. and Reyes, L. (2002) Method and apparatus for wafer-level testing of semiconductor lasers. US Patent 6,448,805; Mooradian, A. (2003) High power laser. US Patent 6,614,827; Martinsen, R. (2003) Method and apparatus for controlling thermal variations in an optical device. US Patent 6,636,539; Harrison, J. and Heald, D.L. (2004) Method and apparatus for wafer-level testing of semiconductor laser. US Patent 6,775,000; Mooradian, A. (2004) Coupled cavity high power semiconductor laser. US Patent 6,778,582.

12 Hadley, M.A. et al. (1993) High single-transverse-mode output from external-cavity surface emitting laser diodes. *Appl. Phys. Lett.*, **63** (12), 1607–1609.

13 Rimington, N.W., Schieffer, S.L., Andreas Schroeder, W., and Brickeen, B.K. (2004) Thermal lens shaping in Brewster gain media: a high-power, diode-pumped Nd:GdVO4 laser. *Opt. Express*, **12** (7), 1426.

14 Siegman, A.E. (1986) *Lasers*, University Science Books, 746.

15 Shchegrov, A.V. (2003) Method for modeling and design of coupled cavity laser devices. US Patent Application 20030120362.

16 Yang, Z.H., Leger, J.R., and Shchegrov, A.V. (2004) Three-mirror resonator with aspheric feedback mirror for laser spatial mode selection and mode shaping. *IEEE J. Quantum Electron.*, **40**, 1258–1269.

17 Martinsen, R. (2003) Method and apparatus for controlling thermal variations in an optical device. US Patent 6,636,539.

18 Watson, J., Mooradian, A., Nabiev, R., Carey, G., Hallstein, S., Cantos, B., Hitchens, W., Jansen, M., Fang, F., and Scholtz, K. (2005) High power high efficiency VCSEL arrays with narrow emission spectrum. Proceedings of ICALEO, Conference paper #405.

19 Seurin, J.-F., Xu, G, Wynn, J.D., Tishinin, D., Wang, Q., Khalfin, V., Miglo, A., Pradham, P., D'Asaro, L.A., and Ghosh, C. (2007) High power vertical cavity surface-emitting laser pump sources. Annual Meeting of IEEE LEOS, Vol. 21, No. 4, pp. 28–32.

20 Mooradian, A. (1992) External cavity semiconductor laser system. US Patent 5,131,002.

21 Hergenhan, G., Lucke, B., and Brauch, U. (2003) Coherent coupling of vertical-cavity surface emitting laser arrays and efficient beam combining by diffractive optical elements: concept and experimental verification. *Appl. Opt.*, **42** (8), 1667–1680.

22 Nabors, C.D. (1992) Coherent coupling of microchip arrays. Proceedings of the Lasers and Electro-Optics Society Annual Meeting, Institute of Electrical and Electronics Engineers, New York, Paper SSLT7.1, pp. 497–498.

23 Takeuchi, T., Chang, Y.-L., Leary, M.H., Mars, D., Song, Y.K., Roh, S.D., Luan, H.-C., Mantese, L., Tandon, A., Twist, R., Belov, S., Bour, D.P., and Tan, M.R.T. (2004) Al contamination in InGaAsN quantum wells grown by metalorganic chemical vapor deposition and 1.3 μm InGaAsN vertical cavity surface emitting lasers. *Jap. J. Appl. Phys.*, **43** (4A), 1260–1263.

24 Takeuchi, T., Chang, Y.L., Leary, M.H., Mars, D.E., Tandon, A., Lin, C.K., Twist, R., Belov, S., Bour, D.P., Tan, M.R.T., Roh, D., Song, Y.K., Mantese, L., and Luan, H.C. (2003) MOCVD growth of InGaAsN QWs and 1. 3 μm VCSELs. IEEE LEOS Annual Meeting – Semiconductor Lasers, October 27, Invited paper #MD1.

25 Chang, Y.L., Takeuchi, T., Leary, M.H., Mars, D.E., Tandon, A., Twist, R., Belov, S., Bour, D.P., Tan, M.R.T., Roh, D., Song, Y.K., Mantese, L., and Luan, H.C. (2003) Development of InGaAsN based 1300 nm VCSELs. 204th Meeting of the Electrochemical Society, State of the Art Program on Compound Semiconductors XXXIX, October 13, #744.

26 Takeuchi, T., Chang, Y.L., Leary, M.H., Tandon, A., Luan, H.C., Bour, D.P., Corzine, S.W., Twist, R., and Tan, M.R.T. (2002) 1.3 µm InGaAsN vertical cavity surface emitting lasers grown by MOCVD. *Electron. Lett.*, **38** (23), 1438–1440.

27 Takeuchi, T., Chang, Y.-L., Tandon, A., Bour, D., Corzine, S., Twist, R., Tan, M., and Luan, H.-C. (2002) Low threshold 1.2 µm InGaAs quantum well lasers grown under low As/III ratio. *Appl. Phys. Lett.*, **80** (14), 2445–2447.

28 Lin, C.-K., Tandon, A., Djordjev, K., Corzine, S.W., and Tan, M.R.T. (2007) High speed 985 nm bottom-emitting VCSEL arrays for chip-to-chip parallel optical Interconnects. *IEEE J. Sel. Top. Quantum Electron.*, **13** (5), 1332–1339.

29 Tandon, A., Lin, C.K., Djordjev, K., Corzine, S.W., and Tan, M.R.T. (2005) High speed 2D VCSEL arrays at 990 nm for short reach interconnects. SPIE-Physics and Simulation of Optoelectronic Devices XII, January 27, Invited paper 5349-28.

30 Tandon, A., Corzine, S.W., Schneider, R.P., and Tan, M.R.T. (2000) Red VCSELs for low cost POF-based data links. SPIE-Vertical Cavity Surface Emitting Lasers IV, January 27, Invited paper 3946-18-02.

31 Tansu, N. and Mawst, L.J. (2001) High-performance, strain compensated InGaAs–GaAsP–GaAs ($\lambda = 1.17$ mm) quantum well diode lasers. *IEEE Photon. Technol. Lett.*, **13** (3), 179–181.

32 Lemoff, B.E., Ali, M.E., Panotopoulos, G., de Groot, E., Flower, G.M., Rankin, G.H., Schmit, A.J., Djordjev, K.D., Tan, M.R.T., Tandon, A., Gong, W., Tella, R.P., Law, B., and Dolfi, D.W. (2005) 500-Gbps parallel-WDM optical interconnect. Proceedings of the Electronics Components and Technology Conference, Session 24 – Optical Transceivers, June 2005.

33 Stringfellow, G.B. (1999) *Organometallic Vapor-Phase Epitaxy: Theory and Practice*, 2nd edn, Academic Press.

34 Much, F., Volkmann, T., Weber, S., and Walther, M. Off-lattice KMC simulations of hetero-epitaxial growth: the formation of nano-structured surface alloys, www.ipam.ucla.edu/publications/maws2/maws2_5441.ppt. 2009.

35 Tandon, A., Bour, D.P., Chang, Y.L., Lin, C.K., Corzine, S.W., and Tan, M.R.T. (2004) High-performance 1300 nm and 1550 nm AlInGaAs lasers grown by MOCVD. SPIE-Physics and Simulation of Optoelectronic Devices XII, January 27, Invited paper 5349-28.

36 Bour, D., Corzine, S., Perez, W., Zhu, J., Tandon, A., Ranganath, R., Lin, C., Twist, R., Martinez, L., Höfler, G., and Tan, M. (2004) Self-aligned, buried heterostructure AlInGaAs laser diodes by micro-selective-area epitaxy. *Appl. Phys. Lett.*, **85** (12), 2184–2186.

37 Robbins, V.M., Lester, S.D., Bour, D.P., Miller, J.N., and Mertz, F. (2004) High-power single-mode 1330- and 1550-nm VCSELs bonded to silicon substrates. Physics and Simulation of Optoelectronic Devices XII, *Proc. SPIE*, **5349**, 366–374.

38 Ramanathan, V., Lee, J., Xu, S., Wang, X., and Reitz, D. (2005) Thermal effects in high repetition rate, high average power chirped pulse amplifiers. Frontiers in Optics 2005 and Laser Science XXI, The 89th OSA Annual Meeting Technical Conference, October 16–20, JTuC44.

39 Jasim, K., Zhang, Q., Nurmikko, A.V., Mooradian, A., Carey, G., Ha, W., and Ippen, E. (2003) passively modelocked vertical extended cavity surface emitting diode laser, *Electron. Lett.*, **39** (4), 373–375.

40 Zhang, Q., Jasim, K., Nurmikko, A.V., Mooradian, A., Carey, G., Ha, W., and Ippen, E. (2004) Operation of a passively mode-locked extended-cavity surface-emitting diode laser in multi-GHz regime. *IEEE Photon. Technol. Lett.*, **16**, 885–887.

41 Stormont, B., Rafailov, E.U., Cormack, I.G., and Sibbett, W. (2004) Extended-cavity surface emitting diode laser as active mirror controlling mode-locked Ti:sapphire laser. *Electron. Lett.*, **40** (12), 732–734.

42 Shchegrov, A.V., Lee, D., Watson, J.P., Umbrasas, A., Strzelecka, E.M., Liebman, M.K., Amsden, C.A., Lewis, A., Doan, V.V., Moran, B.D., McInerney, J.G., and Mooradian, A. (2003) 490-nm coherent emission by intracavity frequency doubling of extended cavity surface-

emitting diode lasers. *Proc. SPIE*, **4994** (1), 197–205.

43 Rafailov, E.U., Sibbett, W., Mooradian, A., McInerney, J.G., Karlsson, H., Wang, S., and Laurell, F. (2003) Efficient frequency double of a vertical-extended-cavity surface-emitting laser diode by use of a periodically poled KTP crystal. *Opt. Lett.*, **28** (21), 2091–2093.

44 Strzelecka, E.M., McInerney, J.G., Mooradian, A., Lewis, A., Shchegrov, A.V., Lee, D., Watson, J.P., Kennedy, K.W., Carey, G.P., Zhou, H., Ha, W., Cantos, B.D., Hitchens, W.R., Heald, D.L., Doan, V.V., and Lear, K.L. (2003) High-power high-brightness 980-nm lasers based on the extended cavity surface emitting lasers concept. *Proc. SPIE*, **4993** (1), 57–67.

45 Boyd, G.D. and Kleinmann, D.A. (1968) Parametric interactions of focused Gaussian light beams. *J. Appl. Phys.*, **39**, 3597–3641.

46 Shchegrov, A.V., Watson, J.P., Lee, D., Umbrasas, A., Hallstein, S., Carey, G.P., Hitchens, W.R., Scholz, K., Cantos, B.D., Niven, G., Jansen, M., Pelaprat, J.-M., and Mooradian, A. (2005) Development of compact blue–green lasers for projection display based on Novalux extended-cavity surface-emitting laser technology. *Proc. SPIE*, **5737** (1), 113–119.

47 Smith, R.G. (1970) Theory of intracavity optical second-harmonic generation. *IEEE J. Quantum Electron.*, **6**, 215–223.

48 Zondy, J.-J. (1991) Comparative theory of walkoff-limited type-II versus type-I second harmonic generation with Gaussian beams. *Opt. Commun.*, **81**, 427–440.

49 Risk, W.P., Gosnell, T.R., and Nurmikko, A.V. (2003) *Compact Blue–Green Lasers*, Cambridge University Press.

50 SNLO Software, http://www.as-photonics.com/SNLO, AS Photonics. 2009.

51 Watson, J.P., Shchegrov, A.V., Umbrasas, A., Lee, D., Amsden, C.A., Ha, W., Carey, G.P., Doan, V.V., Lewis, A., and Mooradian, A. (2004) Laser sources at 460 nm based on intracavity doubling of extended-cavity surface-emitting lasers. *Proc. SPIE*, **5364** (1), 116–121.

52 Shchegrov, A.V., Umbrasas, A., Watson, J.P., Lee, D., Amsden, C.A., Ha, W., Carey, G.P., Doan, V.V., Moran, B., Lewis, A., and Mooradian, A. (2004) 532-nm laser sources based on intracavity frequency doubling of extended-cavity surface-emitting diode lasers. *Proc. SPIE*, **5332** (1), 151–156.

53 Shchegrov, A.V. (2008) Frequency stabilized vertical extended cavity surface emitting lasers. US Patent 7,322,704.

54 Glebov, L. (2005) Optimizing and stabilizing diode laser spectral parameters. *Photon. Spectra*, 90–94.

55 Shchegrov, A.V., Watson, J.P., Lee, D., Umbrasas, A., Dato, R., Green, J., Jansen, M., and Mooradian, A. (2008) Manufacturable vertical extended cavity surface emitting laser arrays. US Patent 7,359,420.

56 Jansen, M., Cantos, B.D., Carey, G.P., Dato, R., Giaretta, G., Hallstein, S., Hitchens, W.R., Lee, D., Mooradian, A., Nabiev, R.F., Niven, G., Shchegrov, A.V., Umbrasas, A., and Watson, J.P. (2006) Visible laser and laser array sources for projection displays. *Proc. SPIE*, **6135** (1), 198–203.

57 Jansen, M., Carey, G.P., Carico, R., Dato, R., Earman, A.M., Finander, M.J., Giaretta, G., Hallstein, S., Hofler, H., Kocot, C.P., Lim, S., Krueger, J., Mooradian, A., Niven, G., Okuno, Y., Patterson, F.G., Tandon, A., and Umbrasas, A. (2007) Visible laser sources for projection displays. *Proc. SPIE*, **6489**, 648908.1–648908.6.

Index

a

absorbers 11
– mirrors 219
– mode locking 246
– saturable 24, 49, 246 ff, 287
– SESAMs 214–222, 233 ff
– external cavities 266, 286
– ultrafast lasers 229
– VECSEL 4–17, 35, 84
absorption coefficient 143, 148, 267
AC-Stark effect 247
active layers 144
– *see also* layers
active mode locking *see* mode locking
active region 75
– GaSb disk lasers 150–166
– MIXSEL 250
– NECSEL 278 ff
AlAs layers 233
AlAs/AlGaAs materials 80, 250
AlAs/GaAs materials 80, 148, 250
AlAsSb materials 80, 148, 164
AlGaAs substrates 280
AlGaAs/AlAs materials 242
AlGaAsSb materials 144, 157
AlGaIn/AsSb materials 179
AlGaInAs/InP materials 43
alloys 13
amplified spontaneous emission (ASE) 31
antimony material system 99, 144 ff
antinodes 20, 32, 173, 198
antireflection (AR) coating 23, 110
– MIXSEL 249
– mode locking 242
– OPS chips 33
antiresonant design 103, 234
aperture 29, 83, 271, 274
applications
– GaSb disk lasers 143
– surface-emitting diode lasers 294 ff
– ultrafast lasers 216 ff
– visible laser sources 287 ff
arrays 50, 274 ff, 297 ff
atomic force microscope (AFM) 131, 237
Auger recombination 17, 223
– external cavities 286
– QD SDLs 197
autocorrelation trace 243

b

BaF_2 76
band edge profiles 144, 153 f
– energy diagrams 12, 75
– external cavities 265
– GaSb disk lasers 144, 158
– SDL 127
bandwidth 103
– GaSb disk lasers 149
– MIXSEL 253
barrier energy 190
barrier layers 153
barrier materials 148
barrier pumping 102, 153, 163
basics
– VECSEL 5–16 f
– GaSb disk lasers 147 ff
BBO crystal 123, 138
beam
– profiles 215, 224
– quality 1–15, 37, 48–54 ff
– splitter 220
bimolecular recombination 17
birefringent filter (BFR) 11, 103, 169
birefringent phase matching 122
Blakeslee thermodynamics 281
blue output 126

Index

blue-induced infrared absorption effects (BLIIRA) 292
Bragg grating 48
Bragg mirrors
– external-cavity 287
– MIXSEL 250
– on-chip 16 ff
– SESAMs 233
– ultrafast lasers 220
– VECSEL 13, 22, 33
Bragg reflectors 79, 121
– see also distributed Bragg reflector
Brewster angle 11, 103, 169
broadening 198
buffer layers 156
– see also layers

c

carbon dioxide/carbonyl sulfide detection 167
carriers
– absorption 7
– confinement 187 f
– density 17, 148, 222
– diffusion 197
– dynamics 286 ff
– excitation 30
– recombination 74
cavity 3 ff, 50–55, 83 ff
– configuration 119
– design 270 ff, 289 ff, 299
– external 263
– format 74
– resonance 161
– ultrafast lasers 213
– V-shaped 134
– Z-type 136
– see also external cavity, intracavity, optical cavity
cavity setup
– GaSb disk lasers 161
– MIXSEL 252
– SESAMs 237
– VECSEL 243 ff
characteristics
– dilute nitride gain media 130 ff
– quantum dot disk lasers 205 ff
charge carriers see carriers
coherence 122, 275
color saturation 218
compact visible sources 297 ff
composition change 194
compound semiconductor materials 12 f
conduction bands 76
– GaSb disk lasers 145

– SDL 130
conductivity 148, 277
confinement factor 17, 20
– in-well-pumped SDLs 155
– QD SDLs 187 f
confinement region 87
confocal fluorescence microscopy 46
construction, VECSEL 16 ff
contact layers 284
continuous-wave (cw) operation
– E-VECSEL 299
– external cavities 274
– ultrafast lasers 213, 231
– visible laser sources 294 ff
copper heat sink 23
coupling effects 77, 138
critical layer thickness 21, 76, 145, 281
– see also thickness
critical phase matching 123
crystals
– BBO 123, 138
– E-VECSELs 289
– GaSb disk lasers 150
– ultrafast lasers 214
current density 188
current injection 1–29, 268, 279 ff

d

de Broglie wavelength 187
defects 36
– dark line 36
– external cavity 266, 283
– QD SDLs 197 ff
– SESAMs 227
delay dispersion 232 ff
delta function quantum dot 188
density functional theory 192
detection sensitivity 53, 167
diamond heat sink 23, 27
diamond heat spreader 88, 132, 205
dielectric mirrors 22
difference frequency generation 44
diffraction 10, 150, 266
diffusion 200
dilute nitrides 120, 127 ff
diode lasers 5 ff
– external-cavity surface-emitting 263–302
– pumped solid-state (DPSS) 6, 29, 214
disk lasers
– quantum dots 200–210
– (3 mm wavelength) 179 ff
– see also semiconductor disk laser
dislocations 156, 156, 283
dispersion 229–236, 243

display applications 119, 218
distributed Bragg reflector (DBR)
– doping levels 277
– dual-band 154
– GaSb disk lasers 144, 147 ff, 166
– mode locking 248
– QD SDLs 198 ff, 205
– SDL 121, 128, 131
– VECSEL 79
divergence 10
doping 7, 11, 91, 265 f
dot-in-a-well (DWELL) structure 195 f
double-heterostructure laser diode 188
dry etching 196
dual wavelength mode 15
dynamic saturation 230

e

edge-emitting lasers 2, 108
– quantum dot 196 ff
– semiconductor 55, 215
efficiency
– external cavities 267
– in-well-pumped SDLs 154
– SDL 122 ff, 138
– VECSEL 17, 29, 75, 100 ff
effusion cells 150
eigenmode radius 273
elastic relaxation *see* relaxation, strain relaxation
electric field
– external-cavities 276
– patterns 79
– SDL 121
– ultrafast lasers 230
electrical carrier injection 108
electrical pumping 29 ff, 240–250, 263–302
electroluminesce 286
electronic constants 148
electronic transitions 153
emission wavelength 109, 202 ff
– external cavities 269
– GaSb disk lasers 144, 150
– InGaAsP-based VECSEL 99
– QD SDLs 192
– temperature dependent 207
– VECSELs 1, 37, 75
– *see also* wavelength 13
energy levels
– charge carriers 189–194
– GaSb disk lasers 158
– in-well-pumped SDLs 153
– splitting 76
enhancement factor 230, 235

epitaxial growth 153 ff, 233, 266
epitaxial structures 21 f, 79 ff, 87, 255
– external cavities 264
– GaSb disk lasers 150
– III-Sb disk lasers 146 ff
EP-VECSEL structures 248
erbium-doped fibers 111, 153
etalon 11, 34
– GaSb disk lasers 170–177
– MIXSEL 252
– mode locking 243
– resonance 84
– SDL 137
excitation bleaching 269
excited states 203
experimentals
– mode locking 242 ff
– single-frequency SDL 176 ff
extended cavity 4
external cavity 2–15, 248 ff, 263–302

f

Fabry–Pérot cavity
– E-VECSEL 294
– GaSb disk lasers 170, 177
– QD SDLs 205
– resonance 80
– SDL 130, 274
femtosecond mode-locked VECSELs 246 ff
Fermi levels 190, 276
fiber lasers 1, 109 ff, 214
fiber mirrors 109
fiber-coupled pumps 31
field enhancement 233 ff, 250
field intensity 267
filters
– GaSb disk lasers 169
– optical 24
– ultrafast lasers 230
flow cytometry 288
flying spot projector 47
Fourier transform spectrometers 53, 151
free spectral range (FSR) 109
free-carrier absorption (FCA) 223
free-running modes 176
frequency comb 213
frequency conversion 39, 291 ff
– E-VECSELs 288
– SDL 119–142
– ultrafast lasers 213 f, 219, 230
Fresnel losses 158
Fresnel reflections 233
fusion interface 22

g

g₁g₂-model 272
GaAlAs material systems 43, 264
GaAs material systems 13, 21, 39 ff, 76
– external cavities 280
– GaSb disk lasers 166
– lattice mismatch 127
– MIXSEL 251
– QD SDLs 201 ff
– SESAMs 233
– tuning ranges 105
– VECSELs 199
GaAs/AlAs DBRs 120, 199
GaAs/AlGaAs 76
GaAs/AlGaAs DBRs 155 ff, 166
GaAsSb barrier layers 158
gain 6–15, 76 ff, 231 ff
– ultrafast lasers 222
gain bandwidth 11
– GaSb disk lasers 172
– QD SDLs 198
– ultrafast lasers 216
gain carriers 32
gain chip 74
– GaSb disk lasers 163
– SDL 136
gain media 13–20, 187 ff
– dilute nitride 130 ff
– GaSb disk lasers 168
– mode locking 240
– SDLs 120–130
– ultrafast lasers 214, 227
gain regions 6–33, 96
– 1:1 mode locking 245 ff
– external cavities 265 ff, 274
– QD SDLs 204
– thermal management 90
gain saturation 229
gain structure 200 ff
– MIXSEL 250
– mode locking 242
– SESAMs 236
GaInAsSb 76, 105, 157 ff
GaInAsSb/AlGaAsSb diode lasers 179
GaInAsSb/GaSb material system 44
GaInNAs/GaInSb material systems 105
GaInP/AlGaInP material systems 126
GaInSb based VECSELs 241
Galilean telescope folded cavity 28
GaN laser diodes 127
GaSb material systems 76, 88, 143–186
GaSb–AlAsSb layer 158
Gaussian mode profiles 92, 224, 248, 289
geometry, optical cavity 24 ff

graded index (GRIN) lens 31
gratings 11
green problem 43
green-induced infrared absorption
   (GRIIRA) 292
ground-state emission 202 f
group delay dispersion (GDD) 53, 228–235,
   251
group III-Sb material system 144 ff
group III–V material systems 13, 74 ff
group V sublattice intermixing 277
growth 81 ff, 128 ff
– GaSb structures 149 ff, 155
– III-Sb structures 146 ff
– interfacial misfit (IMF) 166
– pseudomorphic 145
– QD structures 191, 196
– SESAMs 227
– ultrafast lasers 214

h

harmonic generation 121
– *see also* second harmonic generation
harmonic mode locking 215
heat density 84
heat dissipation 3–9, 53
heat flow modeling 87 ff
heat sink 9
– GaSb disk lasers 151, 158, 164
– MIXSEL 255
– mode locking 243
– QD SDLs 205
– thermal management 88
– VECSELs 33, 23, 85
heat spreaders 11, 22 f, 85
– 1–2 μm approaches 87 ff
– GaSb disk lasers 163
– QD SDLs 205
– Sb-based SDLs 158 ff
– SDL 124, 132
heavy hole (hh) band 144
heterostructures 81
high-average output power 240 ff
high-conductivity crystal 105
high-mode laser sources 294 ff
high-power arrays 274 ff
high-power design 1–72, 157 ff
high-quality beams 96
high-reflectivity gratings (HRGs) 11, 174
high-reflectivity mirrors 79
high-repetition rates 244 ff
high-resolution X-ray diffraction
   (HRXRD) 150
high-speed modulation 1

holes 20, 76 f, 187 ff, 194
hybrid mirrors 149

i
illumination sources 288
imaging projector 47
InAlGaAs/GaAs material systems 76
InAs quantum dots 204, 237
InAs/GaAs submonolayers 190, 194
InAsSb/AlGaAsSb layers 179
incident field intensity 234
index profile *see* field profile 0
indium monolayer coverage 237
InGaAlAs material systems 76
InGaAs material systems 94 ff, 102 f, 126
– external-cavities 281
– GaSb disk lasers 158
– lattice mismatch 127
– quantum dots 120
– quantum wells 203, 242, 247
– external-cavities 276
– thermal management 88
– tuning ranges 105
InGaAs material systems VECSELs 241
InGaAs/GaAs material systems 5, 13–21, 76
InGaAs/GaAsP material system 38, 284
InGaAsN material system 127 ff
InGaAsP material systems 39, 76
InGaN/GaN material systems 76
InGaP material systems 76, 105
injected carriers *see* carriers 0
InP material systems 13, 39, 76, 120
instabilities 231, 244
integrated laser types 73–118, 248 ff
interconnects 217
interfacial misfit (IMF) growth 156, 166
intermediate n-mirror 264
intraband C – X transition 266
intracavity contacts 284
intracavity elements 9, 24 ff
intracavity field 173
intracavity frequency 15, 39
– conversion 119–142, 291 ff
– doubling 48
– selective filters 28
intracavity heat spreaders 124, 158 ff
intracavity laser absorption spectroscopy
  (ICLAS) 11, 53, 57, 168
intracavity light red–orange conversion 136 ff
intracavity losses 266
intracavity pulses 231, 244
intracavity waist 271
in-well pumping 32, 153 ff, 163 ff
island formation 283

j
junctions 29

k
k.p theory 192
Keplerian telescope folded cavity 28
Kerr lens mode locking (KLM) 227
Kerr-effect 219
$KNbO_3$ (potassium niobate) crystals 292
Kramers–Krönig relations 223

l
Langmuir probe 128
laser cavity *see* cavity, intracavity
laser characterization 33 ff
laser functional versatility 11 ff
laser performance 96 ff
laser-excited fluorescence 46
lattice constants 12, 75
– external cavities 280
– SDL 127
lattice matching 75, 145
– GaSb disk lasers 151
– SDL 127
– external-cavities 282
layer thickness 147, 233
layers 12 f
– buffer 156
– GaSb disk lasers 144
LBO (lithium triborate) crystals 123, 292
lead–chalcogenide-based devices 99, 179
lenses 31
light emission 1.2 mm 127 ff
light hole band 145
light output 159
$LiNbO_3$ crystal 123
line scanning projector 47
linear semiconductor disk lasers 133
linewidth enhancement factor 230
liquid-capillary bonding 105
long-cavity solution 272, 294
longitudinal modes 245, 294 f
long-wavelength GaSb disk lasers 143–186
long-wavelength NECSEL 285
losses 101
– doping levels 278
– external cavities 266, 277
– saturation dynamics 53
– ultrafast lasers 221
– *see also* mode locking
low-noise laser sources 294 ff
low-temperature-grown QW-SESAMs 227
luminescence efficiency 75
Lyot filter 169

m

µ-VECSEL   106
marker dyes   46
material systems   2–15, 41, 74 ff
– group III-Sb   144 ff
– QD SDLs   191
– thermal management   93
– ultrafast lasers   222
– *see also* Ga, In etc
Matthew–Blakeslee thermodynamics   21, 281
MEMS-mounted mirror-tunable VECSELs   109
metal-enhanced metamorphic mirrors   22
metal-organic chemical vapor deposition (MOCVD)   82, 263
metal-organic vapor-phase epitaxy (MOVPE)
– dilute nitrides   128
– GaSb disk lasers   149
– QD SDLs   197
– SESAMs   233
– VECSEL   23
metamorphic buffer layers   156
metamorphic semiconductor mirror materials   22
microchip laser configuration   27, 105 f
microsize cavity laser   105
microwave photonics applications   52
mirrors   5, 11
– external-cavities   264
– GaSb disk lasers   160
– hybrid   149
– monolithic   120
– wafer structures   33
– QD SDLs   198
– SDL   130
– SESAMs   214
– thermal management   89
– ultrafast lasers   219
– VECSEL   15, 21 ff, 79 ff, 109
mode control   24 ff, 270 ff
mode converter   6
mode locking
– GaSb disk lasers   161
– high repetition rates   244 ff
– MIXSEL   32, 50, 219, 248 ff
– QD SDLs   200
– Q-switched   231, 244, 263, 269
– regimes   121
– semiconductor disk lasers   213–262
– stability   231 ff
– ultrafast lasers   219
– VECSEL   24, 50 ff, 232–250, 287 ff
mode radius   271
modulation bandwidth   119 f
modulation depth   221
molecular beam epitaxy (MBE)
– external cavities   263
– GaSb disk lasers   149
– QD SDLs   196
– SESAMs   233
– VECSEL   23, 81
– *see also* growth
molecular spectroscopy   53
monolayer coverage   237
monolithic cavities   263
multielement devices   274 ff
multilayer laser Bragg mirror   21 ff
multilayer structures   12, 74, 81
multimode operation   108, 176, 274
multimode pump diodes   9, 30, 51
multiphoton absorption   214
multiple gain chips   39, 48
multiple pulsing   231
multiquantum well structures   5, 20, 76, 168

n

Nd:YAG laser   1, 32
n-distributed Bragg reflector (n-DBR)   277
near-field image, arrays   275
near-infrared laser emission   15
NECSEL
– doping profile   278
– InP-based   285
– mode locking   248
– *see also* electrically pumped VECSEL
n-mirror   264, 277
noise   37, 173, 297
nonlinear crystals   291 ff
nonlinear optical conversion   14 ff, 287 ff
nonlinear optical second harmonic generation   2
n–p–n contact layers   284

o

on-chip beam diameters   9
on-chip mirrors   13–22, 49
operating parameters
– QD SDLs   203
– ultrafast lasers   213, 231
– VECSEL   16 f, 33 ff, 56
optical absorption   8
optical cavity   24 ff, 105
– *see also* cavity
optical characteristics   73, 148
optical clocking   214
optical coherence tomography (OCT)   218

optical conversion 14 ff, 287 ff
optical crystals 13, 24
optical elements 11 ff
optical fibers 1, 264
optical filters 24
optical gain media 187 ff
– see also gain media
optical layer thickness 233
optical links 217
optical losses 266, 278
optical path length 10
optical power 15
optical properties 202
optical pumping
– 1.2 µm SDLs 120
– external cavities 263
– MIXSEL 255
– OPS 2, 24, 33–50
– ultrafast lasers 222
– VECSEL 2 ff, 29 ff, 55, 77
optical rectification 122
optical reflectivity see reflectivity
optoelectronic devices 120
orange–red semiconductor disk
  laser 136
oscillators 143, 214
output characteristics 1 ff
– 50 GHz VECSEL 245
– E-VECSEL 299
– GaSb disk lasers 157 ff, 165, 172
– mode locking 240 ff
– OPS-VECSEL 36
– QD SDLs 206
– SDL 120, 134
– VECSEL 1, 15, 33, 55, 97
output-coupling mirror 198

p

passive mode locking 219, 239 ff
– see also mode locking
PbEuTe/BaF$_2$ material systems 80
PbTe material systems 76, 99, 44
p-distributed Bragg reflector (p-DBR) 277
Peltier device 85
performance
– 1220 nm SDL 132 ff
– external-cavitiess 264 ff
– VECSEL 38 ff, 74 ff
periodically poled KTP (PP-KTP) 289, 292
periodically poled lithium niobate (PPLN)
  292, 299
periodically poled lithium tantalate (PPLT)
  292
permittivity 121

phase changes 232
phase matching 122 f, 265
phase shifts 219
phase-locked arrays 51
photocoagulation treatment 46
photoluminescence 82
– GaSb disk lasers 150, 166
– InGaAsN QWs 128
– OPS chips 35
– QD SDLs 202, 206
– SDL 133
photon energies 8
photopumping 74
photorefractive damage (PRD) 292
piezoelectric potential 192
plane–plane cavity 105, 109
plasma operation 128
plasma-assisted MBE growth 128 ff
p-mirrors 264, 277
p–n junctions 29
polarization
– E-VECSEL 290, 299
– external cavities 273
– SDL 121
– ultrafast lasers 219
post-growth analysis 149 ff
power 1 ff, 51 ff, 214
power scaling
– external-cavities 265, 272
– SDL 124 ff
– VECSEL 3–108, 240 ff
power–current characteristics 297
Poynting vector walk-off 123
profile distortions 10
projection displays 46, 218
propagation distance 122
proton implant 264
pseudomorphic growth 145
pulse/shaping model 221–233, 253, 269
pump absorption 163
pump diodes 30 f
pump integration 9, 107 ff
pump power 17, 37
– see also power
pump spot 10, 26, 94
pumping
– external cavities 267 ff
– ultrafast lasers 222, 230
– VECSEL 4, 8 ff, 29 ff

q

Q-switched mode locking (QML)
– external cavities 263, 269
– instabilities 244

– ultrafast lasers 231
quality beam 1–72
quantum confined compounds 13 f
quantum confinement 187
quantum defects 18, 32, 89, 153, 163
quantum dots 187–212
– absorbers 219
– emission wavelength 202 ff
– gain structures 74
– MIXSEL 249
– SESAMs 233, 236 ff
quantum efficiency 100, 125
– see also efficiency 0
quantum size effect 193
quantum wells 7, 13, 32 f, 187
– absorbers 219
– external cavities 264, 286
– gain structures 16 f, 74 ff, 94
– GaSb disk lasers 143 ff, 171
– MIXSEL 249
– photon-emitting layers 12
– SESAMs 233, 236 ff
– tuning ranges 105
quasiphase matching (QPM) 123
quasisolitons 232, 246
quaternary semiconductors AlGaAsSb 144

r
radio-frequency (RF) plasma 128
radius of curvature (ROC) 106, 133, 252
rapid thermal annealing (RTA) 130
rare-earth-doped fibers 111
recombination 17
– external cavitiess 286
– monomolecular 17
– QD SDLs 189
recovery dynamics 225
red semiconductor disk lasers 119–142
red–green–blue (RGB) wavelengths 46, 119, 297
red–orange emission 127–136
reflectance spectra 151
reflectivity 8
– Bragg Mirror 21
– E-VECSEL 290, 298
– external cavities 264, 270
– GaSb disk lasers 160, 167
– OPS chips 33
– QD SDLs 201
– SDL 133
– SESAMs 233
– ultrafast lasers 220 ff
– VECSEL 16 f, 79
reflector materials 80

refractive index
– external cavities 264, 270
– GaSb disk lasers 149 ff
– mode locking 242
– QD SDLs 188, 201
– SDL 122
– SESAMs 233 f
– ultrafast lasers 223
– VECSEL 21, 35
relative intensity noise (RIN) 51
relaxation 192, 199, 205
repetition rate
– external cavities 287
– GaSb disk lasers 161
– mode locking 240
– SDL 121
– SESAMs 236
– ultrafast lasers 213, 231, 244
– VECSEL 51
resistance 265, 278
resonant design 103
resonant in-well pumping 32
resonant periodic gain (RPG) 7, 77
– GaSb disk lasers 147, 172
– QD SDLs 200
resonant structures
– GaSb disk lasers 161, 171
– QD SDLs 202
– SDL 125
– SESAMs 234
– ultrafast lasers 219
– VECSEL 27, 50, 74, 80
ring laser gyro 52
roll-off power 274

s
sapphire/SiC 76
saturable absorber
– external cavities 268, 287
– MIXSEL 249 f
– mode locking 246
– SESAMs 234
– ultrafast lasers 220 f, 232
Sb material system 143 ff
Sb-based materials 155 ff, 166 f, 177 f
scalable power 3, 54 ff
– see also power scaling
scanning projection applications 119
scanning tunneling micrograph 193
Schottky barriers 265
second harmonic generation (SHG)
– E-VECSELs 289 ff
– external cavities 283
– SDL 121 ff

– VECSEL 2–27
secondary ion mass spectroscopy (SIMS) 150
selective wet chemical etching 23
self-phase modulation (SPM) 53, 228
semiconductor disk lasers (SDLs) 2, 91, 119–142
– quantum dots 187–212
semiconductor gain medium *see* gain media
semiconductor lasers 1 ff
– ultrafast 215 ff
– VECSEL 1 ff
semiconductor material systems 2, 12 ff, 94
semiconductor saturable absorber mirrors (SESAMs) 27, 200–245
semiconductor structures
– GaSb disk lasers 143
– MIXSEL 249
– VECSEL 75
– wafers 22 ff
semiconductor–air interface 34
short pulse generation 38
– mode locking 240
– repetition rates 52
– VECSELs 15
short-cavity solution 272
short-wavelength pump diodes *see* optical –, electrical –, pumping
silica etalon 243
silicon carbide 90
single-array chip 273
single-frequency external cavities 167–176, 273
single-frequency operation 20
single-longitudinal mode 173, 176, 294 f
single-mode operating regime 51, 106
single-sided chip design 264
single-transverse mode optical resonator cavity 27
SiO_2/HfO_2 80
size quantization 187 ff
slope efficiency *see* efficiency 0
slow saturable absorbers 227 f
solitons 228 ff, 246
spacer layers 196, 200 ff
spatial hole burning 20
spectral tuning 104
split-off valence band 145
Stark-SESAMs 228
strain
– compensation 21 ff, 36 f, 49, 56, 76
– critical thickness 21
– epitaxial layers 191
– GaSb disk lasers 144
– lattice-mismatched layers 282

– layers 13, 21, 36 f, 49, 56, 75
– quantum wells material 21, 145
– reducing layer (SRL) 197–203
– relaxation 205, 276, 283
Stranski–Krastanow growth 191–205, 237
structures 12 ff, 73–118
– long-wavelength NECSEL 285
– VECSEL 6 ff
subcavity designs 80 ff
subcavity resonance 202
submonolayers 190, 198–207
submount 89 ff
substrates
– doping 280
– external cavities 266
– GaSb disk lasers 155
– removal technique 149
– VECSEL 21, 41, 76, 86
– *see also* materials 76
surface reflection losses 101
surface-emitting lasers 2–74, 248 ff
– external-cavity/diode 263–302
– quantum dot 198 ff

t
TEM_{00} mode *see* transverse mode
temperature dependence
– doping 281
– emission wavelength 207
– external cavities 275
– gain spectra 74
– GaSb disk lasers 160
– pump diode lasers 32
– SDL 126, 135, 189
– VECSEL 21, 84, 90 f
temperature-tuning coefficient 152
temporal SESAM response 225 ff
tensile strained material 145
terahertz radiation 44, 49, 57
thermal conductivity 22, 86 ff
thermal constants 148
thermal distortions 10
thermal impedance 22 f, 32, 44
thermal lensing 10, 270 ff
thermal load 125
thermal management 73, 84 ff
thermal profiles 270
thermal resistance 87, 265
thermal rollover 8, 124
thermal stability 200
thermalization 74
thickness
– critical 21, 76

– external cavities 281
– SESAMs 233
– ultrafast lasers 222
thin devices 87 ff
thin-disk geometry 91, 124
three-chip InGaAs QW device 97
three-mirror cavities 83, 170
Ti:sapphire laser 52, 101, 214
time constants 227 ff
time–bandwidth product (TBP) 245
titanium–gold ohmic contact 265
top-hat pump profile 91
transfer matrix method 79
transmission 15 f, 296
transmission electron microscopy (TEM), IMF arrays 156
transparency condition 202
transverse beam profile 138
transverse mode 4, 10, 24, 91
– external cavities 263, 269
– mode locking 240
– ultrafast lasers 220, 224
– VECSEL 39, 55
tuning 103 ff, 168 ff, 203 ff
– broad spectral 104
– external cavities 273
– GaSb disk lasers 152, 162
– QD SDLs 187
– SDL 137
tunnel junction 284
tunneling 77
two-photon absorption (TPA) 221, 236

u

ultracompact VECSELs 105
ultrafast lasers 213 ff
ultrashort pulse generation 20
ultrashort pulse VECSELs 103
UV–IR wavelength 1–72

v

valance bands
– VECSEL 76
– external cavities 265, 276
– GaSb disk lasers 144, 148
vertical-cavity surface-emitting lasers (VCSELs) 198
vertical-external cavity surface-emitting laser (VECSEL) 1–120, 187, 215, 240–250
visible laser sources 287 ff
volume Bragg gratings (VBGs) 174, 298

w

wafers
– fusion 22, 120
– growth/structure 7 ff, 21 ff
– MIXSEL 255
– SDL 132
– VECSEL 13
– window-on-substrate 22, 33
wall-plug efficiency 276
wavelength 37 ff, 74
– E-VECSELs 292
– GaSb disk lasers 143
– scaling 48
– SESAMs 235
– versatility 12 ff, 94 ff
wavelength tuning *see* tuning 0
well thickness 36
wet chemical etching 23
wetting layer 190
window layer 198
W-structure 179

y

Yb:YAG absorption 8, 32

z

Z-type cavity 136